METHODS IN MOLECULAR BIOLOGY™

Series Editor
John M. Walker
School of Life Sciences
University of Hertfordshire
Hatfield, Hertfordshire, AL10 9AB, UK

For further volumes:
http://www.springer.com/series/7651

Oxidative Stress and Nanotechnology

Methods and Protocols

Edited by

Donald Armstrong

University at Buffalo, Buffalo, NY, USA;
University of Florida, Gainesville, FL, USA

Dhruba J. Bharali

The Pharmaceutical Research Institute, Albany College of Pharmacy and Health Sciences, Rensselaer, NY, USA

Humana Press

Editors
Donald Armstrong
University at Buffalo
Buffalo, NY, USA

University of Florida
Gainesville, FL, USA

Dhruba J. Bharali
The Pharmaceutical Research Institute
Albany College of Pharmacy and Health Sciences
Rensselaer, NY, USA

ISSN 1064-3745 ISSN 1940-6029 (electronic)
ISBN 978-1-4939-5934-1 ISBN 978-1-62703-475-3 (eBook)
DOI 10.1007/978-1-62703-475-3
Springer New York Heidelberg Dordrecht London

Printed on acid-free paper

Humana Press is a brand of Springer
Springer is part of Springer Science+Business Media (www.springer.com)

Preface

Over the past decade, nanotechnology has been a buzzword that the scientific community has embraced with great compassion. The emergence of nanotechnology has had a profound effect on almost every aspect of the twenty-first century's daily life. It has had a revolutionary impact from stain-resistant clothing and cosmetics to environmental issues, including energy and medicine and even aerospace engineering. This volume, comprised of nine Reviews, nine new Protocols, and one Future Directions chapter, is an endeavor to present various aspects of nanotechnology from the oxidative stress point of view.

The authors reflect the large participation of internationally eminent researchers, who have already distinguished themselves by making a deep impact in the fields of nanotechnology and oxidative stress. Starting from a single concept, nanotechnology has emerged as one of the fastest growing niches in this exciting field. Proponents of this avant-garde technology also describe it as one of the most prevalent engineering innovations in science and society since the Industrial Revolution.

There is an enormous disparity between the amount of funding dedicated to nanoparticle safety studies as compared to its mammoth applications in industrial and consumer products. However, there has been a sustaining interest by the scientific community in the effects of short-term and chronic exposure to nanomaterials on humans and the environment. Succinct nanoparticles are defined as submicron particles, usually <100 nm in size. Materials in this range usually produce a set of distinctive physicochemical properties. Therefore, size, chemical composition, and surface properties of nanomaterials play a crucial role in oxidative stress-induced nanotoxicity. Part I of this volume includes detailed Reviews and characterization of different types of nanomaterials, including polymeric and inorganic nanoparticles, nanotubes, nanogels, and quantum dots along with their adverse effects, particularly nanotoxicity induced by oxidative stress. It also reviews in vivo and in vitro toxic effects caused by oxidative stress as a result of exposure to nanomaterials.

The chapters in Part II provide detailed Protocols for the synthesis of different antioxidant nanoparticles and antioxidant-loaded nanoparticles, as well as their in vitro/in vivo mechanisms of action along with their clinical relevance. The advent of nanotechnology provides unprecedented opportunities for addressing current pitfalls in diagnosis, prevention, and therapy of many diseases. This rapidly developing field of nanotechnology, often referred to as "nanomedicine," has the potential to profoundly impact numerous aspects of healthcare. Of particular interest are the improved pharmacological aspects ensuring that healing antioxidants are delivered to a specific site for enhanced efficacy. The surface-tunable properties of nanoparticles can be manipulated to prevent their opsonization, providing a mechanism for sustained blood circulation, vis-à-vis prolonging the residence time of the antioxidant. These antioxidants, therefore, sit within this tiny "Trojan horse," not only capable of protecting them from external assault but also accumulating at the site of action by active or passive targeting. The main focuses in this part of the volume is utilization of various inorganic and polymeric nanoparticles loaded with antioxidant(s) for diagnosis, prevention, and therapy of various diseases. Many of the chapters in this part also include the findings that some inorganic nanoparticles can be used as potent free radical scavengers and anti-inflammatory agents.

Recently, several intriguing findings by numerous researchers in the field of nanomaterials have sparked a wave—paradigm shift. Part III is an attempt to look into the future and explore a vision of newer technology. We hope this approach will stimulate additional research. Overall, chapters in this volume of Methods in Molecular Biology can be of great help to a wide audience including graduate students in basic sciences and senior scientists or medical residents, which will provide tremendous opportunities to learn techniques of nanotechnology in pathophysiology and protection from the oxidative stress point of view. This volume can also be used as a handbook for applied nanotechnology.

Buffalo, NY, USA *Donald Armstrong*
Rensselaer, NY, USA *Dhruba J. Bharali*

Contents

Contributors

SOOAD AL-DAIHAN • *Biochemistry Department, Science College, King Saud University, Riyadh, Saudi Arabia*
DONALD ARMSTRONG • *University at Buffalo, Buffalo, NY, USA; University of Florida, Gainesville, FL, USA*
ABIR BEN BACHA • *Biochemistry Department, Science College, King Saud University, Riyadh, Saudi Arabia*
RAFAEL BASSO • *Department of Plastic Surgery, School of Medicine, University of Campinas, Campinas, Brazil*
DHRUBA J. BHARALI • *The Pharmaceutical Research Institute, Albany College of Pharmacy and Health Sciences, Rensselaer, NY, USA*
MARTIN J.D. CLIFT • *Adolphe Merkle Institute and Fribourg Center for Nanomaterials, University of Fribourg, Fribourg, Switzerland*
NIRMALENDU DAS • *Biomembrane Division, Indian Institute of Chemical Biology, Kolkata, West Bengal, India*
KRISHNA R. DEVARAKONDA • *Clinica Pharmacology and Pharmacokinetic Division, Covidien/Hazelwood, St. Louis, MO, USA*
ALOK DHAWAN • *Indian Institute of Toxicology Research, Lucknow, Uttar Pradesh, India*
ANCA DINISCHIOTU • *Department of Biochemistry and Molecular Biology, University of Bucharest, Bucharest, Romania*
HUGH DOYLE • *Tyndall National Laboratory, University College Cork, Cork, Ireland*
AFAF EL-ANSARY • *Biochemistry Department, Science College, King Saud University, Riyadh, Saudi Arabia*
DANIELA GRADINARU • *Department of Biochemistry, Faculty of Pharmacology, Barol Davila, University of Medicine and Pharmacy, Bucharest, Romania*
DANIELA IACOPINO • *Tyndall National Laboratory, University College Cork, Cork, Ireland*
RADE INJAC • *Department of Pharmaceutical Biology, Faculty of Pharmacy, University of Ljubljana, Ljubljana, Slovenia*
RIDHIMA JUNEJA • *Department of Chemistry, University of Delhi, Delhi, India*
SARAH KEANE • *Proteomics Research Laboratory, Department of Biochemistry, University College Cork, Cork, Ireland*
GWANGSEONG KIM • *Department of Chemistry, University of Michigan, Ann Arbor, USA*
RAOUL KOPELMAN • *Department of Chemistry, University of Michigan, Ann Arbor, USA*
MALAK KOTB • *Biochemistry Department, Science College, King Saud University, Riyadh, Saudi Arabia; Department of Molecular Genetics, Biochemistry, and Microbiology, College of Medicine, University of Cincinnati, Cincinnati, OH, USA*

PETER KOVACIC • *Department of Chemistry and Biochemistry, San Diego State University, San Diego, CA, USA*
ALEXANDRA KROLL • *Department of Environmental Toxicology, Eawag, Swiss Federal Institute of Aquatic Science and Technology, Dübendorf, Switzerland*
DANA KÜHNEL • *Department of Bioanalytical Ecotoxicology, Helmholtz-Center for Environmental Research - UFZ, Leipzig, Germany*
ASHUTOSH KUMAR • *Institute of Life Sciences, Ahmedabad University, Ahmedabad, Gujarat, India*
RAJIV KUMAR • *Department of Physics, Northeastern University and Dana Farber Cancer Institute, Harvard Medical School, Boston, MA, USA*
SHICHANG LIU • *Nankai University, Ministry of Education, Tianjin, China*
YONG-EUN KOO LEE • *Department of Chemistry, University of Michigan, Ann Arbor, MI, USA*
GERARD A. LUTTY • *Wilmer Ophthalmological Institute, Johns Hopkins Hospital, Baltimore, MD, USA*
ARDHENDU KUMAR MANDAL • *Biomembrane Division, Indian Institute of Chemical Biology, Kolkata, West Bengal, India*
SHAKER A. MOUSA • *The Pharmaceutical Research Institute, Albany College of Pharmacy and Health Sciences, Rensselaer, NY, USA*
DATTATRI K. NAGESHA • *Department of Physics, Northeastern University and Dana Farber Cancer Institute, Harvard Medical School, Boston, MA, USA*
IRENE O'DONOVAN • *Proteomics Research Laboratory, Department of Biochemistry, University College Cork, Cork, Ireland*
SORINA NICOLETA PETRACHE • *Department of Biochemistry and Molecular Biology, University of Bucharest, Bucharest, Romania*
MATEVZ PRIJATELJ • *Department of Pharmaceutical Biology, Faculty of Pharmacy, University of Ljubljana, Ljubljana, Slovenia*
TARL W. PROW • *Dermatology Research Centre, Princess Alexandra Hospital, School of Medicine, The University of Queensland, Brisbane, QLD, Australia*
MIHAELA RADU • *Department of Biochemistry and Molecular Biology, University of Bucharest, Bucharest, Romania*
DEVENDRA KUMAR RAI • *Christopher Bond Life Sciences Center, Department of Molecular Microbiology & Immunology, University of Missouri School of Medicine, Columbia, USA*
PRASHANT KUMAR RAI • *Department of Chemical Technology, University of Johannesburg, Johannesburg, South Africa; Department of NMR, All India Institute of Medical Sciences, New Delhi, India*
RAMA NARSIMHA REDDY ANREDDY • *Department of Pharmacology, Vaageswari College of Pharmacy, Karimnagar, Andhra Pradesh, India*
NARSIMHA REDDY YELLU • *University College of Pharmaceutical Sciences, Kakatiya University, Warangal, Andhra Pradesh, India*
BARBARA ROTHEN-RUTISHAUSER • *Adolphe Merkle Institute and Fribourg Center for Nanomaterials, University of Fribourg, Fribourg, Switzerland*
INDRAJIT ROY • *Department of Chemistry, University of Delhi, Delhi, India*
MARIA HELENA A. SANTANA • *Department of Biotechnological Processes, School of Chemical Engineering, University of Campinas, Campinas, Brazil*
SIBANI SARKAR • *Biomembrane Division, Indian Institute of Chemical Biology, Kolkata, West Bengal, India*

KRISTIN SCHIRMER • *Department of Environmental Toxicology, Eawag, Swiss Federal Institute of Aquatic Science and Technology, Dübendorf, Switzerland*
ANDREEA IREN SERBAN • *Department of Preclinical Sciences, Agronouical Sciences and Vetrinary Medicine*
PATRÍCIA SEVERINO • *Faculty of Health Sciences, Fernando Pessoa University, Porto, Portugal; Department of Biotechnological Processes, School of Chemical Engineering, University of Campinas, Campinas, Brazil*
BECHAN SHARMA • *Department of Biochemistry, University of Allahabad, Allahabad, India*
GAURAV SHARMA • *Department of NMR Institute, All India Institute of Medical Sciences, New Delhi, India*
VYOM SHARMA • *Indian Institute of Toxicology Research, Lucknow, Uttar pradesh, India*
DAVID SHEEHAN • *Proteomics Research Laboratory, Department of Biochemistry, University College Cork, Cork, Ireland*
RATNASAMY SOMANATHAN • *Centro de Graduados e Investigación del Instituto Tecnológico de Tijuana, Tijuana, BC, Mexico*
ELIANA B. SOUTO • *Faculty of Health Sciences, Centre of Genetics and Biotechnology, Institute of Biotechnology and Bioengineering, Fernando Pessoa University, Porto, Portugal*
LOREDANA STANCA • *Department of Biochemistry and Molecular Biology, University of Bucharest, Bucharest, Romania*
BORUT STRUKELJ • *Department of Pharmaceutical Biology, Faculty of Pharmacy, University of Ljubljana, Ljubljana, Slovenia*
DANIEL SUNDH • *Dermatology Research Centre, Princess Alexandra Hospital, School of Medicine, The University of Queensland, Brisbane, QLD, Australia*
SARA TEDESCO • *Proteomics Research Laboratory, Department of Biochemistry, University College Cork, Cork, Ireland*
APARNA WATAL • *Department of Chemical Engineering, Drexel University, Philadelphia, USA; Faculty of Law, Australian National University, Canberra, Australia*
GEETA WATAL • *Medicinal Research Lab, Alternative Therapeutics Unit, Drug Discovery and Development Division, Department of Chemistry, University of Allahabad, Allahabad, India*
ZHUO YANG • *The Key Laboratory of Bioactive Materials, Ministry of Education, College of Medicine, Nankai University, Tianjin, China*

Part I

Reviews

Chapter 1

Nanoscale Biosensor for Detection of Reactive Oxygen Species

Tarl W. Prow, Daniel Sundh, and Gerard A. Lutty

Abstract

Noninvasive detection of biological responses to reactive oxygen species (ROS) in vivo could shed light on mechanisms at work in diverse areas like developmental dynamics, therapeutic effectiveness, drug discovery, pathogenic processes, and disease prevention. Research on ROS is usually dependent on in vitro models without translational relevance. Nanoscale (<100 nm) particulates are attractive carriers and platforms for biosensor technology due to their small size, flexible assembly, and favorable toxicity profiles. Intracellular signalling pathways activated in response to ROS have been well documented and mechanisms elaborated. Likewise, there is a wealth of genetic reporter systems that utilize fluorescent proteins capable of being monitored noninvasively. We combined these elements into a platform technology that utilizes nanoparticle-tethered synthetic genetic elements that respond to cellular response elements to report endogenous responses to oxidative insult through fluorescent gene expression. We envision the future of this technology to play a research role quantifying oxidative stress in vivo and a future clinical role as an automated theragnostic for ROS-related diseases. The production of this nanobiosensor technology utilizes off-the-shelf components and can be carried out in a molecular biology laboratory. Assessment of fluorescent protein expression can be done with noninvasive imaging and quantitative protein expression analysis. This is a flexible nanoparticle-based reporter system for monitoring in vivo responses to ROS.

Key words Magnetic nanoparticles, Reactive oxygen species, Non-invasive imaging, Antioxidant response element

1 Introduction

Some common nanoparticles for human use are ZnO/TiO_2 in sunscreens [1, 2], colloidal silver for wound healing and antimicrobial uses [1], and magnetic nanoparticles (MNPs) used as contrast agents [3]. Clinically, magnetic resonance imaging utilizing MNP relies on the MNP to accelerate the relaxation rate of nearby water molecules after a magnetic field pulse. In the laboratory, MNPs have been developed for cell separation applications. Decades of clinical use show that ultrasmall superparamagnetic iron oxide injected intravenously is safe [4]. Magnetic-activated cell sorting

Donald Armstrong and Dhruba J. Bharali (eds.), *Oxidative Stress and Nanotechnology: Methods and Protocols*, Methods in Molecular Biology, vol. 1028, DOI 10.1007/978-1-62703-475-3_1,

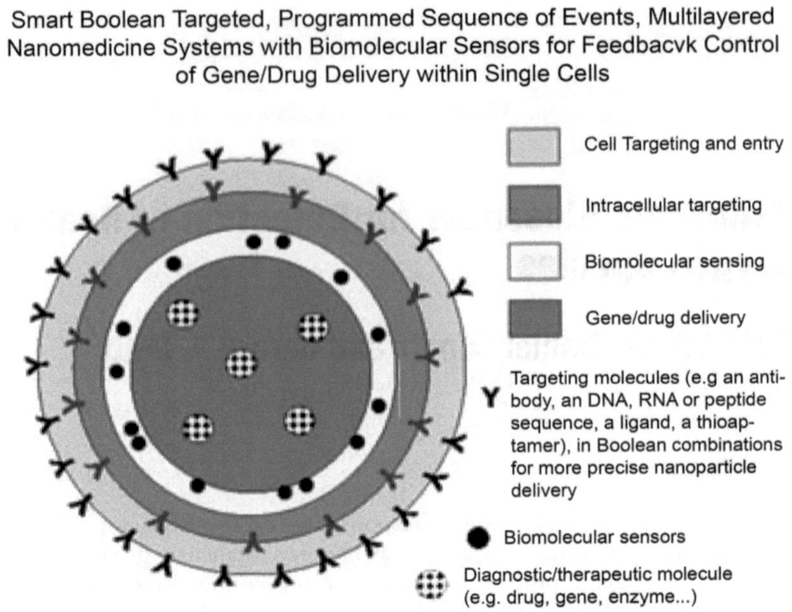

Fig. 1 Schematic image of a smart nanoparticle system that can produce an ordered series of events for drug/gene delivery (Prow et al. (2004) J Mol Histol;35:555–564)

has proved to be nontoxic and had a significant impact on stem cell [5] and immunology [6] research in recent decades. There are a wide variety of MNP products available for purchase. The relative ease of preparation and conjugation of surface molecules has led to the commercialization of MNP with fluorescent, antibody, and/or streptavidin moieties displayed on the MNP surface from a variety of manufactures. The streptavidin and amine surfaces are of direct relevance to our MNP-ROS biosensor application. These MNPs can be coupled to biotin or other labeled PCR products. We started using MNP primarily because of the safety profiles generated by years of contrast agent use. The fact that there are now a selection of sizes and surface chemistries further supports the use of MNP as in vivo research tools. We have successfully used streptavidin-coated nanoparticles (Fig. 3) for in vitro antibody and peptide targeting [7], in vitro gene delivery [8, 9], in vivo toxicology [10, 11] and in vivo biosensor delivery coupled with reactive oxygen species (ROS) monitoring [12] (Figs. 1 and 2). These studies were done with a primary focus on ocular applications, but MNP-tethered antioxidant response elements (ARE) could be used for a variety of models and offer a unique set of advantages that form the framework for this novel biosensor technology.

ROS and resultant oxidative stress has been implicated in a variety of disease states, including diabetes [13–16] and age-related macular degeneration (AMD) [17]. Brownlee has proposed that

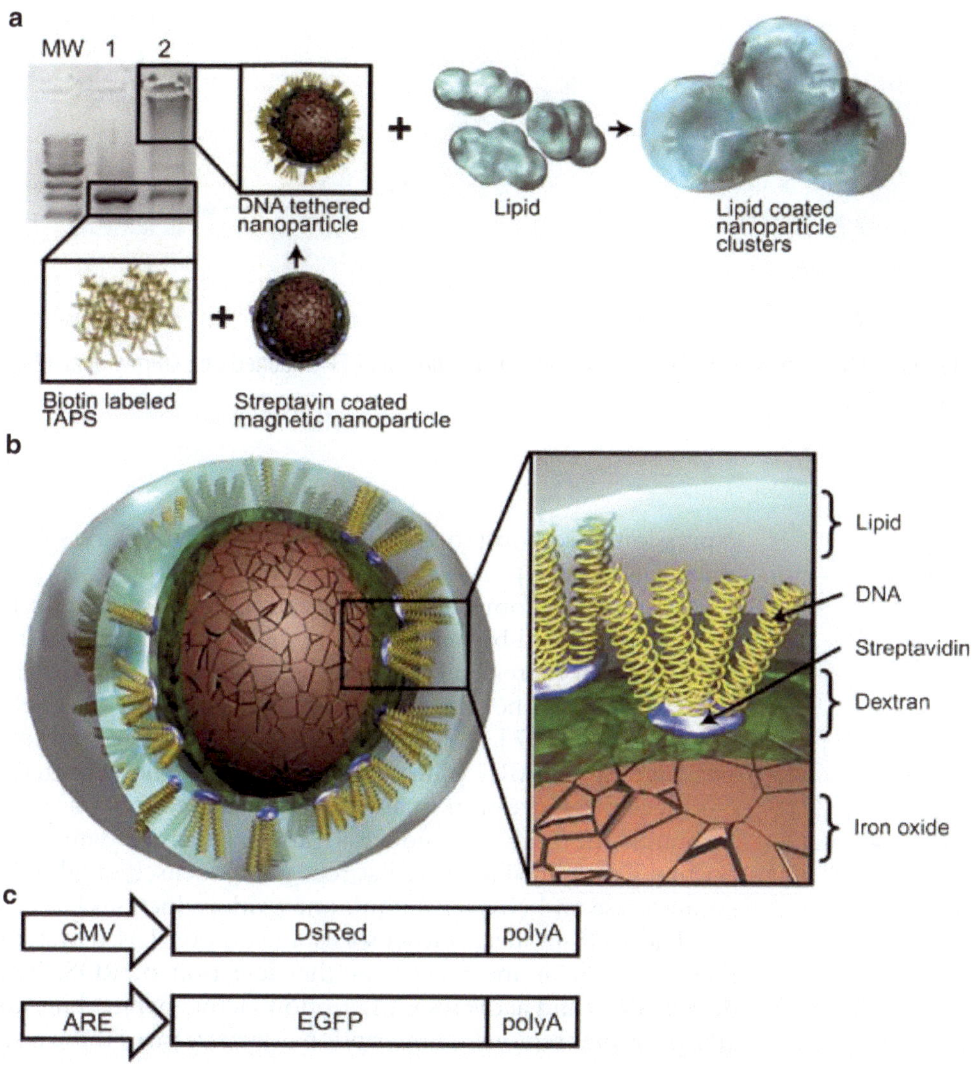

Fig. 2 (**a**) Schematic of the construction of the MNP. (**b**) The layered anatomy of a lipid coated DNA-tethered nanoparticle. (**c**) Schematics of the two DNA constructs used to assess transfection and ARE activity (Prow et al. (2006) Mol Vis;12:616–625)

the major molecular mechanism in diabetic vascular injury is hyperglycemia-induced process of overproduction of the ROS superoxide by the mitochondrial electron transport chain [18]. In AMD, oxidative stress occurs in the retinal pigment epithelium (RPE) resulting in death of RPE and ultimately photoreceptors.

Oxidative stress is an induced state whose levels are communicated through signalling pathways resulting in the induction of antioxidant genes as a protective measure in cells [19–21]. These responses are coordinated through a complex signalling network involving the ARE [22, 23]. Recently, ARE-based sensors have been developed that are capable of initiating reporter gene

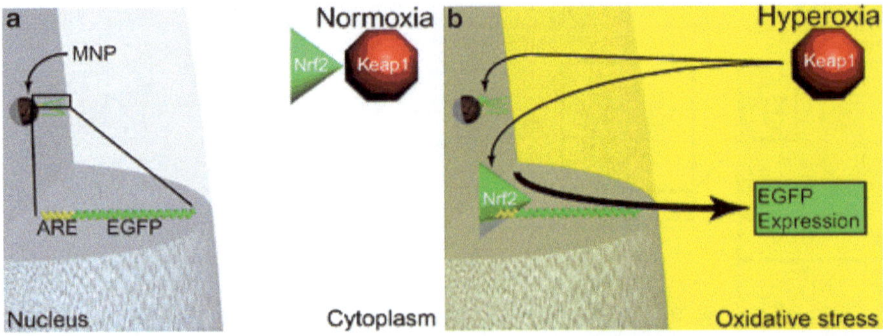

Fig. 3 Normoxia (**a**) and hyperoxia (**b**) influence on transcription of EGFP attached gene (Prow et al. (2006) Mol Vis;12:616–625)

expression in response to the presence of oxidative stress inducing chemicals [22–26].

The ARE is composed of a *cis*-acting enhancer region that is directly controlled by several factors that include the transcription factor Nrf2, a repressor Keap1, and small Maf proteins. Oxidative stress stimulates the dissociation of Keap1 protein from Nrf2 in the cytoplasm [27]. Nrf2 then undergoes nuclear translocation and binding to ARE (illustrated in Fig. 4). The activated ARE enhances the expression of any gene downstream of its sequence. Although the ARE can induce a variety of antioxidant and supporting genes, the literature has repeatedly focused on glutathione s-transferase and NAD(P)H:quinone oxidoreductases [28, 29].

The ROS-induced biosensor is a promoter-based system that relies on cellular machinery for the detection of ROS. The cell detects ROS and activates transcription factors, which bind to specific promoter regions. Transcription downstream of these response elements, in this case a fluorescent reporter gene, is then activated. This construct is composed of a number of ARE repeats followed by a minimal thymidine kinase promoter, ahead of a reporter gene, which enhanced green fluorescent protein or EGFP. The sensitivity of the biosensor is dramatically affected by the number of ARE repeats [24, 25].

The MNP-ARE described herein utilizes a fluorescent reporter protein, EGFP, and the transcriptionally active DNA construct is bound to MNP to detect the cellular response to oxidative stress (Fig. 5). We have used this ARE-tethered nanoparticle to detect oxidative stress from hyperglycemia in vitro and in RPE in the diabetic eyes of rats [12]. This biosensor system has the potential to detect oxidative stress in many disease states and could be used to induce expression of an antioxidant transgene placed in tandem with the ARE sequence when oxidative stress is sensed, i.e., provide antioxidant only when oxidative stress occurs.

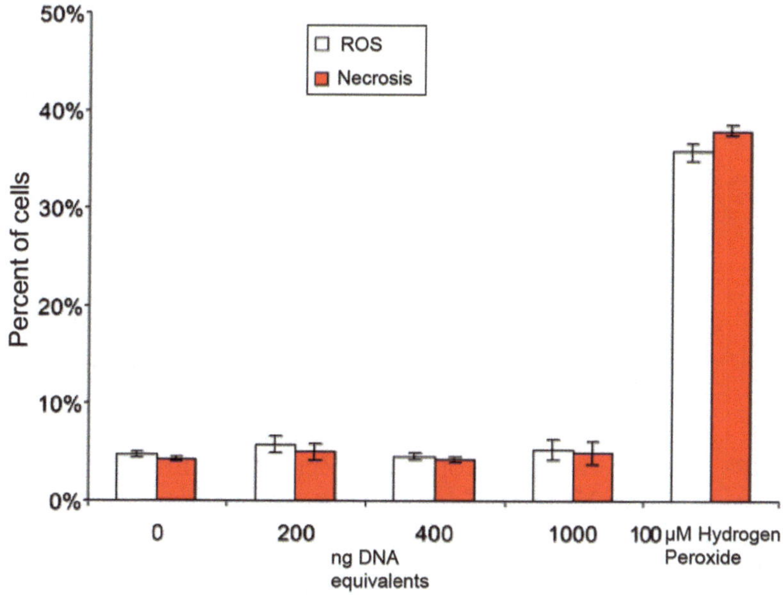

Fig. 4 Uncoated magnetic nanoparticles ability to induce ROS formation or necrosis. Hydrogen peroxide was used as a positive control (Prow et al. (2006) Mol Vis;12:606–615)

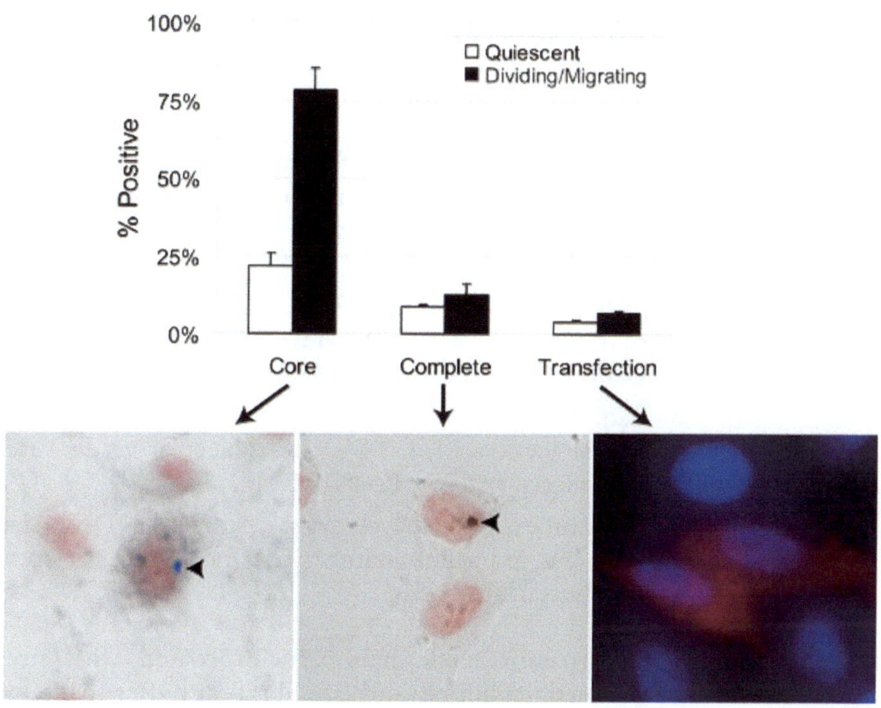

Fig. 5 The presence of magnetic nanoparticle components in transfected cells is indicated by arrowheads (Prow et al. (2006) Mol Vis;12:616–625)

2 Materials

2.1 Equipment

2.1.1 Fluorescence Microscopy

1. Imaging is easiest with a Nikon Eclipse TE2000-U (Nikon, Inc., Melville, NY) inverted fluorescence or a Zeiss 510 META (Carl Zeiss MicroImaging, Inc., Thornwood, NY) confocal microscope. Both UV and 488 nm excitation light are needed for visualizing Hoechst 33342-stained nuclei and EGFP fluorescence, respectively. Emission band-pass filters with 420–480 nm and 510–550 nm ranges will be needed for Hoechst 33342 and EGFP fluorescence imaging, respectively. Light microscopy and fluorescence images can then be overlayed.

2.1.2 Fluorescence Microscopy for ARE-GFP Assessment

1. Confocal analysis can be done with a Zeiss 510 META confocal microscope (Carl Zeiss MicroImaging, Inc., Thornwood, NY).
2. Set the excitation wavelengths to include 405 and 488 nm for Hoechst 33342 and EGFP, respectively.
3. The laser powers are set to 10 % and 5 % for the 405 nm and 488 nm lasers, respectively.
4. Set the emission filters to 420–480 nm emission bandpass for Hoechst 33342 and 510–550 nm emission bandpass for EGFP.
5. The objective can be set to 5, 10, 20, 40, 63, or 100×.
6. Take images in the plane of the nuclear fluorescence from the Hoechst 33342 in order to assure that the images contain cellular bodies, *see* **Note 3**.

2.2 PCR Reagents

1. Oligonucleotide primers for the PCR amplification of the ARE promoter and the EGFP reporter with a biotin label on the 5′ terminus of the forward primer (Integrated DNA Technologies, Coralville, Iowa, USA).
2. pARE-EGFP for the PCR template (William Fahl, University of Wisconsin, Madison, WI, USA and Ming Zhu, Arizona Cancer Center, Tuscon, AZ, USA).
3. Taq, PCR butter, dNTPs, agarose, ethidium bromide, Tris, glacial acetic acid, EDTA, and molecular grade water (Sigma Chemicals, Inc., St. Louis, MO).
4. PCR tubes, thermocycler, gel casting tray, DNA size marker and UV transilluminator (Thermo Fisher Scientific, Inc., Pittsburgh, PA, USA).

2.3 Magnetic Nanoparticles

1. Superparamagnetic nanoparticles coated with streptavidin, magnetic columns, and equilibration buffer (Miltenyi Biotec, Auburn, CA).

2.4 Cell Culture, Transfection, and Fixation Reagents

1. Opti-MEM®, Lipofectamine™ 2000, trypsin, DMEM/F12, fetal bovine serum (FBS) from Invitrogen, Carlsbad, CA, USA.
2. Tissue culture flasks and chamber slides (Thermo Fisher Scientific, Inc., Pittsburgh, PA, USA).
3. Hoechst 33342 is a nuclear counterstain dye (Sigma-Aldrich, St. Louis, MO, USA).
4. Paraformaldehyde, Triton-X, and Tris-buffered saline (TBS) are used to fix, permeabilize, and rinse the treated cells (Sigma-Aldrich, St. Louis, MO, USA).

2.5 Prussian Blue Staining Reagents

Hydrochloric acid and potassium ferrocyanide can both be purchased from Sigma-Aldrich, St. Louis, MO, USA.

3 Methods

3.1 Preparation of Biotin-Labeled Transcriptionally Active PCR Products

1. Combine the following components on ice to make a master PCR mix:
   (a) Forward primer (5′-TAG TTA TTA ATA GTA ATC-3′) at 20 μM or 20 μL.
   (b) Reverse primer (5′-TAC ATT GAT GAG TTT GGA-3′) at 20 μM or 20 μL.
   (c) Template at 10 ng/μL 10 μL.
   (d) Taq at 1.5 units per reaction × 20 reactions 10 μL.
   (e) 10× buffer 100 μL.
   (f) dNTPs 20 μL.
   (g) H_2O 820 μL.
2. Mix the master mix thoroughly and then aliquot the mix into 9 PCR tubes at 100 μL per tube and the balance in a final tube.
3. Cap tubes and put them into the thermocycler to run the PCR.
4. PCR temperature profile:
   (a) Denaturation at 94 °C for 2 min.
   (b) Denaturation at 94 °C for 30 s (35 cycles of **steps b–d**).
   (c) Annealing at 53 °C for 30 s.
   (d) Extension at 72 °C for 2 min.
   (e) Final extension at 72 °C for 6 min.
   (f) Store at 4 °C until ready to run the gel.
5. Weigh out agarose to make a 1 % gel in a total volume of 25 mL (0.25 g) in Tris-acetic acid-EDTA buffer (TAE). Add 25 mL of buffer and heat agarose until dissolved. Let the gel sit until the gel cools slightly. Add ethidium bromide to a final concentration of 0.5 μg/mL and pour the gel.
6. Set up gel in the gel casting apparatus. Place the comb.

7. Pour the gel into the gel tray and ensure there are no air bubbles in the gel or on the comb.
8. Allow to cool (about 10 min), then fill the electrophoresis chamber with TAE.
9. Mix 5 μL of PCR product with 2 μL of gel loading dye.
10. Load 5 μL of PCR product.
11. Load 5 μL of DNA 1 kb ladder.
12. Run gel at 100 V for 30 min.
13. Visualize bands with a UV transilluminator. The expected product size is 2 kb.

3.2 DNA-Tethered Nanoparticle Construction

1. Quantitate the amount of purified PCR product you purified on the previous day using a spectrophotometer.
2. Incubate the magnetic nanoparticles with the biotin-labeled PCR fragments at ~30 ng of PCR product per 1 μL of nanoparticle solution. Use 1 μg of DNA combined with 33 μL of magnetic nanoparticles.
3. Incubate the mixture at room temperature for 30 min.
4. During this time, prepare the magnetic column by equilibrating with 50 μL of equilibration buffer. Add all solutions directly to the center of the column.
5. Wash the column once with 100 μL Opti-MEM®.
6. Load the column with the DNA/nanoparticle mixture.
7. Wash the column three times with 100 μL Opti-MEM®.
8. Elute the nanoparticles by removing the column from the magnet and adding 100 μL of Opti-MEM®. Allow the column to gravity drain into a sterile tube.
9. Coat the DNA-labeled magnetic nanoparticles with lipid (Lipofectamine™ 2000).
   (a) Add 1 μL of Lipofectamine™ 2000 to 100 μL of Opti-MEM®. Gently mix by tapping the side of the tube. Do not vortex. Let the solution stand at room temperature for 5 min, *see* **Note 1**.
   (b) Add the eluted nanoparticles drop by drop to the center of the Opti-MEM® Lipofectamine™ 2000 solution. Gently mix by tapping the side of the tube. Let the solution stand at room temperature for 20 min before transfection onto cell culture (4-well chamber slide).

3.3 Cell Culture for In Vitro Transfection

1. Remove the media from the flask and wash the cells with PBS. Remove the PBS and discard.
2. Add 0.5 mL of trypsin to a T25 flask of cells. Place at 37 °C for 5 min.

3. Dislodge cell with gentle tapping on the side of the flask.
4. Quench the trypsin with 5 mL of DMEM/F12 10 % FBS.
5. Resuspend the cells by gentle pipetting up and down until the mixture is homogenous.
6. Determine cell density with a Neubauer chamber.
7. Prepare a 10 mL cell suspension of 40,000 cells/mL in DMEM/F12 w/10 % FBS.
8. Add 1 mL of cell suspension in a random pattern to each well in a 4-well chamber slide, for example.

3.4 *In Vitro* Transfection

1. Check the chamber slides to ensure the cells are viable and healthy.
2. Label chamber slides with the following:
   (a) Negative control—Opti-MEM® only + Lipofectamine™ 2000.
   (b) Plasmid control—Plasmid only + Lipofectamine™ 2000.
   (c) PCR control—purified biotin-labeled PCR product + Lipofectamine™ 2000.
   (d) Complete magnetic nanoparticles + Lipofectamine™ 2000.
3. Add 400 μL of Opti-MEM® to a microfuge tube plus 4 μL of Lipofectamine™ 2000. Gently mix by tapping the side of the tube. Do not vortex. Let the tube stand at room temperature for 5 min.
4. Prepare transfection solutions for the four wells in sterile tubes:
   (a) 100 μL of Opti-MEM®.
   (b) 100 μL of Plasmid (already diluted so the final concentration added to cells will be 1 μg).
   (c) Dilute purified PCR product in 100 μL, for a total of 1 μg, of Opti-MEM®.
5. After a 5 min incubation, add 100 μL of the Opti-MEM®/Lipofectamine™ 2000 solution to each of the transfection solutions, **steps 2a–c**. Gently mix by tapping the tube. Add the complete MNP solution from Subheading 3.2 to the well for the **step 2d** group.
6. Remove media from all wells of the chamber slide.
7. Wash with 1 mL of Opti-MEM® to remove any traces of antibiotics from the media.
8. Add 200 μL of each transfection solution to the corresponding wells and incubate the chamber slides at 37 °C for 1 h. Gently shake the chamber slides every 15 min for 1 h.
9. Remove the transfection solutions and wash the cells with 1 mL of Opti-MEM®.

10. Remove Opti-MEM® and replace with 1 mL of complete media (DMEM/F12 with 10 % FBS).
11. Incubate chamber slides at 37 °C overnight.

3.5 Prussian Blue Staining to Visualize Iron Nanoparticles

Prussian blue staining is used to detect cells that contain magnetic nanoparticles. This staining protocol detects nanoscale iron oxide crystals particles. Therefore, this procedure does not imply that the cells contain complete magnetic particles, just the iron oxide core of the nanoparticle.

1. The cells are fixed and permeabilized using 2 % paraformaldehyde and permeabilized with 0.25 % Triton-X.
2. Remove media from all wells of the chamber slide.
3. Add 0.5 mL of 2 % paraformaldehyde in Tris-buffered saline pH 7 (TBS) for 10 min at room temperature.
4. Remove paraformaldehyde and discard.
5. Permeabilize cells by the addition of 0.5 mL of 0.25 % Triton-X 100 in TBS for 10 min at room temperature.
6. Remove and discard Triton solution.
7. Counterstain cells with a 20,000 dilution (i.e., 1 μL in 20 mL of PBS) of Hoechst 33342. Add 0.5 mL per well and incubate for 15 min in the dark at room temperature.
8. Remove Hoechst stain and discard. Wash cells with 0.5 mL of PBS, three times.
9. After the final wash, leave the PBS in the wells.
10. Make a fresh staining solution containing equal parts of 20 % hydrochloric acid and 10 % potassium ferrocyanide.
11. Remove the PBS from the cells and add 0.5 mL of the staining solution. Incubate for 20 min at room temperature.
12. Wash cells three times with 0.5 mL of PBS.
13. Leave the PBS in the well and visualize cells, *see* **Note 2**.

4 Notes

1. The amount of lipid needs to be balanced by the toxicity profile in the cells or tissue and the potential for increases in transfection efficiency.
2. Positive reactions generally occur promptly and look like blue spots within the cell. These blue spots are the result of the ferric iron in the nanoparticle combined with the potassium ferrocyanide to form ferric-ferrocyanide, which has a bright blue color. Viewing the cells or tissue section with a bright light for prolonged time can cause a false-positive reaction.

3. The zoom, objective, brightness, contrast, laser power, and gain must remain the same for all groups and images to enable comparisons to be made.

5 Conclusions

The use of a synthetic signalling pathway to noninvasively study the progression of disease and therapeutic interventions could facilitate the development of better therapeutics for a number of diseases. There are a large number of diseases with ROS components, yet studying ROS during disease progression has remained a significant challenge. We developed a magnetic nanoparticle biosensor that can report cellular responses to oxidative stress via fluorescent proteins. This technology enables researchers to monitor, in real time, ROS responses with noninvasive microscopy. This is useful, for example, in monitoring microvascular diseases with ROS components, e.g., diabetic retinopathy (*see* Figs. 4 and 5). Microvascular disease is implicated in the etiology of diabetic complications in most organ systems. It is unclear if loss in viable microvasculature is due to toxicity of the hyperglycemic state or actual occlusive events caused circulating cells or substances. It is likely that both toxicity and occlusions play a role in diabetic complications and these effects are initiated through the metabolic hallmark of diabetes, hyperglycemia. Hyperglycemia is a cause of pancreatic beta cell loss and also activates peripheral blood mononuclear cells (PBMC) and endothelial cells. At the source of these pathogenic processes is hyperglycemia-induced mitochondrial oxidative stress (MOS). To understand how we can treat MOS in diabetes, we must better understand the cellular responses to the hyperglycemic state and define the best strategies for the treatment of MOS. Using systems like our nanobiosensor for ROS can enable researchers to ask the following questions in the context of animal models of diabetes: How does MOS affect the major antioxidant promoters? What are the best mitochondrial-targeted antioxidant therapies for treating MOS? Can synthetic or natural antioxidant promoters that are sensitive to MOS be used as biosensors to switch on and off the expression of mitochondrial-targeted antioxidant enzymes?

References

1. Lin L, Grice JE, Butler MK et al (2011) Time-correlated single photon counting for simultaneous monitoring of zinc oxide nanoparticles and NAD(P)H in intact and barrier disrupted volunteer skin. Pharm Res 28(11):2920–2930
2. Monteiro-Riviere NA, Wiench K, Landsiedel R, Schulte S, Inman AO, Riviere JE (2011) Safety evaluation of sunscreen formulations containing titanium dioxide and zinc oxide nanoparticles in UVB sunburned skin: an in vitro and in vivo study. Toxicol Sci 123(1):264–280
3. Violante MR (1990) Potential of microparticles for diagnostic tracer imaging. Acta Radiol Suppl 374:153–156

4. Bernd H, De Kerviler E, Gaillard S, Bonnemain B (2009) Safety and tolerability of ultrasmall superparamagnetic iron oxide contrast agent: comprehensive analysis of a clinical development program. Invest Radiol 44:336–342
5. Mahmoudi M, Hosseinkhani H, Hosseinkhani M et al (2011) Magnetic resonance imaging tracking of stem cells in vivo using iron oxide nanoparticles as a tool for the advancement of clinical regenerative medicine. Chem Rev 111:253–280
6. Grutzkau A, Radbruch A (2010) Small but mighty: how the MACS-technology based on nanosized superparamagnetic particles has helped to analyze the immune system within the last 20 years. Cytometry A 77:643–647
7. Prow TW, Kotov NA, Lvov YM, Rijnbrand R, Leary JF (2004) Nanoparticles, molecular biosensors, and multispectral confocal microscopy. J Mol Histol 35:555–564
8. Prow T, Grebe R, Merges C et al (2006) Nanoparticle tethered antioxidant response element as a biosensor for oxygen induced toxicity in retinal endothelial cells. Mol Vis 12: 616–625
9. Prow T, Smith JN, Grebe R et al (2006) Construction, gene delivery, and expression of DNA tethered nanoparticles. Mol Vis 12: 606–615
10. Prow TW (2010) Toxicity of nanomaterials to the eye. Wiley Interdiscip Rev Nanomed Nanobiotechnol 2:317–333
11. Prow TW, Bhutto I, Kim SY et al (2008) Ocular nanoparticle toxicity and transfection of the retina and retinal pigment epithelium. Nanomedicine 4:340–349
12. Prow TW, Bhutto I, Grebe R et al (2008) Nanoparticle-delivered biosensor for reactive oxygen species in diabetes. Vision Res 48:478–485
13. Bergua A, Mayer B, Neuhuber WL (1996) Nitrergic and VIPergic neurons in the choroid and ciliary ganglion of the duck Anis carina. Anat Embryol (Berl) 193:239–248
14. Brownlee M (2003) A radical explanation for glucose-induced beta cell dysfunction. J Clin Invest 112:1788–1790
15. Giardino I, Edelstein D, Brownlee M (1996) BCL-2 expression or antioxidants prevent hyperglycemia-induced formation of intracellular advanced glycation endproducts in bovine endothelial cells. J Clin Invest 97:1422–1428
16. Wiernsperger NF (2003) Oxidative stress as a therapeutic target in diabetes: revisiting the controversy. Diabetes Metab 29:579–585
17. Beatty S, Koh H, Phil M, Henson D, Boulton M (2000) The role of oxidative stress in the pathogenesis of age-related macular degeneration. Surv Ophthalmol 45:115–134
18. Brownlee M (2001) Biochemistry and molecular cell biology of diabetic complications. Nature 414:813–820
19. Cory JG, Szentivanyi A (1987) Cancer biology and therapeutics. Plenum, New York
20. Prochaska HJ, De Long MJ, Talalay P (1985) On the mechanisms of induction of cancer-protective enzymes: a unifying proposal. Proc Natl Acad Sci USA 82:8232–8236
21. Prochaska HJ, Talalay P (1988) Regulatory mechanisms of monofunctional and bifunctional anticarcinogenic enzyme inducers in murine liver. Cancer Res 48:4776–4782
22. Friling RS, Bergelson S, Daniel V (1992) Two adjacent AP-1-like binding sites form the electrophile-responsive element of the murine glutathione S-transferase Ya subunit gene. Proc Natl Acad Sci USA 89:668–672
23. Wasserman WW, Fahl WE (1997) Functional antioxidant responsive elements. Proc Natl Acad Sci USA 94:5361–5366
24. Zhu M, Fahl WE (2000) Development of a green fluorescent protein microplate assay for the screening of chemopreventive agents. Anal Biochem 287:210–217
25. Zhu M, Fahl WE (2001) Functional characterization of transcription regulators that interact with the electrophile response element. Biochem Biophys Res Commun 289: 212–219
26. Zhu M, Chapman WG, Oberley MJ, Wasserman WW, Fahl WE (2001) Polymorphic electrophile response elements in the mouse glutathione S-transferase GSTa1 gene that confer increased induction. Cancer Lett 164: 113–118
27. Dinkova-Kostova AT, Holtzclaw WD, Cole RN et al (2002) Direct evidence that sulfhydryl groups of Keap1 are the sensors regulating induction of phase 2 enzymes that protect against carcinogens and oxidants. Proc Natl Acad Sci USA 99:11908–11913
28. Day RM, Suzuki YJ, Fanburg BL (2003) Regulation of glutathione by oxidative stress in bovine pulmonary artery endothelial cells. Antioxid Redox Signal 5:699–704
29. Jaiswal AK (2000) Regulation of genes encoding NAD(P)H:quinone oxidoreductases. Free Radic Biol Med 29:254–262

Chapter 2

Nanoparticles: Toxicity, Radicals, Electron Transfer, and Antioxidants

Peter Kovacic and Ratnasamy Somanathan

Abstract

In recent years, nanoparticles have received increasing attention in research and technology, including a variety of practical applications. The bioactivity appears to be related to the small particle size, in addition to inherent chemical activity as electron transfer (ET) agents, generators of reactive oxygen species (ROS) with subsequent oxidative stress (OS), and as antioxidants (AOs). The mechanism of toxicity, therapeutic action, and AO property is addressed based on the ET-ROS-OS approach. There are several main classes of ET functionalities, namely, quinones (or phenolic precursors), metal compounds, aromatic nitro compounds (or reduction products), and imine or iminium species. Most of the nanospecies fall within the metal category. Cell signaling is also discussed. This review discusses recent developments based on ET-ROS-OS-AO framework.

Key words Nanoparticles, Toxicants, Antioxidants, Therapeutics, Electron transfer, Reactive oxygen species, Mechanisms

1 Introduction

Nanotechnology involves matter on an atomic and molecular scale at dimensions between 1 and 100 nm. Nanoparticles are chemically highly active, in part, because of their large surface area proportional to their volume. This causes formation of agglomerates in which particles are held together by relatively weak forces, including van der Waals forces, electrostatic forces, and surface tension. Nanoparticles may also form a group of strongly associated particles called aggregates. Nanoparticles less than 30 nm have markedly altered properties and are often referred to as "quantum dots" because their size controls the separation of energy levels within them, behaving like a semiconductor. Nanoscale materials find use in a variety of different areas, such as electronic, magnetic, optoelectronic, biomedical, pharmaceutical, cosmetic, energy, environmental, catalytic, and materials

Donald Armstrong and Dhruba J. Bharali (eds.), *Oxidative Stress and Nanotechnology: Methods and Protocols*, Methods in Molecular Biology, vol. 1028, DOI 10.1007/978-1-62703-475-3_2,

applications. Nanotechnology is considered by many as the next step in science, integrating engineering with biology, chemistry, and physics. The potential of this technology has triggered tremendous investigation in the academic and industrial worlds in recent years. The technology has moved from an exotic research pursuit to inclusion in hundreds of mainstream consumer products with nanomaterial markets exceeding billions of dollars annually. Many medical innovations, imaging, drug delivery, nanotherapeutics, and nanophase formulations are in clinical trials. With this nanophase invasion of new materials and products into every aspect of life comes increasing safety and exposure risk. A recent comprehensive review puts focus on biomechanisms [1]. It is recognized that nanoparticles produce reactive oxygen species (ROS) inside and outside the cell and are the key factor in toxicological effects.

Electron transfer (ET) is probably the most important process in chemical transformations. Large impetus was provided to the area by Marcus theory. The preponderance of bioactive substances or their metabolites incorporate ET functionalities, which, we believe, play an important role in physiological responses. The main groups include quinones (or phenolic precursors), metal complexes (or complexors), aromatic nitro compounds (or reduced hydroxylamine and nitroso derivatives), and conjugated imines (or iminium species). In vivo redox cycling with oxygen can occur, giving rise to oxidative stress (OS) through generation of ROS, such as hydrogen peroxide, hydroperoxides, alkyl peroxides, and diverse radicals (hydroxyl, alkoxyl, hydroperoxyl, and superoxide). In some cases, ET results in interference with normal electrical effects (e.g., in respiration or neurochemistry). Generally, active entities possessing ET groups display reduction potentials in the physiologically responsive range (i.e. more positive than −0.5 V). ET, ROS, and OS have been increasingly implicated in the mode of action of drugs and toxins, e.g., anti-infective agents [2], anticancer drugs [3], carcinogens [4], reproductive toxins [5], nephrotoxins [6], hepatotoxins [7], cardiovascular toxins [8], nerve toxins [9], mitochondrial toxins [10], abused drugs [11], ototoxins [12], immunotoxins [13], eye toxins [14], pulmonary toxins [15], dermal toxins [16], gastrointestinal toxins [17], thyroid toxins [18], and various other categories [19].

There is a plethora of experimental evidence supporting the ET-ROS theoretical framework, including generation of the common ROS, lipid peroxidation, degradation products of oxidation, depletion of AOs, effect of exogenous AOs, and DNA oxidation and cleavage products, as well as electrochemical data. This comprehensive, unifying mechanism is in keeping with the frequent observations that many ET substances display a variety of activities (e.g., multiple-drug properties) as well as toxic effects.

This review, an update of a prior one [1], provides a unifying mechanism for the physiological action of nanoparticles, involving toxicity, antioxidant effects, and therapeutic use. However, it is important to recognize that biological action is often complex, entailing a variety of factors. Literature references are mainly representative with original ones present in prior reviews in some cases.

1.1 Carbon Nanotubes

Carbon nanotubes (CNTs) have been the object of much attention for application in tissue engineering scaffolds [20]. Although they possess promising properties, they are associated with hazardous effects in human health. For example, CNTs are able to produce toxicity in lungs; the carcinogenicity is ambiguous. Adverse effects on various other organs are not definitively established. More studies are needed to obtain clearer insight, including mechanistic aspects. A recent review deals with mechanisms underlying the toxicity of CNTs [21]. These materials can exert an oxidative response that stimulates inflammatory, genotoxic, and cytotoxic consequences. A common mechanism may be responsible for the toxicity, despite the fact that CNTs are a diverse population of materials. Authors of the present review suggest that ROS-OS may play an important role in the common mechanism. Implications of similarity in the structure of asbestos to that of CNTs are discussed. An evaluation was made of the toxic potential of CNTs in relation to various variables [22]. Genotoxicity might be induced through OS or inflammation. At the present, there is no global agreement about the risk of CNTs on human health [23]. Evidence indicates that CNTs and asbestos share comparable physical properties. Hence, CNTs might display toxic effects, and the extent is specifically directed to lung and pleura. These effects are related to various properties, including extent of oxidation. An overview of toxicity issues takes into consideration the well-known asbestos-induced mechanism of biodamage.

Investigations are reported involving multi-walled carbon nanotubes (MWCNTs) and single-walled carbon nanotubes (SWCNTs). Apoptotic changes in human lung epithelium cells were studied with MWCNTs [24]. ROS and malondialdehyde, a product of OS, were produced, together with significant decrease in the activity of AO agents, catalase, and GSH. Apoptosis was accompanied by OS. Recent reports suggest that ROS induced in mammalian cells exposed to MWCNTs could mediate the cytotoxicity [25]. Superoxide levels were significantly increased in blood macrophages after treatment by the nanotubes. Activation of NADPH oxidases may contribute to generation of ROS. Rats were exposed to MWCNTs with focus on respiratory tract and systemic toxicity [26]. An etiopathological sequence of inflammatory events resulted. A review addresses involvement of ROS-OS in inflammation [27]. Rat lungs were injected with MWCNTs [28]. The exposure produced a significant dose-dependent reduction of blood AO capacity (GSH, SOD, and catalase) and increased lipid peroxidation product, namely, malondialdehyde. Induction of OS was achieved by reduction of total AO capacity. The findings point to possible occupational health hazard in chronic exposures. Toxicity and mechanism were investigated with human embryonic kidney cells treated with MWCNTs [29]. Findings showed decreased GSH levels and increased TBARS, both of which result in pathogenicity in the lung [30]. Asbestos is known to induce ROS-OS [31].

The literature also contains material on SWCNTs. This type has been proposed for various medical applications [32]. However, the safety aspect has not yet been fully demonstrated. In vitro experiments have pointed to OS as a mechanism involved in their cytotoxicity. A report examined the effects of pulmonary exposure to SWCNTs [33]. The nanotubes can exacerbate murine allergic airway inflammation via increased OS. This class of NTs with its unique physicochemical and mechanical properties has many potential new applications in medicine and industry [34]. There has been concern regarding toxicity, pathogenicity, and biopersistance. Exposure of mesothelial cells to SWCNTs induced ROS generation, enhanced DNA damage and MAPKs activation, and increased cell death.

Mice exposed to MWCNATs induced cytokine, chemokine, and proteases production that promoted the inflammatory and fibric changes in the lung [35]. Inhalation was more effective than aspiration in causing inflammatory response, OS, collagen deposition, and fibrosis as well mutation of K-ras gene in the lung tissue of mice exposed to SWCNTs [36]. A similar study with mice showed that inhalation of nanotubes induces inflammation, fibrosis, alteration of oxidant and AO levels, and induction of apoptosis-related proteins in the lung tissues, triggering cell death [37]. A report showed OS and DNA damage responses in rat and mouse lung to inhaled carbon nanoparticles [38].

Topical exposure of mice skin to SWCNTs caused OS, depletion of glutathione, oxidation of protein thiols and carbonyls, elevated myloperoxidase activity, increase of dermal cell numbers, and skin thickening, resulting from accumulation of polymorphonuclear leukocyte and mast cells [39].

1.2 Engineered Nanoparticles

The chemical and catalytic activity of nanoparticles have strongly contributed to the current, tremendous interest in engineered nanomaterials and often serve as a guiding principle for the design of functional materials [40]. Since these active materials can enter into cells or organisms, a study was made of the level of intracellular oxidation after exposure to Fe, Co, Mn, and titania containing silica nanoparticles and the corresponding oxides in vitro. The resulting OS was measured as the release of ROS. Results indicate that the particles can enter the cells efficiently by a Trojan-horse-type mechanism which provoked an up to eight times OS in the case of Co and Mn as compared to controls. A systematic investigation of Fe-containing nanoparticles, as used in industry, showed that the catalytic activity could strongly alter the damaging action of a nanomaterial. The results indicate that a risk assessment of nanomaterials should consider chemical and catalytic properties beyond a mere focus on physical properties, such as size, shape, and degree of agglomeration. OS is known to play important roles in engineered nanomaterial-induced cellular toxicity [41]. However, the proteins and signaling pathways associated with engineered nanomaterial-mediated OS and toxicity are largely unknown. In order to identify the toxicity pathways and networks associated with the engineered materials, a proteomic study was conducted using human bronchial cells. The phosphorylation profiles of upstream signaling pathways were analyzed in response to the TiO_2 treatment and the induced OS. The study provided novel insights into the biological responses and the potential toxicity and detoxification pathways of TiO_2. Engineered nano-sized Cu oxide particles are extensively used in diverse applications [42]. Because aquatic environments are the ultimate "sink" for all contaminants, it is expected that nanoparticles will follow the same fate [42]. Two marine invertebrates were chosen as ecotoxicological models. Biomarkers of OS were enhanced in both species. Newly emerged scientific areas of research involve interactions between NPs and biomolecules in body fluids, cellular microenvironment, intracellular components, or secreted cellular proteins, such as cytokines, growth factors, and enzymes, and use of engineered NPs to target various signal transduction pathways in cancer therapy [43]. OS has largely been implicated in NP-induced toxicity. It could activate a wide variety of cellular events, such as cell cycle arrest, apoptosis, inflammation, and induction of AO enzymes. These responses occur after the activation of different cellular pathways. In this context, these groups of MAP kinase cascades (ERK, p38 protein kinase, and JNK), as well as redox-sensitive transcription factors, were investigated. The ability of NPs to interact with these signaling pathways could potentially explain their cytotoxicity. The induction of apoptosis is also closely related to the modulation of signaling pathways induced by NPs. Evidence supports involvement of ROS-OS in apoptosis [44]. Metal-containing engineered NPs, an important group, are often made of one metal, e.g., Cu or Ag, metal oxides, e.g., ZnO or TiO_2, or metal composites [45]. Evidence suggests that nanometals can cause a range of sublethal effects in various organs involving common responses, including OS. The nano forms of some metals may be more toxic to embryo or juveniles than the corresponding compound. Genotoxicity of some engineered nanomaterials, e.g., metal oxides, carbon black, or nanotubes, has commonly been related to OS and subsequent inflammation [46]. Polymer nanomaterials have raised concern because of their toxic responses. As common engineered nanomaterials, TiO_2 nanoparticles are usually perceived as nontoxic and are widely used in many products and applications [47]. However, recent information indicates a serious risk of reproductive impairment of environments contaminated with low levels of the metal oxide. Reviews document involvement of ET-ROS-OS in reproductive toxicity [5, 48–50]. Pulmonary exposure to engineered NPs generates vasomotor dysfunction [51]. Small doses are associated with a small increase in risk of cardiovascular disease, possibly through OS and inflammatory pathways. To unravel toxicity mechanisms, physicochemical properties of engineered NPs relevant to toxicity, e.g., of genes, need to be assessed, typically involving the release or quenching of ROS [52].

1.3 Combustion

Ultrafine particles or nanoparticles are emerging as the most abundant species of this class in urban and industrial areas [53]. Their exposure has increased dramatically from sources such as the internal combustion engine, power plants, incinerates, and other types of thermo-degradation. The current scientific literature is examined concerning the health effects. Combustion-derived and manufactured nanoparticles are known to provoke responses in human lung cells [54]. Thus, they play an important role in the development of adverse health effects. The production of ROS was induced by all particle types in most cell cultures. The interplay of different cell types seems to substantially modulate the OS and the inflammatory responses after exposure to nanoparticles.

Relevant material is also presented in other sections.

1.4 Diesel Exhaust Particulate

Studies with animal models indicate that various types of particles, including those from combustion processes, such as from diesel oil, produce elevated levels of lipid peroxidation products and oxidatively damaged DNA [55]. Particles of this type generate ROS, deplete AOs, alter mitochondrial function, and produce oxidative damage to lipids and DNA. Results indicate that OS and elevated levels of oxidatively altered biomolecules are important intermediates that may be useful markers in hazard characterization of particulates. In another study, DEPs were found to promote both skin and gut cell damage [56]. Researchers determined the effects of administration of DEP on blood pressure, inflammation, oxidative stress, and morphological alterations of organs [57]. DEP caused acute effects involving inflammation and OS. Toxicity mainly was directed toward lung tissue. Ultrafine particles from diesel vehicle emissions induce vascular inflammatory responses [58]. Epidemiological evidence supports the association of ambient particulate matter with cardiovascular diseases, such as atherosclerosis. These particles gave rise to OS via c-Jun N-terminal kinases activation in endothelial cells.

1.5 Asbestos

A report deals with the iron-related molecular toxicity mechanism of synthetic asbestos nanofibers [59]. Asbestos shares with CNTs some morphological and physicochemical features. An asbestos-like behavior for CNTs has been reported. The effect of iron loading on asbestos toxicity was examined. A linear relationship between the formation of radicals and the amount of iron sites was found. Even the smallest iron contamination imparts radical reactivity, hence toxicity. Different conventional pathogenic particle types, including asbestos and quartz, caused diverse pathological effects which have been reported to cause very similar OS effects in cultured cells [60]. Likely explanation and implications are discussed. An earlier review addresses the mechanistic features [4]. Evidence strongly supports involvement of ROS-OS. The content of iron is directly related to the fibers ability to generate ROS. Redox cycling of the surface metal seems to be occurring. Copper may play a similar role. Asbestos-induced formation of ROS also causes lipid peroxidation. Formation of 8-OHdG, attributed to ROS, has also been noted. Cells with the genotype for speedy restoration of the AO GSH were nearly immune to the toxic effects of ROS.

1.6 Silica

Quartz dust (silica) has also been cited as a source of cancer [61]. Silica damages tissues and DNA by creating ROS.

Silica nanoparticles are being increasingly used in diagnosis, imaging, and drug delivery for the central nervous system. Little is known concerning the potential adverse effects on the brain associated with exposure to silica nanoparticles. A report showed that these nanoparticles entered the brain via intranasal instillation and deposited in the striatum inducing oxidative damage, increased inflammatory response in the striatum, and apoptosis [62].

Workers exposed to silica nanoparticles showed damage mainly in macrophages, pulmonary microvessels, vascular endothelial cells, microlymphatic vessels, pleural effusions, and in the alveolar epithelial cells. These silica nanoparticles may have contributed in part to the illness reported in these workers through ROS and OS [63]. Positively charged silica nanoparticles proved to be more cytotoxic in reducing mitochondrial metabolic activity and effects on phagocytosis than neutral silica nanoparticles [64]. Human bronchial epithelial cells exposed to silica nanoparticles showed toxicity via OS [65]. OS contributes to silica nanoparticle-induced cytotoxicity in human embryonic kidney cells [66].

The study of nanotechnology has emerged in the last few decades due to advances in the field of engineering, which has facilitated rapid and mass production of engineered nanoparticles (NP). In addition, ultrafine particles include NPs that are components of emission from combustion and dust-generating industrial processes. These particles, defined as having at least one dimension in the nanoscale, are so small that they can easily enter or diffuse through pores of membranes. There are three main routes of entry of the NPs into the body. First, NPs are inhaled into the body from atmosphere air via the upper respiratory tract. Oral ingestion and entry via the dermal route absorption through skin pores are other methods of entry [67].The human lung is a vulnerable organ for NP invasion as it offers a large surface area, which is in contact with the environmental atmosphere. The large surface area leads to increased possibility for the formation of free radicals, such as ROS, superoxide anions, and hydroxyl radicals, which consequently lead to OS. There is increasing evidence linking the NPs with human health problems. It has been shown that inhaled NPs possess the potential to aggravate existing respiratory disorders, such as asthma or bronchitis [68]. Translocation of NPs from lungs towards other organs has been demonstrated and possible consequences include inflammation, heart rate, function anomalies, homeostatic disturbances, and OS [68].

2 Metal Compounds

In relation to the metal category, reduction by the metal compound comprises electron donation, whereas oxidation consists of electron acceptance. Certain valences of metal compounds could assume both roles. Conditions play an important role.

2.1 Titanium Dioxide (TiO_2)

Titanium dioxide NPs have increasingly been used in the fields of paints, waste water treatment, sterilization, cosmetics, food additives, biomedical ceramic, and implanted biomaterials. Exposure of human epithelial tissues to MWCNTs resulted in clastogenic and aneugenic effects [69]. Carbon and titanium dioxide induce apoptosis by a ROS-dependent mitochondrial pathway, whereas titanium dioxide NPs induce cell death through lysosomal membrane destabilization and lipid peroxidation. Low intrapulmonary doses of titanium dioxide and gold NPs aggravate pulmonary inflammation and asthmatic response in mouse model of diisocyanate-induced asthma [70].

Titanium dioxide injected at the abdominal cavity was shown to be translocated into the brain causing OS, lipid peroxidation, decrease in antioxidative enzymes, excessive release of nitric oxide, reduction of glutamic acid, and the downregulated level of acetylcholinesterase activities [71].

A report showed titanium dioxide nanoparticle-induced genotoxicity, oxidative DNA damage, and inflammation in a mice model [72]. Results showed TiO_2 NPs induced 8-hydroxy-2′-deoxyguanosine, γ-H2AX foci, micronuclei, and DNA double-strand breaks. A related report showed titanium dioxide nanoparticles induce oxidative stress and DNA-adduct formation in human lung cells [73]. TiO_2 nanoparticles in the absence of photoactivation induced oxidative DNA damage, lipid peroxidation, and micronuclei formation in bronchial epithelial cell lines [74].

Exposure of brain microglia to titanium dioxide NPs caused an increase in the production of ROS through oxidative burst and through interference with mitochondrial electron transfer chains [75]. ROS stimulates apoptotic pathways leading to neuronal damage, suggesting nanoparticle-induced neurotoxicity. A study revealed exposure of mice to TiO_2 nanoparticles led to identifying different proteins in the brain tissue, suggesting TiO_2 nanoparticle-induced change in protein expression in mouse brain [76].

A novel type of platinum-modified titanium dioxide nanoparticles was synthesized and tested for OS and mitochondrial membrane potential in rat cardiac cells [77]. Results showed increase in OS and greater decrease in mitochondrial membrane potential, suggesting Pt-modified TiO_2 is more cytotoxic than their commercial counterparts.

A study showed that ultraviolet irradiation or sunlight system combined with TiO_2 nanoparticles could efficiently inactivate microorganisms by ROS generation [78].

Titanium dioxide and zinc oxide nanoparticles are being increasingly formulated in sunscreens. While the same compounds, in larger particle form, work by reflecting UV radiation, in nanoparticle form, they absorb UV radiation, resulting in photocatalysis, releasing ROS. These ROS are known to have capability to alter DNA [79]. Data also suggest TiO_2 may accelerate the lipid peroxidation process through production of ROS.

2.2 Silver

Silver nanoparticles are among the most commercialized nanoparticles worldwide. They can be found in many diverse products, mostly because of their antibacterial properties. Silver impregnated catheters and wound dressings are used in therapeutic applications. Data showed even low doses of silver nanoparticles exerted adverse effects in human macrophages [80].

A study involving starch-coated silver nanoparticles using normal lung fibroblast cells and human glioblastoma cells showed cell cycle arrest, possibly due to repair of DNA damage [81]. Transmission electron microscopy indicated the presence of silver nanoparticles inside the mitochondria and nucleus, implicating their direct involvement in the mitochondrial toxicity and DNA damage. A possible mechanism is proposed involving the disruption of the mitochondrial respiratory chain by the silver nanoparticles, leading to the production of ROS, and interruption of ATP synthesis which in turn cause DNA damage.

Results demonstrate that silver nanoparticles may inhibit the vascular endothelial growth factor and interleukin-1 beta-induced permeability through inactivation of Src kinase pathway and provide a potential therapeutic target to inhibit ocular disease, such as diabetic retinopathy [82].

Mitochondrial and cell membrane exposed to silver nanoparticles showed a size-dependent toxicity, and the predominant mechanism of toxicity was found to be largely mediated through OS [83].

2.3 Cerium

Cerium is a rare-earth element of the lanthanide series. The oxide form CeO_2 has been used in polishing glass, but current research is focused on use of cerium oxide nanoparticles in catalytic converters for automobile exhaust systems, oxygen sensors, as electrolyte for solid oxide fuel cells, or as ultraviolet absorbent. Exposure of cultured human lung cancer cells to cerium oxide nanoparticles produced significant OS in the cells, as reflected by reduced glutathione and α-tocopherol levels [84]. The toxic effects of CeO_2 are dose dependent; elevated OS increases the production of malondialdehyde and lactate dehydrogenase, which are indicators of lipid peroxidation and cell membrane damage, respectively.

It has been reported that CeO_2 nanoparticles have a dual natured property as AO and as agents responsible for OS [85]. Data suggest that CeO_2 nanoparticles have a potential to induce intracellular OS and increase the intracellular Ca^{2+} level in culture cells [86]. Exposure of cultured human lung epithelial cells to cerium oxide nanoparticles resulted in cell death, increase in ROS, decrease in GSH, and the induction of OS-related genes, such as heme oxygenase-1, catalase, glutathione-transferase, and thioredoxin reductase [87].

Normal and cancer cell linings were exposed to cerium oxide nanoparticles with different surface charge (positive, negative, and neutral) [88]. The results showed that cerium oxide with positive or neutral charge enters most of the cell lines, while the negatively charged CeO_2 internalizes mostly in the cancer cell lines. The internalization and subcellular localization (cytoplasm and lysosomes) of CeO_2 play a key role in the nanoparticles cytotoxicity, exhibiting significant toxicity when they localize in the lysosomes of the cancer cells.

Cerium oxide nanoparticles activate mast cells contributing to pulmonary inflammation, impairment of vascular relaxation, and exacerbation of myocardial ischemia/reperfusion injury [89].

2.4 Zinc Oxide

Exposure of mesangial and epithelial human kidney cell lines to ZnO and CdS nanoparticles significantly increased cell mortality through ROS and OS, indicating their potential nephrotoxicity [90]. Zinc oxide nanoparticles induce oxidative stress and genotoxicity in human liver cells (HepG2) [91]. Exposure of human bronchial epithelial cells to zinc oxide showed increased levels of OS [92].

Exposure to ZnO nanoparticles killed zebrafish embryos, retarded the embryo hatching, reduced body length of larvae, and caused tail malformation [93]. A report revealed the genotoxic potential of ZnO nanoparticles and their ability to perturb the mitochondrial membrane potential, causing nanoparticle-induced DNA damage in human lymphocytes possibly through OS [94]. A related work showed zinc oxide nanoparticles damaging DNA on human epidermal cells [95]. Exposure of RAW and BEAS-2B cell lines to ZnO nanoparticles leads to generation of ROS, oxidant injury, excitation inflammation, and cell death [96]. CeO_2, on the other hand, suppressed ROS production and induced cellular resistance to an exogenous source of OS. Zinc oxide nanoparticles showed antibacterial activity against *Campylobacter jejuni*. The mechanism most likely involves disruption of the cell membrane and OS [97].

2.5 Other Metal Compounds

Tungsten carbide-cobalt nanoparticles exert hypoxia-like effects on gene expression in human keratinocytes [98]. A report showed that exposure of human cell lines to engineered cobalt oxide nanoparticles causes rapid induction of ROS. The material enters the cell and remains confined to vesicles inside the cytoplasm [99].

Exposure of human lung carcinoma A549 and keratinocyte cells to chromium (III) oxide nanoparticles showed an increase in intracellular ROS, activation of antioxidant defense systems, and apoptosis [100]. A study involving exposure of suckling rats to chromium (VI) nanoparticles showed disorders in the bone mineralization, decrease in calcium and phosphorous levels, increase in malondialdehyde levels, and decrease in glutathione, nonprotein thiol, and vitamin C [101].

Manganese toxicity in humans has been observed as manganism, a disease that resembles Parkinson's disease. Exposure to Mn oxide nanoparticles leads to increased ROS generation and cell death through apoptosis [102]. Increased levels of NF-kB, which mediates the cellular inflammatory response, were also observed. A related study showed exposure of alveolar epithelial cells to manganese oxide nanoparticles led to increased OS and cell apoptosis [103].

Nickel ferrite nanoparticles are being utilized in many applications, including magnetic resonance imaging, drug delivery, and hyperthermia. Exposure of cultured A549 cells to nickel ferrite nanoparticles induced cell apoptosis through ROS and OS [104]. Data suggested rats exposed to nickel oxide nanoparticles have persistent inflammatory effects, and the transient increase in cytokine expression and persistent increase in chemokine were involved in the persistent pulmonary inflammation [105].

Gold nanoparticles are chemically stable, electron dense, and possess an affinity to biomolecules, such as amino acids, protein, and DNA, making them suitable as drug carriers and imaging reagents. Rats exposed to gold nanoparticles showed histological alterations, suggesting that gold nanoparticles interact with proteins and enzymes of the hepatic tissue interfering with the antioxidant defense mechanism and leading to ROS generation which in turn may induce hepatocytes to undergo necrosis by OS [106]. Antimicrobial properties of gold nanoparticles are also attracting the attention of the food industry. Recent studies have shown that gold nanoparticles have been proposed as components of food containers, due to their antibacterial, antiseptic, deodorant, and catalytic properties. Exposure of *Mytilus edulis* to gold particles leads to OS and toxicity [107]. Exposure of human MRC-5 lung fibroblasts in vitro to gold nanoparticles induces autophagy and OS [108]. A study showed gold particles trigger necrosis by OS and mitochondrial damage [109].

Maternal mouse exposure to platinum nanoparticles led to increased pup mortality and decreased infant growth during the lactation period [110]. Colloidal platinum nanoparticles used in the preparation of dental fillings increase mitochondrial stress induced by resin composite components [111].

A report suggested that exposure to Fe_3O_4 nanoparticles could cause oxidative damage on mouse liver in vitro, with increase in ROS generation and decrease in glutathione content, and might have toxic effects on human body under high concentrations [112]. A related report deals with concentration-dependent toxicity of iron oxide nanoparticles mediated by increased OS [113]. A study revealed zerovalent and Fe(II) nanoparticles induce OS in human bronchial epithelial cells [114]. Exposure of MRC-5 fibroblast cells to Fe_2O_3 nanoparticles induce OS, with decrease in glutathione and increased lipid peroxidation [115]. Data suggested that high levels of perinuclear localized iron oxide nanoparticles diminish the efficiency of protein expression and sterically hinder the mature actin fibers and could have detrimental effects on cell migration and differentiation [116].

3 AOs

Generation of ROS can be promoted by large surface area, which is one of the main characteristics of nanoparticles. Consequently, the smaller the nanoparticles, the greater the OS they can induce. In addition, with their high surface area, compounds of metals, such as titanium, cadmium, chromium, cobalt, iron, nickel, zinc, gold, and silver, have the potential to cause the conversion of cellular oxygen metabolic products, such as H_2O_2, superoxide, and hydroxyl radicals, which interact with membranes, DNA, and other vital tissues. Interestingly, many of these metal compound nanoparticles can also act as scavengers of free radicals and act as AOs. Rationale for the oxidant-AO property for other substances is presented in a review [117]. In application to metal compounds, certain valences could act both as oxidants (electron acceptors) and as AOs (electron donors). A recent review addresses the implications of OS mechanisms due to nanoparticles [118] and another deals with antioxidant nanoparticles for control of infectious diseases [119].

Of all the AO metal oxides, cerium oxide takes the leading position. Cerium oxide nanoparticles (nanoceria), well known as catalysts, possess an astonishing pharmacological potential due to their AO properties, deriving from the fraction of Ce^{3+} and Ce^{4+} ions present in CeO_2. Reactions involving redox cycles between the Ce^{3+} and Ce^{4+} oxidation states allow nanoceria to react catalytically with superoxide and hydrogen peroxide, mimicking the behavior of two key AO enzymes, superoxide dismutase, and catalase, potentially abating all noxious intracellular ROS. Hence, nanoceria, apparently well tolerated by the organism, might fight chronic inflammation and the pathologies associated with OS [120]. Cerium oxide acts as an AO against hepatic oxidative damage induced by monocrotaline in rat model [121]. A report showed cerium oxide nanoparticles inhibit cigarette smoke-induced cell damage via inhibition of ROS generation, NFkB activation, inflammatory gene expression, and AO depletion and may have a great potential for treatment of smoking-related diseases [122]. Cerium oxide and platinum nanoparticles protect cells from oxidant-mediated apoptosis [123]. Cerium oxide administration to mice with induced liver toxicity (by CCl_4) showed similar findings to mice treated with N-acetyl cysteine, a common therapeutic to reduce OS [124]. Cerium oxide nanoparticles were reported to act as an AO against radiation-induced pneumonitis [125]. The antioxidant efficacy was enhanced by dissolving zirconium oxide in the CeO_2 lattice [126]. CeO_2 nanoparticles protect against the progression of cardiac dysfunction and remodeling by attenuation of myocardial OS and the inflammatory processes through their autoregenerative AO properties [127]. A report showed that cerium and yttrium oxide nanoparticles act as direct AOs and protect nerve cells from OS [128].

Platinum nanoparticles synthesized within cavities of hollow protein nanospheres act as AOs against externally induced stress [129]. Data showed platinum nanoparticles act as AOs in inhibiting pulmonary inflammation induced by acute cigarette smoking [130]. A study showed platinum nanoparticles have an AO property similar to NADH/ubiquinone oxidoreductase activity, a potential medicinal substance for OS diseases [131]. A related work showed platinum nanoparticles could behave as superoxide dismutase/catalase mimetics and also have an activity similar to that of mitochondrial electron transport complex [132].

A report showed that gold nanoparticles act as AOs by inhibiting lipid peroxidation and ROS generation during hyperglycemia in diabetic mice [133]. Gold nanoparticles inhibit the receptor activator of NF-kB ligand RANKL-induced osteoclast formation by acting as AOs [134].

The delayed wound healing is a well-known problem with type 2 diabetic patients. Nanoparticle metal oxides, such Al_2O_3, CeO_2, Y_2O_3, acting as ROS scavengers help to cure impaired healing of diabetic foot wounds [135].

A report revealed the scavenging of ROS and potential cell protection by functionalized fullerene materials [136].

4 Nanomedicine

Nanomedicine, the application of nanotechnology to health care, holds great promise for revolutionizing medical treatments and therapies in various areas, such as imaging, faster diagnosis, drug delivery, as well as development of new medical products. Devices of nanometric dimensions are already approved for clinical use and numerous products are being evaluated in clinical trials [137–140].

Among the many types of metal nanomaterials developed for nanomedicine applications, gold nanoparticles play a major role in cancer therapy and a wide variety of rheumatic diseases, including arthritis, and discoid lupus erythematosus [141–145]. Gold nanoparticles bound to thioglucose enhance radio-cytotoxic targeting of ovarian cancer [146].

Like gold, silver nanoparticles also have wide therapeutic application. A review addresses the role of silver nanoparticle application to human health [147].

A report showed the antiviral action of silver nanoparticles against HIV-1 strain [148]. They exhibited antitumor activity in lymphoma tumor models [149]. Magnetic nanoparticles derived from cobalt, nickel, and iron are being increasingly used as magnetically targeted nanoparticles for brain tumor therapy [150]. A review deals with nanotechnology-based drug delivery including anticancer drugs, such as paclitaxel, doxorubicin, 5-fluorouracil, and dexamethasone [151]. A report showed that polymeric nanoparticle formulation of curcumin inhibits growth, clonogenicity, and stemlike fractions in malignant brain [152].

In addition to the reports presented, there is much related literature on drug delivery.

Acknowledgment

Editorial assistance by Thelma Chavez is acknowledged.

References

1. Kovacic P, Somanathan R (2010) Biomechanisms of nanoparticles (toxicants, antioxidants and therapeutics): electron transfer and reactive oxygen species. J Nanosci Nanotechnol 10:7919–7930
2. Kovacic P, Becvar LE (2000) Mode of action of anti-infective agents: emphasis on oxidative stress and electron transfer. Curr Pharm Des 6:143–167
3. Kovacic P, Osuna JA (2000) Mechanisms of anticancer agents: emphasis on oxidative stress and electron transfer. Curr Pharm Des 6:277–309
4. Kovacic P, Jacintho JD (2001) Mechanism of carcinogenesis: focus on oxidative stress and electron transfer. Curr Med Chem 8:773–796
5. Kovacic P, Jacintho JD (2001) Reproductive toxins: pervasive theme of oxidative stress and electron transfer. Curr Med Chem 8:863–892
6. Kovacic P, Sacman A, Wu-Weis M (2002) Nephrotoxins: widespread role of oxidative

stress and electron transfer. Curr Med Chem 9:823–847

7. Poli G, Cheeseman GH, Dianzani MU, Slater TF (1989) Free radicals in the pathogenesis of liver injury. Pergamon Pree, Oxford, pp 1–330
8. Kovacic P, Thurn LA (2005) Cardiovascular toxins from the perspective of oxidative stress and electron transfer. Curr Vasc Pharmacol 3:107–117
9. Kovacic P, Somanathan R (2005) Neurotoxicity: the broad framework of electron transfer, oxidative stress and protection by antioxidants. Curr Med Chem-CNS Agents 5:249–258
10. Kovacic P, Pozos RS, Somanathan R, Shangari N, O'Brien PG (2005) Mechanism of mitochondrial uncouplers, inhibitors, and toxins: focus on electron transfer, free radicals, and structure-activity relationships. Curr Med Chem 5:2601–2623
11. Kovacic P, Cooksy AL (2005) Unifying mechanism for toxicity and addiction by abused drugs: electron transfer and reactive oxygen species. Med Hypotheses 64:357–366
12. Kovacic P, Somanathan R (2008) Ototoxicity and noise trauma: electron transfer, reactive oxygen species, cell signaling, electrical effects, and protection by antioxidants: practical medical aspect. Med Hypotheses 70: 914–923
13. Kovacic P, Somanathan R (2008) Integrated approach to immunotoxicity: electron transfer, reactive oxygen species, antioxidants, cell signaling, and receptors. J Recept Signal Transduct 28:323–346
14. Kovacic P, Somanathan R (2008) Unifying mechanism for eye toxicity: electron transfer, reactive oxygen species, antioxidant benefits, cell signaling and cell membranes. Cell Membr Free Rad Res 1:56–69
15. Kovacic P, Somanathan R (2009) Pulmonary toxicity and environmental contamination: radicals, electron transfer, and protection by antioxidants. Rev Environ Contam Toxicol 201:41–69
16. Kovacic P, Somanathan R (2010) Dermal toxicity and environmental contamination: electron transfer, reactive oxygen species, oxidative stress, cell signaling, and protection by antioxidants. Rev Environ Contam Toxicol 203:119–138
17. Kourounakis PN, Tsiakitzis K, Kourounakis AP, Galanakis D (2000) Reduction of gastrointestinal toxicity of NSAIDs via molecular modifications leading to antioxidant anti-inflammatory drugs. Toxicology 144:205–210
18. Kovacic P, Edwards C (2010) Integrated approach to the mechanisms of thyroid toxins: electron transfer, reactive oxygen species, oxidative stress, cell signaling, receptors, and antioxidants. J Recept Signal Transduct 30:133–142
19. Halliwell B, Gutteridge JMC (1999) Free radicals in biology and medicine. Oxford University Press, New York, pp 1–936
20. van der Zande M, Junker R, Walboomers XF, Jansen JA (2010) Carbon nanotubes in animal models: a systemic review on toxic potential. Tissue Eng Part B Rev 17:58–69
21. Johnston HJ, Hutchinson GR, Christensen FM, Peters S, Hankin S, Aschberger K, Stone V (2010) A critical review of the biological mechanisms underlying the in vivo and in vitro toxicity of carbon nanotubes: the contribution of physico-chemical characteristics. Nanotoxicology 4:207–246
22. Kim JS, Lee K, Lee YH, Cho HS, Kim KH, Choi KH, Lee SH, Song KS, Kang CS, Yu IJ (2011) Aspect ratio has no effect on genotoxicity of multi-walled nanotubes. Arch Toxicol 85:775–786
23. Stella GM (2011) Carbon nanotubes and pleural damage: perspectives of nanosafety in the light of asbestos experience. Biointerphases 6:1–17
24. Srivastava RK, Pant AB, Kashyap MP, Lohani M, Jonas L, Rahman Q (2011) Multi-walled carbon nanotubes induce oxidative stress and apoptosis in human lung cancer cells line-A549. Nanotoxicology 5:195–207
25. Ye S, Wang Y, Jiao F, Zhang H, Lin C, Wu Y, Zhang Q (2011) The role of NADPH oxidase in multi-walled carbon nanotubes-induced oxidative stress and cytotoxicity in human macrophages. J Nanosci Nanotechnol 11:3773–3781
26. Pauluhn J (2010) Subchronic 13-week inhalation exposure of rats to multiwalled carbon nanotubes: toxic effects are determined by density of agglomerate structures, not fibrillar structures. Toxicol Sci 113:226–242
27. Kovacic P, Somanathan R (2011) Inflammation and anti-inflammatory agent: reactive oxygen species and toxicity. In: Laher I (ed) Systems biology of free radicals and anti-oxidants. Springer, New York
28. Reddy AR, Rao MV, Krishna DR, Himabindu V, Reddy YN (2011) Evaluation of oxidative stress and anti-oxidant status in rat serum following exposure of carbon nanotubes. Regul Toxicol Pharmacol 59:251–257
29. Reddy AR, Reddy YN, Krishna DR, Himabindu V (2010) Multi wall carbon

nanotubes induce oxidative stress and cytotoxicity in human embryonic kidney (HEK 293) cells. Toxicology 272:11–16

30. Takaya M, Serita F, Yamazaki K, Aiso S, Kubota H, Asalura M, Ikawa N, Nagano K, Arito H, Fukushima S (2010) Characteristics of multiwalled nanotubes for an intratracheal installation study with rats. Ind Health 48: 452–459
31. Halliwell B, Gutteridge JMC (1999) Free radicals in biology and medicine. Oxford University Press, New York, p 464
32. Clichici S, Mocan T, Filip A, Biris A, Simon S, Daicoviciu D, Decea N, Parvu A, Moldovan R, Muresan A (2011) Blood oxidative stress generation after interperitonial administration of functionalized single walled-carbon nanotubes in rats. Acta Physiol Hung 98: 231–241
33. Inoue K, Yanagisawa R, Koike E, Nishkawa M, Takano H (2010) Repeated pulmonary exposure to single-walled carbon nanotubes exacerbates allergic inflammation of the airway: possible role of oxidative stress. Free Rad Biol Med 48:924–934
34. Pacurari M, Yin XJ, Zhao J, Ding M, Leonard SS, Schwegler-Berry D, Ductman BS, Sbarra D, Hoover MD, Castranova V, Vallyathan V (2008) Raw single-wall carbon nanotubes induce oxidative stress and activate MAPKs, AP-1, NF-kappaB, and Akt in normal and malignant human mesothelial cells. Environ Health Perspect 116:1211–1217
35. Wang X, Katwa P, Podila R, Chen P, Ke PC, Rao AM, Walters DM, Wingard CJ, Brown JM (2011) Multi-walled carbon nanotube instillation impairs pulmonary function in C57BL/6 mice. Particle Fibre Toxicol 8:24
36. Shevedova AA, Kisin E, Murray AR, Johnson VJ, Gorlik O, Arepalli S, Hubbs AF, Mercer RR, Keohavong P, Sussman N, Jin J, Stone S, Chen BT, Deye G, Maynard A, Castranova V, Baron PA, Kagan VE (2008) Inhalation vs. aspiration of single-walled carbon nanotubes in C57BL/mice: inflammation, fibrosis, oxidative stress, and mutagenesis. Am J Physiol Lung Cell Mol Physiol 295:L552–L565
37. Ravichandran P, Baluchamy S, Gopikrishnan R, Biradar S, Ramesh V, Goornavar V, Thomas R, Wilson BL, Jeffers R, Hall JC, Ramesh GT (2011) Pulmonary biocompatibility assessment of inhaled single-wall and multiwall carbon nanotubes in BALB/c mice. J Biol Chem 286:29725–29733
38. Wessels A, van Berlo D, Boots AW, Gerloff K, Scherbart AM, Casseee FR, Gerlofs-Niijland ME, van Schooten F-J, Albrecht C, Schins RPF (2011) Oxidative stress and DNA damage responses in rat and mice lung to inhaled carbon nanoparticles. Nanotoxicology 5: 66–78
39. Murray AR, Kisin E, Leonard SS, Yoiung SH, Kommineni C, Kagan VE, Castronova V, Shvedova AA (2009) Oxidative stress and inflammatory response in dermal toxicity of single-walled carbon nanotubes. Toxicology 257:161–171
40. Limbach LK, Wick P, Grass RN, Bruinink A, Stark WJ (2007) Exposure of engineered nanoparticles to human lung epithelial cells: influence of chemical composition and catalytic activity on oxidative stress. Environ Sci Technol 41:4158–4163
41. By GY, Brumno M, Wallace K, Winnik W, Prasad RY (2011) Proteome profiling reveals potential toxicity and detoxification pathways following exposure of BEAS-2B cells to engineered nanoparticle titanium dioxide. Proteomics 11:2406–2422
42. Buffet PE, Tankoua OF, Pan JF, Berhanu D, Herrenknecht C, Poirier L, Amiard-Triquet C, Amiard JC, Bérard JB, Risso C, Guibbolini M, Roméo M, Reip P, Valsami-Jones E, Mouneyrac C (2011) Behavioral and biochemical responses of two marine invertebrates Scrobicularia plana and Hediste diversicolor to copper oxide nanoparticle. Chemosphere 84:166–174
43. Marano F, Hussain S, Rodrigues-Lima F, Baeza-Squiban A, Boland S (2011) Nanoparticles: molecular targets and cell signaling. Arch Toxicol 85:733–741
44. Halliwell B, Gutteridge JMC (2000) Free radicals in biology and medicine. New York, Oxford, p 340
45. Shaw BJ, Handy RD (2011) Physiological effects of nanoparticles on fish: a comparison of nanometals versus metal ions. Environ Int 37:1083–1097
46. Beyerle A, Long AS, White PA, Kissel T, Stoeger T (2011) Poly(ethylene imine) nanocarriers do not induce mutations nor oxidative stress DNA damage in vitro in MutaMouse FE1 cells. Mol Pharm 8:976–981
47. Wang J, Zhu X, Zhang X, Zhao Z, Liu H, George R, Wilson-Rawls J, Chang Y, Chen Y (2011) Disruption of zebrafish (*Danio rerio*) reproduction upon chronic exposure to TiO_2 nanoparticles. Chemosphere 83:461–467
48. Sikka SC (2001) Relative impact of oxidative stress on male reproductive function. Curr Med Chem 8:851–862
49. Kovacic P, Somanathan R (2006) Mechanisms of teratogenesis: electron transfer, reactive

oxygen species and antioxidants. Birth Def Res 78:308–325

50. Kovacic P, Somanathan R (2011) Free radicals and teratogenesis. In: Laher I (ed) Systems biology of free radicals and antioxidants. Springer, New York
51. Møller P, Mikkelsen L, Vesterdal LK, Folkmann JK, Forchhammer L, Roursgaard M, Danielsen PH, Loft S (2011) Hazard identification of particulate matter on vasomotor dysfunction and progression of atherosclerosis. Crit Rev Toxicol 41:339–368
52. Fubini B, Chaiazza M, Fenoglio I (2010) Physico-chemical features of engineered nanoparticles relevant to their toxicity. Nanotoxicology 4:347–363
53. Terzano C, Di Stefano F, Conti V, Graziani E, Petroianni A (2010) Air pollution ultrafine particles: toxicity beyond the lung. Eur Rev Med Pharmacol Sci 14:809–821
54. Müller L, Riediker M, Wick P, Mohr M, Gehr P, Rothen-Rutishauser B (2010) Oxidative stress and inflammation response after nanoparticle exposure: differences between human lung cell monocultures and an advanced three-dimensional model of the human epithelial airways. J R Soc Interface 7:S27–S40
55. Møller P, Jacobsen NR, Folkmann JK, Danielsen PH, Mikkelsen L, Hemmingsen JG, Vesterdal LK, Forchhammer L, Wallin H, Loft S (2010) Role of oxidative damage in toxicity of particulates. Free Rad Res 44:1–46
56. McLeish JA, Chico TJ, Taylor HB, Tucker C, Donaldson K, Brown SB (2010) Skin exposure to micro-and nano-particles can cause haemostasis in zebrafish larvae. Thromb Haemost 103:797–807
57. Nemmar A, Al-Salam S, Zia S, Dhanasekaran S, Shudadevi M, Ali BH (2010) Time-course effects of systematically administered diesel exhaust particle in rats. Toxicol Lett 194: 58–65
58. Li R, Ning Z, Majumdar R, Cui J, Takabe W, Jen N, Sioutas C, Hsiai T (2010) Ultrafine particles from diesel vehicle emissions at different driving cycles induce differential vascular pro-inflammatory responses: implication of chemical components and NF-kappaB signaling. Part Fibre Toxicol 7:6
59. Turci F, Tomatis M, Lesci IG, Roveri N, Fubini B (2011) The iron-related molecular toxicity mechanism of synthetic asbestos nanofibres: a model study for high-aspect-ratio nanoparticles. Chemistry 17:350–358
60. Donaldson K, Borm PJ, Castronova V, Gulumian M (2009) The limits of testing particle-mediated oxidative stress in vitro in predicting diverse pathologies; relevance for testing of nanoparticles. Part Fibre Toxicol 6:13
61. Brown T (2009) Silica exposure, smoking, silicosis and lung cancer-complex interactions. Occup Med 59:89–95
62. Wu J, Wang C, Sun J, Xue Y (2011) Neurotoxicity of silica nanoparticles: brain localization and dopaminergic neurons damage pathways. ACS Nano 5:4476–4489
63. Song Y, Li X, Wang L, Rojanaskul Y, Castranova V, Li H, Ma J (2011) Nanomaterial in humans: identification, characteristics, and potential damage. Toxicol Pathol 39:841–849
64. Bhattacharjee S, de Haan LHJ, Evers NM, Jiang X, Marcelis ATM, Zuihof H, Rietjens IMCM, Alink GM (2010) Role of surface charge and oxidative stress in cytotoxicity of organic monolayer-coated silicon nanoparticles towards macrophage NR8383 cells. Particle Fibre Toxicol 7:25
65. Eom H-J, Choi J (2009) Oxidative stress of silica nanoparticles in human bronchial epithelial cell, Beas-2B. Toxicol Vitro 23:1326–1332
66. Wang F, Gao F, Lan M, Yuan H, Huang Y, Liu J (2009) Oxidative stress contributes to silica nanoparticle-induced cytotoxicity in human embryonic kidney cells. Toxicol Vitro 23:808–815
67. Li JJ, Muralikrishnan S, Ng C-T, Yung L-YL, Bay B-H (2010) Nanoparticles-induced pulmonary toxicity. Exp Biol Med 235: 1025–1033
68. Hussain S, Thomassen LCJ, Ferecatu I, Borot M-C, Andreau K, Martens JA, Fleury J, Baeza-Squiban A, Marano F, Boland S (2010) Carbon black and titanium dioxide nanoparticles elicit distinct apoptotic pathways in bronchial epithelial cells. Particle Fiber Toxicol 7:10
69. Muller J, Decordier I, Hoet PH, Lombaert FH, Lison D, Kirsch-Volders M (2008) Clastogenic and aneugenic effects of multiwall carbon nanotubes in epithelial cells. Carcinogenesis 29:427–433
70. Hussain S, Vanoirbeek JAJ, Luyts K, De Vooght V, Verbeken E, Thomassen LCJ, Martens JA, Dinsdale D, Boland S, Marano F, Nemery B, Hoet PHM (2011) Lung exposure to nanoparticles modulates an asthma response in mouse model. Eurp Respir J 37: 299–309
71. Ma L, Liu J, Li N, Wang J, Duan Y, Yan J, Liu H, Wang H, Hong F (2010) Oxidative stress in the brain of mice caused by translocated nanoparticulate TiO_2 delivered to the abdominal cavity. Biomaterials 31:99–105

72. Trouiller B, Reliene R, Westbrook A, Solaimani P, Schiestl RH (2009) Titanium dioxide nanoparticles induce DNA damage and genetic instability in vivo in mice. Cancer Res 69:8784–8794
73. Bhattacharya K, Davoren M, Boertz J, Schins RPF, Hoffmann E, Dopp E (2009) Titanium dioxide nanoparticles induce oxidative stress and DNA-adduct formation but not DNA-breakage in human lung cells. Particle Fibre Toxicol 6:17
74. Gurr J-R, Wang ASS, Chen C-H, jan K-Y (2005) Ultrafine titanium dioxide particles in the absence of photoactivation can induce oxidative damage to human bronchial epithelial cells. Toxicology 213:66–73
75. Long TC, Saleh N, Tilton RD, Lowry GV, Veronesi B (2006) Titanium dioxide (P25) produces reactive oxygen species in immortalized brain microglia (BV2): implications for nanoparticle neurotoxicity. Environ Sci Technol 40:4346–4352
76. Jeon Y-M, park S-K, Lee M-Y (2011) Toxicoproteomic identification of TiO_2 nanoparticle-induced protein expression changes in mouse brain. Animal Cells Systems 15:107–114
77. Mallik A, Bryan S, Puukila S, Chen A, Khaper N (2011) Efficacy of Pt-modified TiO_2 nanoparticles in cardiac cells. Exp Clin Cardiol 16:6–10
78. Choi O, Hu Z (2009) Reactive oxygen species in determining nitrification inhibition by metallic/oxide nanoparticles. J Environ Eng 135:1365–1370
79. Tran DT, Salmon R (2010) Potential photocarcinogenic effects of nanoparticle sunscreens. Australas J Dermatol. doi: 10.1111/j.1440-0960.2010.00677.x
80. Haase A, Tentscher J, Jungnickel H, Graf P, Mantion A, Draude F, Plendl J, Goetz ME, Galla S, Mašić A, Thuenemann AF, Taubert A, Arlinghaus HF, Luch A (2011) Toxicity of silver nanoparticles in human macrophages: uptake, intracellular distribution and cellular responses. J Phys Conf Series 304(012030):1–15. doi:10.1088/ 1742-6596/304/1/ 012030
81. AshaRani PV, Mun GLK, Hande MP, Valiyaveettil S (2009) Cytotoxicity and genotoxicity of silver nanoparticles in human cells. ACSNano 3:279–290
82. Sheikpranabu S, Kalishwaralal K, Venkataraman D, Ecom SH, Park J, Gurunathan S (2009) Silver nanoparticles inhibit VEGF-and IL-I-induced vascular permeability via dependent pathway in porcine retinal endothelial cells. J Nanotechnol 7:8. doi:10.1186/1477-3155-7-8
83. Carlson C, Hussain SM, Schrand AM, Braydich-Stolle LK, Hess KL, Jones RL, Schlager JJ (2008) Unique cellular interaction of silver nanoparticles: size-dependent generation of reactive oxygen species. J Phys Chem B 112:13608–13619
84. Lin W, Huang Y-W, Zhou X-D, Ma Y (2006) Toxicity of cerium oxide nanoparticles in human lung cancer cells. Int J Toxicol 25:451–457
85. Horie M, Fujita K (2011) Toxicity of metal oxides nanoparticles. In: Fishbein JC (ed) Advances in molecular toxicology. Elsevier, Oxford, UK, Chapter 4
86. Horie M, Nishio K, Kato H, Fujita K, Endoh S, Nakamura A, Miyauchi A, Kinugasa S, Yamamoto K, Niki E, Yoshida Y, Hagihara Y, Iwahashi H (2011) Cellular responses induced by cerium oxide nanoparticles: induction of intracellular calcium level and oxidative stress on cultured cells. J Biolchem doi: 10.1093/jb/mvr081
87. Park E-J, Choi J, Park Y-K, Park K (2008) Oxidative stress induced by cerium oxide nanoparticles in cultured BEAS-2B cells. Toxicology 245:90–100
88. Asati A, Santra S, Kaittanis C, Perez JM (2010) Surface-charge-dependent cell localization and cytotoxicity of cerium oxide nanoparticles. ACS Nano 4:5321–5331
89. Wingard CJ, Walters DM, Cathey BL, Hildrbrand SC, Katwa P, Lin S, Ke PC, Podila R, Rao A, Lust RM, Brown JM (2010) Mast cells contribute to altered vascular reactivity and ischemia-reperfusion injury following cerium oxide nanoparticle instillation. Nanotoxicology 4:1–15
90. Pujalté I, Passagne I, Brouillaud B, Tréguer M, Durand E, Ohayon-Courtés C (2011) Cytotoxicity and oxidative stress induced by different metallic nanoparticles on human kidney cells. Particle Fibre Toxicol 8:10
91. Sharma V, Anderson D, Dhawan A (2011) Zinc oxide nanoparticles induce oxidative stress and genotoxicity in human liver cells (HepG2). J Biomed Nanotechnol 7:98–99
92. Heng BC, Zhao X, Xiong S, Ng KW, Boey FY-C, Loo JC-C (2010) Toxicity of zinc oxide (ZnO) nanoparticles on human bronchial epithelial cells (BEAS-2B) is accentuated by oxidative stress. Food Chemical Toxicol 48:1762–1766
93. Bai W, Zhang Z, Tian W, He X, Ma Y, Zhao Y, Chai Z (2010) Toxicity of zinc oxide nanoparticles to zebrafish embryo: a physiochemical study of toxicity mechanism. J Nanoparticle Res 12:1645–1654
94. Musarrat J, Saquib Q, Azam A, Naqvi AH (2009) Zinc oxide nanoparticles-induced

DNA damage in human lymphocytes. Int J Nanoparticles 2:401–415

95. Sharma V, Shukla RK, Saxena N, Parmar D, das M, Dhawan A (2009) DNA damaging potential of zinc oxide nanoparticles in human epidermal cells. Toxicol Lett 185:211–218
96. Xia T, Kovochich M, Liong M, Mädler L, Gilbert B, Shi H, Yeh JI, Zink JI, Nel AE (2008) Comparison of the mechanism of toxicity of zinc oxide and cerium oxide nanoparticles based on dissolution and oxidative stress properties. ACS Nano 2:2121–2134
97. Xie Y, He Y, Irwin PL, Jin T, Shi X (2011) Antibacterial activity and mechanism of action of zinc oxide nanoparticles against *Campylobacter jejuni*. Appl Environ Microbiol 77:2325–2331
98. Busch W, Kühnel D, Schirmer K, Scholz S (2010) Tungsten carbide cobalt nanoparticles exert hypoxia-like effects on gene expression level in human keratinocytes. BMC Genomics 11:65
99. Papis E, Rossi F, Raspanti M, Dalle-Donne I, Colombo G, Milzani A, Bernardini G, Gornati R (2009) Engineered cobalt oxide nanoparticles readily enter cells. Toxicol Lett 189:253–259
100. Horie M, Nishio K, Endoh S, Kato H, Fujita K, Miyauchi A, Nakamura S, Yamamoto K, Niki E, Yoshida Y, Iwahashi H (2011) Chromium (III) oxide nanoparticles induced remarkable oxidative stress and apoptosis on culture cells. Environ Toxicol doi:10.1002/tox.20695
101. Soudani N, Amara IB, Troudi A, Bouaziz H, Boudawara T, Zeghal N (2011) Oxidative stress induced by chromium (VI) in bone of suckling rats. Toxicol Ind Health doi: 10.1177/0748233710395992
102. Stefanescu DM, Khoshnan A, Patterson PH, Hering JG (2011) Neurotoxicity of manganese oxide nanoparticles. J Nanoparticle Res 11:1957–1969
103. Frick R, Müller-Edenborn B, Schlicker A, Rothen-Rutishauser B, Raemy DO, Günther D, Hattendorf B, Stark W, Beck-Schimmer B (2011) Comparison of manganese oxide nanoparticles and manganese sulfate with regard to oxidative stress, uptake and apoptosis in alveolar epithelial cells. Toxicol Lett 205:163–172
104. Ahamed M, Akgtar MJ, Siddiqui MA, Ahmad J, Musarrat J, Al-Khedhairy AA, AlSalhi MS, Alrokayan SA (2011) Oxidative stress mediated apoptosis induced by nickel ferrite nanoparticles in cultured A549 cells. Toxicology 283: 101–108
105. Morimoto Y, Ogami A, Todoroki M, Yamamoto M, Murakami M, Hirohashi M, Oyabu T, Myojo T, Nishi K-I, Kadoya C, Yamasaki S, Nagatomo H, Fujita K, Endoh S, Uchida K, Yamamoto K, Kobayashi N, Nakanishi J, Tanaka I (2010) Expression of inflammation-related cytokines following intratracheal instillation of nickel oxide nanoparticles. Nanotoxicology 4:161–176
106. Abedelhalim MAK, Jarrar BM (2011) Gold nanoparticles administration induced prominent inflammatory, central vein intima disruption, fatty change and Kupffer cells hyperplasia. Lipid Health Disease 10:133
107. Tefesco S, Doyle H, Blasco J, Redmond G, Sheehan D (2010) Oxidative stress and toxicity of gold nanoparticles in *Mytilus edulis*. Aquatic Toxicol 100:178–186
108. Li JJ, Hartono D, Ong CN, Bay BH, Yung LY (2010) Autophagy and oxidative stress associated with gold nanoparticles. Biomaterials 31:5996–6003
109. Pan Y, Leifert A, Ruau D, Neuss S, Bornemann J, Schmid G, Brandau W, Simon U, Jahnen-Dechent W (2009) Gold nanoparticles of diameter 1.4 nm trigger necrosis by oxidative stress and mitochondrial damage. Small 5:2067–2076
110. Park E-J, Kim H, Kim Y, Park K (2010) Effects of platinum nanoparticles on the postnatal development of mouse pups by maternal exposure. Environ Health Toxicol 25:279–286
111. Kachi H, Noda M, Wataha JC, Nakaoki Y, Sano H (2011) Colloidal platinum nanoparticles increase mitochondrial stress induced by resin composite components. J Biomed Mater Res B Appl Biomater 96:193–198
112. Yuan Y, Tang J, Wei C, Han B, Yang X (2011) Oxidative damage of Fe_3O_4 nanoparticles on mouse hepatic cells in vitro. Bioinfor Biomed Eng 1–4, 5th international conference. doi:10.1109/icbbe.2011.5781375
113. Naqvi S, Abdin MZ, Ahmed FJ, Maitra AN, Prashant CK, Dinda AK (2010) Concentration-dependent toxicity of iron oxide nanoparticles mediated by increased oxidative stress. Int J Nanomed 5:983–989
114. Keenan CR, Goth-Golsrein R, Lucas D, Sedlak DL (2009) Oxidative stress induced by zero-valent iron nanoparticles and Fe(II) in human bronchial epithelial cells. Environ Sci Technol 43:4555–4560
115. Radu M, Munteanu MC, Petrache S, Serban AI, Dinu D, Hermenean A, Sima C, Dinischiotu A (2010) Depletion of intracellular glutathione and increased lipid peroxidation mediate cytotoxicity of hematite

nanoparticles in MRC-5 cells. Acta Biochim Polon 57:355–360

116. Soenen SJH, Nuytten N, De Meyer SF, De Emedt SC, De Cuyper M (2010) High intracellular iron oxide nanoparticle concentrations affect cellular cytoskeleton and focal adhesion kinase-mediated signaling. Small 6:832–842
117. Kovacic P, Somanathan R (2006) Beneficial effects of antioxidants in relation to carcinogen, toxins and various illnesses. In: Panglossi HV (ed) Frontiers in antioxidants research. Nova Science, New York, pp 1–38, Chapter 1
118. Mocan T, Clichici S, Agoşton-Colea L, Mocan L, Şimon Ş, Ilie IR, Biriş AR (2010) Implications of oxidative stress mechanisms in toxicity of nanoparticles. Acta Physiol Hung 97:247–255
119. Elswaifi SF, Palmieri JR, Hockey KS, Rzigalinski BA (2009) Antioxidant nanoparticles for control of infectious disease. Infec Disord Drug Targets 9:445–452
120. Celardo I, Pederson JZ, Traversa E, Ghibell L (2011) Pharmacological potential of cerium oxide nanoparticles. Nanoscale 3:1411–1420
121. Amin KA, Hassan MS, Awad E-ST, Hashem KS (2011) The protective effects of cerium oxide nanoparticles against hepatic oxidative damage induced by monocrotaline. Int J Nanomed 6:143–149
122. Niu J, Wang K, Kolattukudy PE (2011) Cerium oxide nanoparticles inhibit oxidative stress and nuclear factor-kB activation in H9c2 cardiomyocytes exposed to cigarette smoke extract. J Pharmacol Exp Ther 338: 53–61
123. Clark A, Zhu A, Sun K, Petty HR (2011) Cerium oxide and platinum nanoparticles protect cells from oxidant-mediated apoptosis. J Nanopart Res 13(10):5547–5555. doi: 10.1007/s11051-011-0544-3
124. Hirst SM, Karakoti A, Singh S, Self W, Tyler R, Seal S, Reilly CM (2011) Bio-distribution and in vivo antioxidant effects of cerium oxide nanoparticles in mice. Environ Toxicol doi:10.1002/tox.20704
125. Colon J, Herrera L, Smith J, Patil S, Komanski C, Kupelian P, Seal S, Jenkins DW, Baker CH (2009) Protection from radiation-induced pneumonitis using cerium oxide nanoparticles. Nanomed Nanotechnol Biol Med 5:225–231
126. Tsai Y-Y, Oca-Cossio J, Lin S-M, Woan K, Yu P-C, Sigmund W (2008) Reactive oxygen species scavenging properties of ZrO_2–CeO_2 solid solution nanoparticles. Nanomedicine 3:637–645
127. Niu J, Azfer A, Rogers LM, Wang X, Kolattukudy PE (2007) Cardioprotective effects of cerium oxide nanoparticles in a transgenic murine model of cardiomyopathy. Cardiovasc Res 73:549–559
128. Schbert D, Dargusch R, Raitano J, Chan S-W (2006) Cerium and yttrium oxide nanoparticles are neuroprotective. Biochem Biophys Res Commun 342:86–91
129. Zhang L, Laug L, Münchgesang W, Pippel E, Gösele U, Brandsch M, Knez M (2010) Reducing stress on cells with apoferritin-encapsulated platinum nanoparticles. Nano Lett 10:219–223
130. Onizawa S, Aoshiba K, Kajita M, Miyamoto Y, Nagai A (2009) Platinum nanoparticle antioxidants inhibit pulmonary inflammation in mice exposed to cigarette smoke. Pulmonary Pharmacol Therap 22:340–349
131. Hikosaka K, kim J, Kajita M, Kanayama A, Miyamoto Y (2008) Platinum nanoparticles have an activity similar to mitochondrial NADH:ubiquinone oxidoreductase. Colloids Surf B: Biointerfaces 66:195–200
132. Sakaue Y, kim J, Miyamoto Y (2010) Effects of TAT-conjugated platinum nanoparticles on lifespan of mitochondrial electron transport complex I-deficient *Caenorhabditis elegans, nuo-1*. Int J Nanomed 5:687–695
133. Barath Mani Kanth S, Kalishwaralal K, Sriram M, Pandian SRK, Youn H-S, Eom S, Gurunathan S (2010) Anti-oxidant effect of gold nanoparticles restrains hyperglycemic conditions in diabetic mice. J Nanobiotechnol 8:16
134. Sul O-J, Kim J-C, Kyung T-W, Kim H-J, Kim Y-Y, Kim S-H, Kim J-S, Choi H-S (2010) Gold nanoparticles inhibited the receptor activator of nuclear factor-kB ligand (RANKL)-induced osteoclast formation by acting as an antioxidant. Biosci Biotechnol Biochem 74:2209–2213
135. Mohammad G, Mishra V, Pandey HP (2008) Antioxidant properties of some nanoparticles may enhance wound healing in T2DM patient. Digest J Nanomat Biostruct 3:159–162
136. Yin J-J, Lao F, Fu P, Wamer WG, Zhao Y, Wang PC, Qiu Y, Sun B, Xing G, Dong J, Liang X-J, Chen C (2009) The scavenging of reactive oxygen species and the potential for cell protection by functionalized fullerene materials. Biomaterial 30:611–621
137. Bawa R (2008) Nanoparticle-based therapeutics in humans; a survey. Nanotechnol Law Business 2008(5):135–155
138. Sanvicens N, Marco MP (2008) Multifunctional nanoparticles-properties and prospects for their use in human medicine. Trends Biotechnol 26:425–433

139. De Jong WH, Borm PJA (2008) Drug delivery and nanoparticles: applications and hazards. Int J Nanomed 3:133–149
140. Nair HB, Sung B, Yadev VR, Kannappan R, Chaturvedi MM, Aggarwal BB (2010) Delivery of anti-inflammatory nutraceuticals by nanoparticles for prevention and treatment of cancer. Biochem Pharmacol 80: 1833–1843
141. Lim Z-ZJ, Li J-EJ, Ng C-T, Yung A-YL, Bay B-H (2011) Gold nanoparticles in cancer therapy. Acta Pharmacol Sinica 32:983–990
142. Thakor AS, Jokerst J, Zavaleta C, Massoud TF, Gambhir SS (2011) Gold nanoparticles a revival in precious metal administration to patients. Nano Lett 11(10):4029–4036
143. Oo MKK, Yang X, Du H, Wang H (2008) Aminolevulinic acid-conjugated gold nanoparticles for photodynamic therapy of cancer. Nanomedicine 3:777–786
144. Kim D, Jeong YY, Jon S (2010) Drug-loaded aptamer-gold nanoparticle bioconjugated for combined CT imaging and therapy of prostate cancer. ACS Nano 4:3689–3696
145. Reddy AK, Rathinaraj BS, Prathyusha P, BHargav TA, Rajani T, Rajamanickam V, Krishna Reddy P, Arul Selvam M (2011) Emerging trends of nanotechnology in cancer therapy. Int J Pharamc Biolo Arch 2:1
146. Geng F, Song K, Xing JZ, Yuan C, Yan S, Yang Q, Chen J, Kong B (2011) Thio-glucose bound gold nanoparticles enhance radio-cytotoxic targeting of ovarian cancer. Nanotechnology 22(28):285101
147. Ahamed M, Sldalhi MS, Siddiqui MKJ (2010) Silver nanoparticle applications and human health. Clinica Chimica Acta 411:841–1848
148. Lara HH, Ayala-Nuñez N, Ixtepan-Turrent L, Rodriguez-Padilla C (2010) Mode of antiviral action of silver nanoparticles against HIV-1. J Nanobiotechnol 8:1
149. Sriram MI, Mani Kanth SB, Kalishwaralal K (2010) Antitumor activity of silver nanoparticles in Dalton's lymphoma ascites tumor model. Int J Nanomedicine 5: 753–762
150. David AE, Cole AJ, Yang VC (2011) Magnetically targeted nanoparticles for brain tumor therapy; what does the future hold? Nanomedicine 6:1133–1135
151. Suri SS, Fenniri H, Singh B (2007) Nanotechnology-based drug delivery systems. J Occupational Med Toxicol 2:16
152. Lim KJ, Bisht S, Bar EE, Maitra A, Eberhart CG (2011) A polymeric nanoparticle formulation of curcumin inhibits growth, clonogenicity and stem-like fraction in malignant brain tumors. Cancer Biol Therp 11:1–10

Chapter 3

Encapsulation of Antioxidants in Gastrointestinal-Resistant Nanoparticulate Carriers

Eliana B. Souto, Patrícia Severino, Rafael Basso, and Maria Helena A. Santana

Abstract

Reactive oxygen species (ROS) are known to cause several human pathologies. For this reason, antioxidants have gained utmost importance because of their potential as prophylactic and therapeutic agents in many diseases. Examples of their application include their use in diabetic patients, as aging drugs, in cancer diseases, Parkinson's, Alzheimer's, autoimmune disorders, and also in inflammation. Antioxidants have limited absorption profiles, therefore low bioavailability and low concentrations at the target site. Efforts have been done towards loading antioxidant molecules in advanced nanoparticulate carriers, e.g., liposomes, polymeric nanoparticles, solid lipid nanoparticles, self-emulsifying drug delivery system. Examples of successful achievements include the encapsulation of drugs and other active ingredients, e.g., coenzyme Q10, vitamin E and vitamin A, resveratrol and polyphenols, curcumin, lycopene, silymarin, and superoxide dismutase. This review focuses on the comprehensive analysis of using nanoparticulate carriers for loading these molecules for oral administration.

Key words Antioxidants, Nanoparticles, Oral administration, Disease, Bioavailability

1 Introduction

Reactive oxygen species (ROS) such as superoxide anion (O_2^-), hydrogen peroxide (H_2O_2), and hydroxyl radicals (·OH) are generated as byproducts of cellular metabolism, and if remaining in the human body, it would result in a variety of human diseases [1, 2]. Oxidative stress is the result of an imbalance between oxidants and antioxidants in favor of the former. Increased production of oxidants, and/or decreased availability of antioxidants, triggers a cascade of oxidative reactions damaging lipids, proteins, enzymes, carbohydrates, and DNA, ultimately leading to cell death induced by DNA fragmentation and lipid peroxidation [3–5]. To approach this concern, antioxidants have received particular attention

Donald Armstrong and Dhruba J. Bharali (eds.), *Oxidative Stress and Nanotechnology: Methods and Protocols*, Methods in Molecular Biology, vol. 1028, DOI 10.1007/978-1-62703-475-3_3, © Springer Science+Business Media New York 2013

because of their potential as prophylactic and therapeutic agents in many diseases [6]. Examples of such oxidative stress-related disease include atherosclerosis [7], myocardial infarction [8], hypertension [9], diabetes [10], aging [11], cancer [12], Parkinson's and Alzheimer's diseases [13], Huntington's disease [14], autoimmune disorders [15], age-related macular degeneration, acute lung injury, acute respiratory distress syndrome, inflammation [16], and hyperoxia [17].

Due to the limited absorption of the antioxidant molecules and their low bioavailability, this results in subtherapeutic concentrations at the target site. To overcome the bioavailability issues, advanced nanoparticulate carriers, designed to provide localized or targeted delivery of these agents, may represent a more viable therapeutic option. Several carriers, e.g., liposomes [18, 19], polymeric nanoparticles [20, 21], solid lipid nanoparticles [22], self-emulsifying drug delivery system [23], have shown to significantly enhance the preventive/therapeutic efficacy of many antioxidants by increasing their bioavailability and target ability.

A recent increased interest in nanoparticulate carriers represents a new era in the bound of nanotechnology and biotechnology. These studies have revealed a growing realization of the potential utility of novel environmentally benign technologies in diagnosis and therapeutic use in biological systems [24]. The exact nature of the interactions between nanoparticles and cells depends on the nanoparticle exposure route, as this will determine the point of contact between human tissue and the nanomaterial [25].

Despite innovations being released in the field of drug delivery, in addition to new knowledge about promising alternative administration routes, it is estimated that over 90 % of all medicines comprise the oral formulations [26]. For oral use, nanoparticulates reach the gastrointestinal tract and remain in contact with the intestinal epithelium for about 4–8 h. In the gastrointestinal tract, epithelial cells form a monolayer lining the intestinal lumen, and this epithelium forms a barrier that separates the sterile body interior from the harsh conditions of the lumen [27, 28].

Oral nanoparticulate carriers are interesting because a considerable number of newly developed drug candidates and marketed substances belong to the class II and IV of the Biopharmaceutics Classification System (BCS); as a common characteristic, they suffer from insufficient bioavailability due to the low solubility. Nanosizing an oral drug candidate, it may clearly improve its dissolution profile [26, 29]. In addition, it is established that nanoparticulate carriers enhance the oral bioavailability by facilitating drug uptake via a transcellular or paracellular pathway [30].

2 Role of Antioxidants in Oxidative Stress

Antioxidants are substances which counteract free radicals and prevent the damage caused by them [6, 31]. With respect to their activity, antioxidants may be classified as enzymatic or nonenzymatic type, according to Fig. 1.

The enzymatic type comprises a class of antioxidants that has efficient and specific biocatalysts in nature, playing an important role in human beings, because they participate in cellular metabolic processes by enhancing the rates of reactions between biomolecules [32]. In biological organisms, enzymes contribute dominatingly to enhance cellular antioxidative defense against oxidative stress [33–35]. The most well-known example are superoxide bismutase catalyzes dismutation of superoxide radicals to oxygen and hydrogen peroxide superoxide dismutase catalyzes the disproportionation of superoxide (O_2^-) to molecular oxygen and peroxide and thus has a key role in protecting the cell against the toxic products of aerobic respiration [33, 36]. And, catalase catalyzes the decomposition of hydrogen peroxide to water and is one of the fastest biocatalysts known, with the highest turnover numbers of all known enzymes (40,000,000 molecules/s) [33].

Nonenzymatic type is a class of antioxidant depending on dietary supplementation and lifestyle [37]. Human antioxidant defense system is incomplete without dietary supplied vitamins. At some point in the evolutionary process, uric acid replaced the

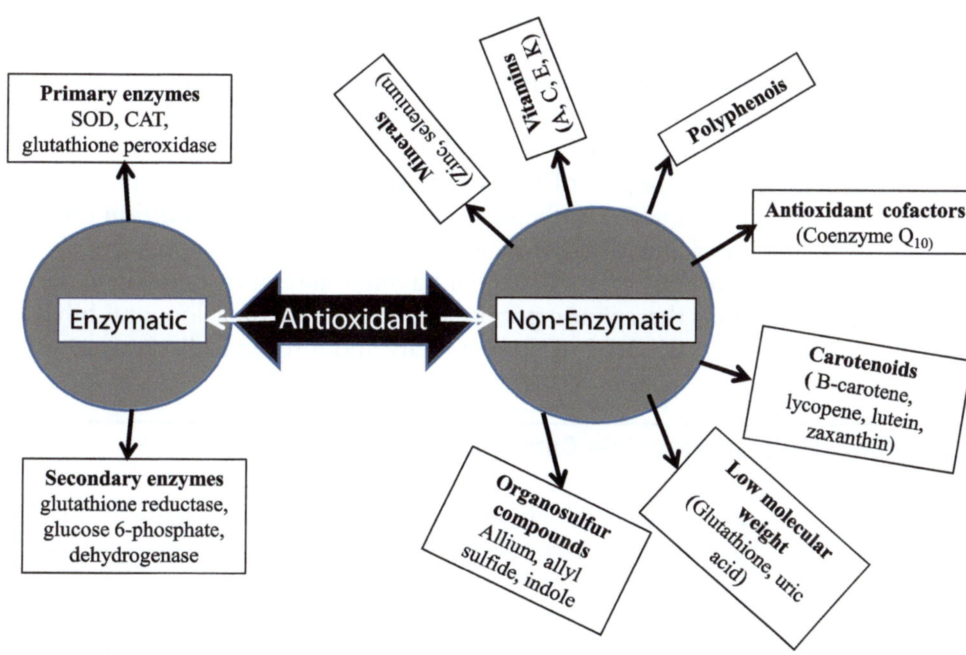

Fig. 1 Classification of anti-oxidants may be classified as enzymatic or nonenzymatic type

ascorbic acid as the major water-soluble antioxidant in human biology. Humans lack in the ability to synthesize ascorbic acid endogenously, for which the organism has absolute requirement. Apart from ascorbic acid, other antioxidants, like vitamin E and vitamin A [37], coenzyme Q_{10} [38], resveratrol [39], and polyphenols [40], are obtained from external sources and play an important role in maintaining human health [41].

Studies in clinical and animal models have already shown a decrease in nonenzymatic antioxidants during aging [42]. Human studies on the role of dietary antioxidants on healthy aging process are not well established. Consequently, most of the studies with significant findings are from animal trials [37]. Examples of pharmacological action of nonenzymatic antioxidants are vitamin A, which exhibits important biological functions as protecting the vision, growth, reproduction, and the differentiation; immune system; photo-protective effects in skin; potent chemotherapeutic agent for treatment of acute promyelocytic leukemia; and for epithelial tumors [43, 44].

Another example is coenzyme Q_{10}, which is found in all cells of the human body, as heart, liver, and skeletal muscles. It functions as a coenzyme in the energy-producing metabolic pathways of every cell of the body. As antioxidant molecule, it scavenges free radicals and inhibits lipid peroxidation [38]. Several studies have provided evidence of the potential of coenzyme Q_{10} in prophylaxis and therapy of various disorders related to oxidative stress [45, 46], e.g., neurodegenerative disorders, diabetes, cancer, and muscular and cardiovascular [47]. Coenzyme Q_{10} is practically insoluble even in the presence of 5 % sodium lauryl sulfate in water and poorly absorbed from the gastrointestinal tract. The slow absorption of coenzyme Q_{10} (T_{max} 2–10 h) from the gastrointestinal tract is attributed to its high molecular weight and poor water solubility [48].

Examples of flavonoids are the catechins, found in green tea, which have also being received considerable attention due to their medicinal attributes, e.g., cardio- and neuro-protection, as anticancers, antioxidants, antiagings and antiviral drugs. However, their therapeutic potential is limited by their low oral bioavailability, attributed to poor stability and intestinal absorption [40].

3 Applications of Nanoparticulate Carriers in Oxidative Stress

Pharmaceutical nanoparticulate carriers have changed the pathway of disease treatment [49]. The therapeutic efficiency of oral administration qualifies this route as the most convenient for delivering drugs [50]. To make oral route delivery possible, the risk of physicochemical degradation of antioxidants in the gastrointestinal tract and its absorption hindrance [41] need to be circumvented.

Advantages of using nanoparticulate carriers to overcome these obstacles are their ability to manipulate the release profile and promoting revolution the conventional drug delivery system. Various types of materials have been employed to produce nanoparticulate carriers, both from natural and synthetic sources [49].

3.1 Liposomes

Liposomes have been employed as drug delivery vehicles for several decades. It is the spherical bilayer vesicles with an aqueous interior formed by the self-association behavior of amphiphilic phospholipids with cholesterol molecules [51, 52]. Advantages of this carrier are ability to interact with cells, produced in a wide range of structural and compositional configurations and capacity to load biopharmaceutical drugs [53].

There are various types of antioxidants formulated in liposomes, e.g., quercetin [54], xanthii fructus extract [19], curcumin [55], lycopene [18], silymarin [56, 57], vitamin E [58], superoxide dismutase [59, 60], and resveratrol [39].

Wang et al. [39] investigated the protective effect of resveratrol derived from *Polygonum cuspidatum* and its liposomal form on nigral cells of Parkinsons disease-bearing rats. *Polygonum cuspidatum* and its liposomal form were administered orally. Results showed that after 14 days of oral treatment with resveratrol or resveratrol liposome (20 mg/kg per day), the abnormal rotational behavior, the loss and apoptosis of nigral cells, and the levels of total ROS were markedly decreased and the total antioxidant capability of nigral tissues was significantly improved. Resveratrol liposome showed more profound effects than free resveratrol. Due to the increased bioavailability, resveratrol liposomes were more potent and may become a promising alternative for the treatment of Parkinsons disease.

3.2 Solid Lipid Nanoparticles

The current increasing interest in lipid nanoparticles (SLN) relies on the advantages of lipid materials in comparison to others, as biocompatibility, biodegradation, controlled release, easy scale up, and low cost [61]. Furthermore, these nanoparticles show excellent physical and chemical stability providing greater protection against degradation of drugs and ability to load hydrophobic, hydrophilic, and amphiphilic biomacromolecules [62].

Quercetin is the major representative of the flavonoid class. It has been shown a variety of biological activities and pharmacological actions, such as dilating coronary arteries, decreasing blood lipid, antiplatelet aggregation, anticancer, antioxidation, anti-anemia, anti-inflammation, and antianaphylaxis effects [63]. Li et al. [22] prepared SLN by an emulsification and low-temperature solidification method to load quercetin (QT-SLN). QT-SLN depicted a mean average size of ca. 155 nm, high zeta potential values (>30 mV) and were spherically shaped. The encapsulation efficiency and loading capacity reached values of above 90 and

13 %. The absorption of QT-SLN in the gastrointestinal tract was studied in rats using an in situ perfusion method. The pharmacokinetic study was conducted in rats after oral administration of quercetin (50 mg/kg) in the form of either QT-SLN or as suspension. QT-SLN enhanced the absorption compared to quercetin suspension.

In another study Wang et al. [64] developed SLN containing total flavones of *Hippophae rhamnoides* (TFH-SLN) by hot or cold high-pressure homogenization (HPH). The influence of process parameters, i.e., size distribution, zeta potential, encapsulation efficiency, crystal form, and in vitro release profile was investigated. Results showed that TFH-SLN improved oral bioavailability of TFH and prolonged the mean retention time and drug release.

3.3 Polysaccharide Nanoparticles

Among several nanoparticulate carriers, the use of ionic interaction between positively and negatively charged polysaccharides to produce biocompatible nanoparticles is being currently exploited. Due to the exceptional characteristics of the positively charged chitosan (e.g., biocompatible, nontoxic, and mucoadhesive polymer), many researchers selected this polysaccharide as potential material for oral drug delivery [51, 65, 66].

Dube et al. [20] encapsulated (+)-catechin (C) and (−)-epigallocatechin gallate (EGCg) in chitosan nanoparticles as a means of enhancing their intestinal absorption. Using excised mouse jejunum, encapsulation significantly enhanced ($p<0.05$) intestinal absorption. The cumulative amounts transported after encapsulation were significantly higher for chitosan nanoparticles. The mechanism by which absorption was enhanced was not through an effect of chitosan nanoparticle on intestinal paracellular or passive transcellular transport processes or an effect on efflux proteins, but it was likely due to stabilization of catechins after encapsulation. Encapsulation of catechins in chitosan nanoparticle enhances their intestinal absorption and is a promising strategy for improving their bioavailability.

3.4 Self-Emulsifying Drug Delivery Systems

Self-emulsifying drug delivery systems (SEDDS) can be used to improve the oral absorption of lipophilic drugs [23, 67]. SEDDS are isotropic mixtures of oils and surfactants that form fine oil-in-water emulsions upon mild agitation in aqueous media such as gastrointestinal tract fluids [23, 67]. Following oral administration, SEDDS were reported to provide a twofold increase in the bioavailability compared to a powder formulation [68]. SEDDS present the solubilized drugs in an oil-in-water emulsion of small droplet size (nanometers), and this therefore ultimately results in increased drug absorption [23, 67].

Balakrishnan et al. [69] formulated SEDDS composed of oil, surfactant, and cosurfactant for oral administration of coenzyme Q_{10} to enhance the solubility and bioavailability. The formulations were prepared using two oils (Labrafil® M 1944 and Labrafil® M 2125), surfactant (Labrasol®), and cosurfactant (Lauroglycol® FCC and Capryol® 90). The level of coenzyme Q_{10} was fixed at 6 % (w/v) of the vehicle for all formulations, which were characterized by solubility of the drug in the vehicle, particle size of the dispersed emulsion, zeta potential, and drug release profile. Ternary phase diagrams were used to evaluate the emulsification domain. The self-emulsification time following introduction into an aqueous medium under gentle agitation was evaluated. The optimized SEDDS showed mean droplet size ca. 240 nm and optimal drug release profile in water. The pharmacokinetic study in rats for the optimized formulation was performed and compared to powder formulation. SEDDS have significantly increased the C_{max} and area under the curve (AUC) of coenzyme Q_{10} compared to powder ($p < 0.05$). Thus, this self-microemulsifying drug delivery system should be an effective oral dosage form for improving oral bioavailability of coenzyme Q_{10}.

In another study, Taha et al. [43] compared the bioavailability of vitamin A SEDDS, vitamin A optimized in capsule, and free vitamin A. Bioavailability was assessed after a single oral dose. The bioavailability increased almost twofold and 1.4-fold for SEDDS and optimized formulation, respectively. The study showed that SEDDS formulations increased the rate and extent of drug absorption compared to the free vitamin A.

3.5 Cerium Oxide Nanoparticles (Nanoceria)

Nanoceria are another type of nanoparticulate carriers, based on cerium oxide nanoparticles. These particles are known as catalysts, revealing high pharmacological potential as antioxidants [70]. Nanoceria catalytically scavenge ROS by mimicking the activities of superoxide dismutase and catalase [71]. These defects, compensated by oxygen vacancies, are enriched at the surface and therefore in nanosized particles. Reactions involving redox cycles between the Ce(3+) and Ce(4+) oxidation states allow nanoceria to react catalytically with superoxide and hydrogen peroxide, mimicking the behavior of two key anti-oxidant enzymes, superoxide dismutase, and catalase, potentially abating all noxious intracellular ROS via a self-regenerating mechanism [70]. These particles are well tolerated by the organism, might fight against chronic inflammation and the pathologies associated with oxidative stress, which include cancer and neurodegeneration.

4 Future Perspectives

ROS are widely believed to cause or aggravate several human pathologies. Antioxidants are assumed to counteract the harmful effects of ROS and therefore prevent or treat oxidative stress-related diseases. Pharmaceutical research has been employing nanoparticulate carriers to improve bioavailability and low absorption and increase the stability of antioxidants in gastrointestinal tract. This area is still relatively new, but antioxidants as flavonoids, coenzyme Q_{10}, quercetin, curcumin, lycopene, silymarin, alpha-tocopherol, superoxide dismutase, resveratrol have found accepted clinical use and are being showing promising results.

References

1. Huang X, Liu Y, Liang K, Tang Y, Liu J (2008) Construction of the active site of glutathione peroxidase on polymer-based nanoparticles. Biomacromolecules 9:1467–1473
2. Naik E, Dixit VM (2011) Mitochondrial reactive oxygen species drive proinflammatory cytokine production. J Exp Med 208:417–420
3. Nur E, Biemond BJ, Otten HM, Brandjes DP, Schnog JJ (2011) Oxidative stress in sickle cell disease; pathophysiology and potential implications for disease management. Am J Hematol 86:484–489
4. Beckman KB, Ames BN (1998) The free radical theory of aging matures. Physiol Rev 78: 547–581
5. Halliwell B, Gutteridge JM (1984) Lipid peroxidation, oxygen radicals, cell damage, and antioxidant therapy. Lancet 1:1396–1397
6. Ratnam DV, Ankola DD, Bhardwaj V, Sahana DK, Kumar MN (2006) Role of antioxidants in prophylaxis and therapy: a pharmaceutical perspective. J Control Release 113:189–207
7. Park JG, Yoo JY, Jeong SJ, Choi JH, Lee MR, Lee MN, Lee JH, Kim HC, Jo H, Yu DY, Kang SW, Rhee SG, Lee MH, Oh GT (2011) Peroxiredoxin 2 deficiency exacerbates atherosclerosis in apolipoprotein E-deficient mice. Circ Res 109(7):739–749
8. Houston MC (2011) Role of mercury toxicity in hypertension, cardiovascular disease, and stroke. J Clin Hypertens (Greenwich) 13:621–627
9. Drummond GR, Selemidis S, Griendling KK, Sobey CG (2011) Combating oxidative stress in vascular disease: NADPH oxidases as therapeutic targets. Nat Rev Drug Discov 10: 453–471
10. Yang H, Jin X, Wai Kei Lam C, Yan SK (2011) Review: oxidative stress and diabetes mellitus. Clin Chem Lab Med 49(11):1773–1782
11. Farboud ES, Nasrollahi SA, Tabbakhi Z (2011) Novel formulation and evaluation of a Q10-loaded solid lipid nanoparticle cream: in vitro and in vivo studies. Int J Nanomedicine 6: 611–617
12. Riemann A, Schneider B, Ihling A, Nowak M, Sauvant C, Thews O, Gekle M (2011) Acidic environment leads to ROS-induced MAPK signaling in cancer cells. PLoS One 6:e22445
13. Klyachko NL, Manickam DS, Brynskikh AM, Uglanova SV, Li S, Higginbotham SM, Bronich TK, Batrakova EV, Kabanov AV (2012) Cross-linked antioxidant nanozymes for improved delivery to CNS. Nanomedicine 8:119–129.
14. Tasset I, Pontes AJ, Hinojosa AJ, de la Torre R, Tunez I (2011) Olive oil reduces oxidative damage in a 3-nitropropionic acid-induced Huntington's disease-like rat model. Nutr Neurosci 14:106–111
15. Caccamo D, Curro M, Ferlazzo N, Condello S, Ientile R (2012) Monitoring of transglutaminase2 under different oxidative stress conditions. Amino Acids 42:1037–1043.
16. Lee DM, Jackson KW, Knowlton N, Wages J, Alaupovic P, Samuelsson O, Saeed A, Centola M, Attman PO (2011) Oxidative stress and inflammation in renal patients and healthy subjects. PLoS One 6:e22360
17. Gao Z, Spilk S, Momen A, Muller MD, Leuenberger UA, Sinoway LI (2012) Vitamin C prevents hyperoxia-mediated coronary vasoconstriction and impairment of myocardial function in healthy subjects. Eur J Appl Physiol 112:483–492.
18. Tian YY, Ge L, Duan XL, Gao ZQ, Chang YZ (2007) Lycopene liposomes: lycopene release in vitro and pharmaceutical behaviors and antioxidation in vivo. Yao Xue Xue Bao 42: 1107–1111
19. Huang MH, Wang BS, Chiu CS, Amagaya S, Hsieh WT, Huang SS, Shie PH, Huang GJ (2011) Antioxidant, antinociceptive, and anti-inflammatory activities of Xanthii Fructus extract. J Ethnopharmacol 135:545–552
20. Dube A, Nicolazzo JA, Larson I (2010) Chitosan nanoparticles enhance the intestinal absorption of the green tea catechins (+)-catechin and (−)-epigallocatechin gallate. Eur J Pharm Sci 41:219–225
21. Ratnam DV, Chandraiah G, Meena AK, Ramarao P, Kumar MN (2009) The co-encapsulated antioxidant nanoparticles of ellagic acid and coenzyme Q10 ameliorates

hyperlipidemia in high fat diet fed rats. J Nanosci Nanotechnol 9:6741–6746

22. Li H, Zhao X, Ma Y, Zhai G, Li L, Lou H (2009) Enhancement of gastrointestinal absorption of quercetin by solid lipid nanoparticles. J Control Release 133:238–244
23. Setthacheewakul S, Kedjinda W, Maneenuan D, Wiwattanapatapee R (2010) Controlled release of oral tetrahydrocurcumin from a novel self-emulsifying floating drug delivery system (SEFDDS). AAPS PharmSciTech 12:152–164
24. Amin KA, Hassan MS, Awad el ST, Hashem KS (2011) The protective effects of cerium oxide nanoparticles against hepatic oxidative damage induced by monocrotaline. Int J Nanomedicine 6:143–149
25. Bimbo LM, Makila E, Laaksonen T, Lehto VP, Salonen J, Hirvonen J, Santos HA (2011) Drug permeation across intestinal epithelial cells using porous silicon nanoparticles. Biomaterials 32:2625–2633
26. Gabor F, Fillafer C, Neutsch L, Ratzinger G, Wirth M (2008) Improving oral delivery. Handb Exp Pharmacol 197:345–398
27. Wilfart A, Montagne L, Simmins H, Noblet J, Milgen J (2007) Digesta transit in different segments of the gastrointestinal tract of pigs as affected by insoluble fibre supplied by wheat bran. Br J Nutr 98:54–62
28. Thubagere A, Reinhard BM (2010) Nanoparticle-induced apoptosis propagates through hydrogen-peroxide-mediated bystander killing: insights from a human intestinal epithelium in vitro model. ACS Nano 4: 3611–3622
29. Kesisoglou F, Panmai S, Wu Y (2007) Nanosizing-oral formulation development and biopharmaceutical evaluation. Adv Drug Deliv Rev 59:631–644
30. Ramesan RM, Sharma CP (2009) Challenges and advances in nanoparticle-based oral insulin delivery. Expert Rev Med Devices 6:665–676
31. Azzi A, Davies KJ, Kelly F (2004) Free radical biology—terminology and critical thinking. FEBS Lett 558:3–6
32. Yin Y, Huang X, Lv C, Wang L, Yu S, Luo Q, Xu J, Liu J (2010) Construction of an artificial glutathione peroxidase active site on copolymer vesicles. Macromol Biosci 10:1505–1516
33. McCord JM, Fridovich I (1969) Superoxide dismutase. An enzymic function for erythrocuprein (hemocuprein). J Biol Chem 244: 6049–6055
34. Miriyala S, Holley AK, St Clair DK (2011) Mitochondrial superoxide dismutase-signals of distinction. Anticancer Agents Med Chem 11:181–190
35. Iborra M, Moret I, Rausell F, Bastida G, Aguas M, Cerrillo E, Nos P, Beltran B (2011) Role of oxidative stress and antioxidant enzymes in Crohn's disease. Biochem Soc Trans 39:1102–1106
36. Bertini I, Mangani S, Viezzoli MS (1998) Structure and properties of copper-zinc superoxide dismutases. Adv Inorg Chem 45(45): 127–250
37. Chong-Han K (2011) Dietary lipophilic antioxidants: implications and significance in the aging process. Crit Rev Food Sci Nutr 50: 931–937
38. Aberg F, Appelkvist EL, Dallner G, Ernster L (1992) Distribution and redox state of ubiquinones in rat and human tissues. Arch Biochem Biophys 295:230–234
39. Wang Y, Xu H, Fu Q, Ma R, Xiang J (2011) Protective effect of resveratrol derived from Polygonum cuspidatum and its liposomal form on nigral cells in Parkinsonian rats. J Neurol Sci 304:29–34
40. Saito ST, Gosmann G, Pungartnik C, Brendel M (2009) Green tea extract-patents and diversity of uses. Recent Pat Food Nutr Agric 1:203–215
41. Firuzi O, Miri R, Tavakkoli M, Saso L (2011) Antioxidant therapy: current status and future prospects. Curr Med Chem 18:3871–3888
42. Van der Loo B, Bachschmid M, Spitzer V, Brey L, Ullrich V, Luscher TF (2003) Decreased plasma and tissue levels of vitamin C in a rat model of aging: implications for antioxidative defense. Biochem Biophys Res Commun 303:483–487
43. Taha E, Ghorab D, Zaghloul AA (2007) Bioavailability assessment of vitamin A self-nanoemulsified drug delivery systems in rats: a comparative study. Med Princ Pract 16: 355–359
44. D'Ambrosio DN, Clugston RD, Blaner WS (2011) Vitamin A metabolism: an update. Nutrients 3:63–103
45. Orsucci D, Mancuso M, Ienco EC, Logerfo A, Siciliano G (2011) Targeting mitochondrial dysfunction and neurodegeneration by means of coenzyme Q10 and its analogues. Curr Med Chem 18(26):4053–4064
46. Kalayci M, Unal MM, Gul S, Acikgoz S, Kandemir N, Hanci V, Edebali N, Acikgoz B (2011) The effect of coenzyme Q10 on ischemia and neuronal damage in an experimental traumatic brain injury model in rats. BMC Neurosci 12:75
47. Villalba JM, Parrado C, Santos-Gonzalez M, Alcain FJ (2010) Therapeutic use of coenzyme Q10 and coenzyme Q10-related compounds

and formulations. Expert Opin Investig Drugs 19:535–554
48. Seo DW, Kang MJ, Sohn Y, Lee J (2009) Self-microemulsifying formulation-based oral solution of coenzyme Q10. Yakugaku Zasshi 129:1559–1563
49. Lopes CM, Martins-Lopes P, Souto EB (2010) Nanoparticulate carriers (NPC) for oral pharmaceutics and nutraceutics. Pharmazie 65:75–82
50. Harde H, Das M, Jain S (2011) Solid lipid nanoparticles: an oral bioavailability enhancer vehicle. Expert Opin Drug Deliv 8(11):1407–1424
51. Bansal SS, Goel M, Aqil F, Vadhanam MV, Gupta RC (2011) Advanced drug delivery systems of curcumin for cancer chemoprevention. Cancer Prev Res (Phila) 4:1158–1171
52. Bawarski WE, Chidlowsky E, Bharali DJ, Mousa SA (2008) Emerging nanopharmaceuticals. Nanomedicine 4:273–282
53. Barea MJ, Jenkins MJ, Gaber MH, Bridson RH (2010) Evaluation of liposomes coated with a pH responsive polymer. Int J Pharm 402:89–94
54. Ghosh A, Mandal AK, Sarkar S, Das N (2011) Hepatoprotective and neuroprotective activity of liposomal quercetin in combating chronic arsenic induced oxidative damage in liver and brain of rats. Drug Deliv 18:451–459
55. Takahashi M, Uechi S, Takara K, Asikin Y, Wada K (2009) Evaluation of an oral carrier system in rats: bioavailability and antioxidant properties of liposome-encapsulated curcumin. J Agric Food Chem 57:9141–9146
56. Wagoner J, Morishima C, Graf TN, Oberlies NH, Teissier E, Pecheur EI, Tavis JE, Polyak SJ (2011) Differential in vitro effects of intravenous versus oral formulations of silibinin on the HCV life cycle and inflammation. PLoS One 6:e16464
57. Yan-yu X, Yun-mei S, Zhi-peng C, Qi-neng P (2006) Preparation of silymarin proliposome: a new way to increase oral bioavailability of silymarin in beagle dogs. Int J Pharm 319:162–168
58. Nacka F, Cansell M, Meleard P, Combe N (2001) Incorporation of alpha-tocopherol in marine lipid-based liposomes: in vitro and in vivo studies. Lipids 36:1313–1320
59. Regnault C, Soursac M, Roch-Arveiller M, Postaire E, Hazebroucq G (1996) Pharmacokinetics of superoxide dismutase in rats after oral administration. Biopharm Drug Dispos 17:165–174
60. Regnault C, Roch-Arveiller M, Tissot M, Sarfati G, Giroud JP, Postaire E, Hazebroucq G (1995) Effect of encapsulation on the anti-inflammatory properties of superoxide dismutase after oral administration. Clin Chim Acta 240:117–127
61. Souto EB, Muller RH (2010) Lipid nanoparticles: effect on bioavailability and pharmacokinetic changes. Handb Exp Pharmacol 197: 115–141
62. Severino P, Pinho SC, Souto EB, Santana MH (2011) Polymorphism, crystallinity and hydrophilic-lipophilic balance of stearic acid and stearic acid-capric/caprylic triglyceride matrices for production of stable nanoparticles. Colloids Surf B Biointerfaces 86:125–130
63. Hollman PC, Katan MB (1999) Dietary flavonoids: intake, health effects and bioavailability. Food Chem Toxicol 37:937–942
64. Wang D, Zhao P, Cuia F, Li X (2007) Preparation and characterization of solid lipid nanoparticles loaded with total flavones of Hippophae rhamnoides (TFH). PDA J Pharm Sci Technol 61:110–120
65. Nagamoto T, Hattori Y, Takayama K, Maitani Y (2004) Novel chitosan particles and chitosan-coated emulsions inducing immune response via intranasal vaccine delivery. Pharm Res 21:671–674
66. Da Silva CF, Severino P, Martins F, Chaud MV, Santana MH (2009) The intestinal permeation of didanosine from granules containing microspheres using the everted gut sac model. J Microencapsul 26:523–528
67. Constantinides PP (1995) Lipid microemulsions for improving drug dissolution and oral absorption: physical and biopharmaceutical aspects. Pharm Res 12:1561–1572
68. Ljusberg-Wahren H, Seier Nielsen F, Brogard M, Troedsson E, Mullertz A (2005) Enzymatic characterization of lipid-based drug delivery systems. Int J Pharm 298:328–332
69. Balakrishnan P, Lee BJ, Oh DH, Kim JO, Lee YI, Kim DD, Jee JP, Lee YB, Woo JS, Yong CS, Choi HG (2009) Enhanced oral bioavailability of coenzyme Q10 by self-emulsifying drug delivery systems. Int J Pharm 374:66–72
70. Celardo I, Pedersen JZ, Traversa E, Ghibelli L (2011) Pharmacological potential of cerium oxide nanoparticles. Nanoscale 3:1411–1420
71. Zhou X, Wong LL, Karakoti AS, Seal S, McGinnis JF (2011) Nanoceria inhibit the development and promote the regression of pathologic retinal neovascularization in the Vldlr knockout mouse. PLoS One 6:e16733

Chapter 4

Toxicity of Novel Nanosized Formulations Used in Medicine

Afaf El-Ansary, Sooad Al-Daihan, Abir Ben Bacha, and Malak Kotb

Abstract

Nanotechnology involves the creation and manipulation of materials at nanoscale levels (1–100 nm) to create products that exhibit novel properties. While this motivation has driven nanoscience and technology in physics and engineering, it is not the main reason that nanoparticles are useful for systemic applications in the human body. The application of nanotechnology to medicine, known as nanomedicine, concerns the use of precisely engineered materials at this length scale to develop novel therapeutic and diagnostic modalities. A number of nanotherapeutic formulations are already approved for medical use and more are in the approval pipeline currently. This chapter is intended to provide an overview of the toxicity of these therapeutic nanoparticles and to summarize the current state of the field. We begin with background on the sources of exposure to nanoparticles, followed by reviewing different forms of nanosized therapeutic tools as quantum dots, nanoshells, nanocapsules, echogenic bubble, and "nanoshuttles." Moreover, cytotoxic effects of nanoparticles on cell membrane, mitochondrial function, prooxidant/antioxidant status, enzyme leakage, DNA, and other biochemical endpoints were elucidated. We highlight the need for caution during the use and disposal of such manufactured nanomaterials to prevent unintended environmental impacts. Moreover, different strategies which could be used to minimize or eliminate nanotoxicity were also discussed in detail. Understanding of how to tune size and surface properties to provide safety will permit the creation of new, more effective nanomedicines for systemic use.

Key words Nanoparticles, Nanomedicine, Quantum dots, Nanoshells, Nanocapsules, Nanoshuttles, Nanotoxicity

1 Introduction

Nanotechnology is an emerging science involving manipulation of materials at the nanometer scale (1–100 nm). There are several exciting prospects for the application of engineered nanomaterials in medicine. However, concerns over adverse and unanticipated effects on human health have also been raised. In fact, the same properties that make engineered nanomaterials attractive from a technological and biomedical perspective could also make these novel materials harmful to human health and the environment [1]. The application of nanotechnology to medicine, known as

Donald Armstrong and Dhruba J. Bharali (eds.), *Oxidative Stress and Nanotechnology: Methods and Protocols*, Methods in Molecular Biology, vol. 1028, DOI 10.1007/978-1-62703-475-3_4, © Springer Science+Business Media New York 2013

nanomedicine, concerns the use of precisely engineered materials at this length scale to develop novel therapeutic and diagnostic modalities [2]. Nanomaterials have unique physicochemical properties, such as ultra small size, large surface-area-to-mass ratio, and high reactivity, which are different from bulk materials (in microscale) of the same composition. These properties can be used to overcome some of the limitations found in traditional therapeutic and diagnostic agents. Nanotoxicology is emerging as an important subdiscipline of nanotechnology. Nanotoxicology refers to the study of the interactions of nanostructures with biological systems with an emphasis on elucidating the relationship between the physical and chemical properties (e.g., size, shape, surface chemistry, composition, and aggregation) of nanostructures with induction of toxic biological responses [3, 4].

1.1 Sources of Exposure to Nanoparticles

Most of the manufactured nanoparticles (NPs) currently used are made from metal oxides, silicon, and carbon [5]. So far the majority of nanosized approved drug delivery systems are lipid-, liposomal-, and polyethylene glycol-based [6]. Potential exposure to manufactured NPs may increase dramatically in the future [5]. Research on toxicity of NPs is still limited, although toxicity relevant to humans and other mammals have, however, been studied. Much of this research has been done with experimental mammals, though there is also epidemiological research on the effects on humans of manufactured ultrafine carbon particles (carbon black) and non-manufactured NPs. Animal experiments can identify hazards; they cannot be the basis for precise quantitative estimates regarding actual risk for humans because of differences between experimental animals and humans that make extrapolations uncertain [7, 8].

A subject that has attracted considerable interest concerns the inhalation of solid, ultrafine particles. Inhaled NPs have high deposition efficiency in the pulmonary region, which is increased in people with asthma or chronic obstructive lung disease [9]. Inflammation of the lung is often seen as a response to the inhalation of NPs [7, 10, 11].

Apart from exposure due to inhalation, dermal penetration of NPs is a matter of interest for humans. The issue of dermal penetration is highlighted by the application of TiO_2 and ZnO NPs in sunscreens, currently the most important use of ultrafine metal oxide particles in personal care products [12]. Such particles may be coated, for instance, with silicones, Al_2O_3, or SiO_2 [12, 13]. TiO_2 and ZnO particles sized w15e w50 nm can be photocatalytically active on exposure to sunlight [14, 15]. According to a study of Dunford et al. [13], this also holds for coated particles that are actually applied in sunscreens. Menzel et al. [16] demonstrated in experiments with pig skins that TiO_2 NPs can penetrate through the stratum corneum into the underlying stratum granulosum within the first 8 h after application. And studies with ZnO suggest that ZnO NPs may penetrate deep into the rat and rabbit skins [13]. Dermal exposure of humans may also be important in handling NPs in laboratories or industries.

Another pathway of exposure that may be relevant to environmental pollution by NPs involves the digestive tract. NPs may enter the digestive tract by swallowing inhaled particles and by their presence in food. In experiments with TiO_2 particles of on average 200 nm, Powell et al. [17] found that these particles were immunologically active and may be important adjuncts in overcoming normal gut cell hyporesponsiveness to endogenous molecules. Such 200-nm particles are used as food additives, for instance, as anticaking agents. Powell et al. [17] suggest that the immune potentiation found may apply to ultrafine particles in general. This may be particularly relevant to humans with inflammatory bowel disease, such as Crohn's disease, that have abnormal intestinal permeability. For this disease, food contamination with toxic nanoparticles is a putative risk factor [18].

Limited research has been done on the impact on the human body of submicron particles that originate in wear and corrosion of implants such as hip prostheses and of dental work such as bridges. This may be relevant to the effect of NPs generated within the body or ultrafine particles that are ingested. Submicron debris from implants was shown to be disseminated by the lymphatic system and possibly via blood vessels and may end up in the spleen and liver [19]. Submicron debris from dental work may pass through the intestinal barrier [20]. From the research done so far, there is some evidence that submicron particles derived from implants and dental work may be associated with phenomena such as inflammation and granulomatosis [20, 21].

A number of ultrafine particles have been studied regarding their application in drug delivery systems. A common side effect of drugs using such systems is a hypersensitivity reaction after intravenous administration, possibly due to complement activation. This occurs in a considerable number of recipients [6, 22]. Other immune reactions to NPs used in drug delivery have also been shown [6].

Another application of NPs is for biological labeling or imaging as an alternative to fluorescent dyes. *A relatively well-studied example is the use of semiconductor quantum dots made from CdSe. These have been found to exhibit cytotoxicity correlating with the liberation of Cd_2C* [23].

1.2 Nanomedicine

Nanomedicine is referred as the application of nanotechnology to disease treatment, diagnosis, and monitoring and to the control of biological systems at the level of single molecules or molecular assemblies [24]. The major goal in this area is to design rational delivery and targeting of pharmaceutical, therapeutic, and diagnostic agents. Compared with conventional drugs, nanomedicine usually exhibits different response to light, magnetic, or electronic irritation, and its physicochemical properties such as pH and temperature sensitivity also change obviously.

Over the last decade, nanotechnology-based formulations have received great interest due to their ability to cross various biological membranes of the body [25–27]. In recent years, the focus is in fabricating polymeric NPs as oral drug delivery vehicles due to numerous advantages offered by them from the pharmaceutical perspective over conventional approaches. By virtue of their absorption properties through gastrointestinal tract (GIT), polymer NPs offer a possibility to deliver various drugs falling under different classes of BCS [28]. After oral administration, polymeric NPs can protect the incorporated drug molecules from GI degradation as well as gut wall metabolism, thereby ensuring the stability of the drug in GI tract before absorption. Also, these nanocarriers have the ability to increase the oral bioavailability by preventing the first-pass metabolism of the drug through their specialized lymphatic uptake mechanism. Therefore, nanomedicine shows numerous advantages in the biological characteristics for targeted drug delivery and therapeutics, which overcome the limitations of molecular imaging, and gene/drug delivery commonly arose in recent years. For example, it will protect drugs against degradation and enhances drug stability, prolong the circulation and target searching time, and reduce the side effect and improve the distribution and metabolic process in tissues [29].

Our body is constructed from nanoscale building blocks such as DNA and proteins, which have long been targeted by the pharmaceutical industry long before the emergence of nanotechnology [30]. This category of drugs includes aspirin, cisplatin, and other anticancer agents as well as much more complex molecules like beta-blockers and anti-inflammatory agents [31]. The difference between nanomedicine and conventional drugs is that nanomedicine is entirely based on small molecule chemistry. It not only covers the therapeutic agents themselves but also promises to combine the abilities to deliver those agents to specific regions or tissues in the body, to specific cells, and perhaps to a specific location within a cell and also to make release of the therapeutic nanoparticle responsive to a physiological condition and perform a specific task [32]. The increased biological activity of NPs can be either positive or desirable (e.g., antioxidant activity, carrier capacity for therapeutic penetration of blood–brain barrier, and the stomach wall or tumor pores), dispersed throughout the whole body including entering the central nervous system, or negative and undesirable (e.g., toxicity, induction of oxidative stress, or cellular dysfunction) or a mix of both [3].

NPs have been found to be distributed to the colon, lungs, bone marrow, liver, spleen, and the lymphatics after intravenous

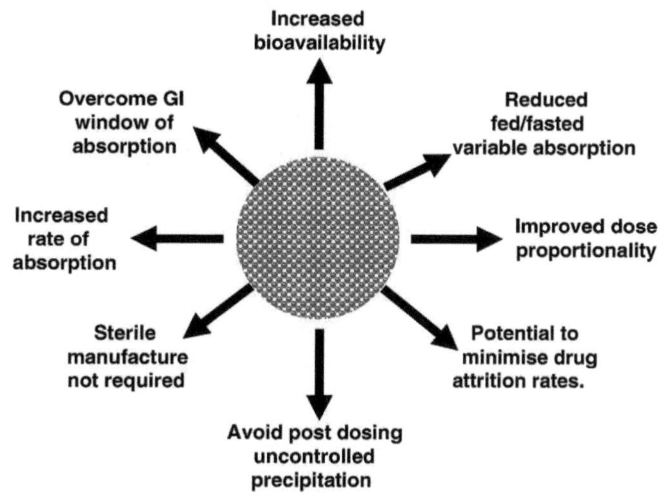

Fig. 1 Possible advantages of oral nanoparticulates

injection [33]. Distribution is followed by rapid clearance from the systemic circulation, predominantly by action of the liver and splenic macrophages [34]. Clearance and opsonization, the process that prepares foreign materials to be more efficiently engulfed by macrophages, occur under certain conditions for NPs depending on size and surface characteristics [34].

When inhaled, NPs are found to be distributed to the lungs, liver, heart, spleen, and brain [33]. NPs are cleared in the alveolar region via phagocytosis by macrophages facilitated by chemotactic attraction of alveolar macrophages to the deposition site [35]. The average half-life ($t_{1/2}$) for NPs in the respiratory tract is ~700 days in humans [3].

After intraperitoneal injection, NPs have been found to cross the transplacental membrane or cross the peritoneal cavity into uterus. This affected the embryo's cranial development and even caused embryo death [36].

After oral exposure, NPs distribute to the kidneys, liver, spleen, lungs, brain, and the gastrointestinal (GI) tract [33]. Some NPs can pass through the GI tract and are rapidly eliminated in feces and in urine, indicating that they can be absorbed across the GI barrier and into the systemic circulation [33]. However, some NP systems can accumulate in the liver during the first-pass metabolism [3] (Fig. 1).

Nano-preparations are currently under investigation as novel treatments of acne vulgaris, recurrent condyloma acuminata, atopic dermatitis, hyperpigmented skin lesions, and other non-dermatologic diseases [37].

1.3 Nanosized Therapeutic Tools

Many approaches have been developed to use NPs in the area of biomedical imaging and drug delivery. Quantum dots (QDs), one of the most well studied, are nanocrystals that fluoresce in different colors depending on their sizes (e.g., cadmium selenide) [38]. They typically have a core made of inorganic element but are generally coated with organic materials such as polyethylene glycol to enhance their biocompatibility or attach them to specific target molecule like proteins or DNA strands [39]. If the target molecule is an indicator of disease, detection of that molecule may indicate a higher propensity for disease. An example is to use NPs to bind to blood clots and to help make clots more visible by ultrasound [40].

Nanoshells, another nanodelivery system that is composed of copolymers, are used in combination with specific wavelengths of light and heat technology for cancer therapy. Gold nanoshells (GNSs) are particularly suitable for use in the surgical arena as their outer shell is composed of a commonly used reduced inert gold. When activated by near-infrared light, GNS can raise surrounding temperatures to levels sufficient for cellular ablation. This strategy was recently used by Stern and Cadeddu, (2008), for the therapeutic ablation of urologic malignancies [41].

Incorporation of cancer-killer genes into nanocapsules is being carried out. One of the genes being investigated is the gene elaborating tumor necrosis factor, a protein that is toxic not only to cancer but also to healthy cells when injected in large doses. To avoid damage to normal tissue, the nanocapsule is coated with sensors that target only on tumor cells.

Delivery of drugs to the brain has always been a challenge. The use of NPs to deliver drug to the brain using the pathfinder technology is being investigated. This technology uses nanoparticulate drug carriers in combination with the novel targeting principle of "differential protein adsorption" to cross the blood–brain barrier [42]. As the NPs are not efficiently scavenged by macrophages, the resulting increase in blood circulation time and hence bioavailability are expected to extend the duration of controlled system drug delivery or to improve the prospects for NPs to reach target sites by extravasation [43].

The delivery of magnetic NPs (MNPs) to or into various cell types has become an area of increasing interest in the biomedical sciences [44, 45]. Targeted delivery is used to deliver drugs or genes by attaching them to MNPs and locally concentrating the resulting complexes in vivo to the desired locale [46]. Similarly, magnetic hyperthermia, the local concentration of MNPs and subsequent heating via magnetic fields, has shown promise as a potentially viable cancer therapy [47].

Nanosized calcium fluoride (CaF_2) that could be used as F reservoir for more effective F regimens was recently prepared [48]. The nano-CaF_2 can be used as an effective anticaries agent in increasing the labile F concentration in oral fluid and thus enhances the tooth remineralization. It can also be very useful in the treatment for the reduction of dentin permeability.

Among the many novel nanocarriers being developed, echogenic bubble formulations have been gaining lots of attention in recent decades [49]. Bubbles are filled with gas, spherically shaped and stable in aqueous solution [50]. Compared to other particles, bubbles have the special properties of being "explosive" under ultrasound-energy illumination, prompting the destruction of bubbles and cellular membrane permeability changes. Various bubble formulations were used for local drug delivery [51], for gene delivery [52], and for targeting drug delivery [53]. They were developed based on micron-scale (μm) bubbles commercially available as ultrasound contrast agent for imaging diagnosis. Besides drug delivery, bubbles have also attracted investigators involved in ultrasonic imaging.

An example of third-generation NPs is biologically active molecular networks called "nanoshuttles," which are self-assemblies of gold NPs within a bacteriophage matrix. Nanoshuttles combine the hyperthermic response of the gold to near-infrared external energy with the biological targeting capabilities of the 4C-RGD sequence presented by the phage [54, 55]. These nanoshuttles collectively accommodate enhanced fluorescence, dark-field microscopy, and surface-enhanced Raman scattering detection. Another example of third-generation nanovectors is the "nanocell," based on a disease-inspired approach to therapy [56]. Utilizing the combinatorial chemotherapy approach, researchers have developed a nested NP construct that comprises a lipid-based NP enveloping a polymeric NP core called a "nanocell." A conventional chemotherapeutic drug, doxorubicin, is conjugated to a polymer core and an antiangiogenic agent, combretastatin, is then trapped within the lipid envelope. When nanocells accumulate within the tumor through the EPR effect, the sequential time release of the antiangiogenic agent, and then the cytotoxic drug, causes an initial disruption of tumor vascular growth and effectively traps the drug conjugated NP core within the tumor to allow eventual delivery of the cancer cell killing agent. However, the challenge in terms of utilizing NPs for targeted drug delivery in cancer patients is thus, in principle, how to target NPs exclusively to cancer cells, allowing the NPs to release their payload, and then become excreted or, alternatively, biodegraded in a programmable manner, while avoiding the uptake and accumulation of NPs in bystander cells, including cells of the immune system. Garcia-Bennett et al. [57] have highlighted recent developments of folate-mediated uptake of NPs for several main classes of NPs as well as noted other tumor and tumor vasculature targeting strategies. There are several advantages of using NPs conjugated to folate to target specific cancer cells that express the folate receptor, rather than using the therapeutic drugs directly conjugated to folate. First, the potential to deliver large quantities of drugs through individual receptor molecules and the utilization of the very effective folate receptor-mediated endocytosis pathway is especially feasible when combined with porous NPs or such particles that contain internal voids allowing large amounts of cargo to be transported.

Iron oxide NPs with unique magnetic properties have a high potential for use in several biomedical, bioengineering, and in vivo applications, including tissue repair, magnetic resonance imaging, immunoassay, drug delivery, detoxification of biologic fluids, cell sorting, and hyperthermia [58].

RNA interference (RNAi)-mediated knockdown of genes implicated in disease is the new paradigm in gene medicine. The clinical potential of RNAi-based therapeutics, however, is restricted by inadequate delivery to the extracellular and intracellular target sites [59, 60]. The poor pharmacokinetic profile, as a consequence of the susceptibility of RNA molecules to serum nucleases, renal clearance, and nonspecific biodistribution, restricts systemic applications. Furthermore, the polyanionic nature of nucleic acids that reduces interaction with cellular membranes further compromises the efficacy of naked siRNA duplexes. The silencing activity of RNA-based therapeutics is dependent on molecular interactions at specific intracellular sites. The combined requirement for modulation of the pharmacokinetics and intracellular trafficking is a major challenge for improvement in the therapeutic efficacy of RNAi therapeutics [61]. The use of polycationic NPs is a method used to improve the therapeutic potential of siRNA. The approach utilizes synthetic or natural cationic polymers or cationic species to form siRNA-containing NPs (termed polyplexes) to overcome extracellular and intracellular barriers to gene delivery [62] facilitating site-specific delivery, cellular uptake, and intracellular target interactions.

Polyplexes formed by the self-assembly of siRNA with synthetic and natural polycations that include polyethylenimine PEI, chitosan, atelocollagen, and cyclodextrin have been developed to maximize the RNA target payload and gene silencing in vivo. The entropy-driven formation of submicron spherical particles results from ionic interaction between the cationic amines or other positively charged groups and RNA-bearing anionic phosphates referred to as the N:P ratio. The physicochemical properties of positive surface charge at N:PN1 and nanoscale particle form facilitate electrostatic cellular membrane interaction and uptake and nuclease degradation protection [63] and permit size-mediated extravasation across vasculature endothelial barriers into tissue [64].

1.4 Nanotoxicology

While the small size of particles is what makes nanotechnology so useful in medicine and industry, it is also one of the main factors that might make them potentially dangerous to human health.

Research is now showing that even the harmless bulk materials can be harmful in a nanoparticle form, and in many cases the smaller the particles the more reactive and toxic are their effects. It is because any intrinsic properties of particles will likely be emphasized with the increase in surface area per unit mass [65, 66].

The potential risks inherent to any new technology are recognized. However, the special concern with nanotechnology is the unique type of toxicity due to surface modification. Enhanced endocytosis including a potential for inflammatory and prooxidant activity are shown to be largely dependent on NPs' surface chemistry (coating) and in vivo surface modifications [3, 65]. Increase in pulmonary toxicity (e.g., inflammation, granuloma formation) of carbon nanotubes when compared with that of the carbon black and carbonyl iron particles was seen in mice and rats [3, 67]. In vivo studies on the toxicity of single-wall nanotubes (SWCNTs) in mouse and rat models have shown various chronic inflammatory responses [68–70]. Moreover, multi-walled carbon nanotubes (MWCNTs) administered to mice via inhalation induce short-term pulmonary and systemic immune responses [71]. Intratracheal instillation of MWCNT in rat lung induces a significant dose-dependent increase in micronucleated pneumocytes, both centromere positive and negative, indicating the induction of both clastogenic and aneugenic events [72]. These results demonstrate that exposure to (CNTs) induces severe toxicity, but the dose, the degree of agglomeration, and the route of administration have to be considered carefully. Recently, Di Giorgio et al. [73] have investigated the inflammatory response, the release of tumor necrosis factor-α TNF-α, the production of intracellular reactive oxygen species (ROS), cell death (both through necrosis and apoptosis), chromosomal aberrations, and ultrastructural cellular alterations caused by carbon black (CB), MWCNTs. and SWCNTs. Their data confirm that both CNTs and CB are cytotoxic and genotoxic to RAW 264.7 mouse macrophages. Exposure to CNTs induced release of ROS, necrosis, and chromosomal aberrations, but did not cause an inflammatory response. In addition, CNTs induced ultrastructural damage and apoptosis. CNTs penetrate the cell membrane and individual MWCNTs are observed to be associated with the nuclear envelope.

There is an initial finding that indicated that gold NPs might move through a mother's placenta to the fetus [68]. Efficient uptake of NPs via the gastrointestinal tract has also been well documented in oral feeding studies and gavage studies [74]. All these findings indicate that NPs may potentially present problems with body burdens and it is hypothesized that NPs, because of their long retention in the body tissues, might repeat their highly catalytic activity with the host in cascade [36].

Though nanosilver-based dressing and surgical sutures have received approval for clinical application and good control of wound infection is achieved, their dermal toxicity is still a topic of concern. Despite laboratory and clinical studies confirming the dermal biocompatibility of nanosilver-based dressings, several other researchers have demonstrated the cytotoxicity of these materials [3, 75–79]. Paddle-Ledinek et al. [80] exposed cultured keratinocytes to extracts of several types of silver-containing dressings. The results showed that extracts of nanocrystalline-coated dressings are among those, which are the most cytotoxic. Keratinocyte proliferation was significantly inhibited, and cell morphology was affected [80].

Ag NP is now able to access human reproductive system via a variety of commercial products such as contraceptive devices and feminine hygiene products. Braydich-Stolle et al. [81] have shown that Ag NP caused toxicity to germ line stem cells via reduction in mitochondrial function and induction of membrane leakage and apoptosis. Other studies also examined the DNA damage response to polysaccharide surface functionalized (coated) and nonfunctionalized (uncoated) Ag NP in two murine mammalian cells: embryonic stem cells and embryonic fibroblasts [82]. Ahamed et al. [83] found that both types of Ag NP stimulated apoptosis via upregulation of cell cycle checkpoint protein p53, DNA damage repair proteins Rad51, and phosphorylated-H_2AX expression. These results suggested that the surface chemistry of Ag NP induces different DNA damage responses, i.e., coated Ag NP induced more severe damage versus uncoated Ag NP. Teodoro et al. [84] evaluated the toxicity of silver NPs (40 and 80 nm) and their effects on rat liver mitochondria bioenergetics. Wistar rat liver mitochondria demonstrate alterations in respiration and membrane potential capacities in the presence of either 40- or 80-nm silver NPs. Their data demonstrated a statistically significant decrease in mitochondrial membrane potential (Fig. 2), ADP-induced depolarization, and respiratory control ratio (RCR) upon exposure to silver NPs. They showed that silver NPs cause impairment of mitochondrial function, due mainly to alterations of mitochondrial membrane permeability which results in an uncoupling effect on the oxidative phosphorylation system. The increase in the lag phase observed in mitochondria incubated with either 40- or 80-nm silver NPs indicates an uncoupling between the oxidation of reduced substrates by the electron transport chain and the phosphorylation of ADP to ATP by ATP synthase. Mitochondrial permeability transition (MPT) induction disrupts the permeability barrier of the inner membrane, thus dissipating the membrane potential and pH gradient that together drive ATP synthesis through oxidative phosphorylation, causing impairment of mitochondrial function (Fig. 3).

Both comet assay and chromosomal aberration test showed DNA damage after 1, 3, and 24 h at 0.1 μg/mL Ag NP of human mesenchymal stem cells (hMSC). A significant increase of IL-6,

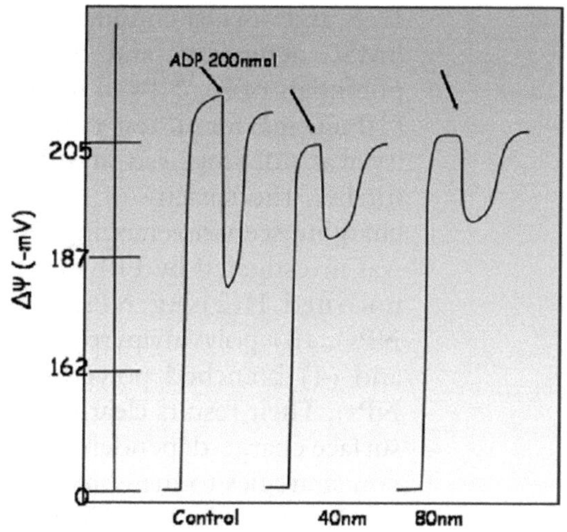

Fig. 2 Mitochondrial transmembrane potential ($\Delta\Psi$) in succinate-energized liver mitochondria, upon a 10-min incubation with silver nanoparticles. $\Delta\Psi$ was measured with a TPP^+-selective electrode. Reactions were carried out in 1 mL of reaction medium, supplemented with 3 μM Rotenone and 1 mg of freshly isolated mitochondria, as described in materials and methods. Energization was achieved with 5 mM succinate and phosphorylation induced by 200-nmol ADP. The traces represent typical direct recordings of experiments performed with ten different mitochondrial preparations

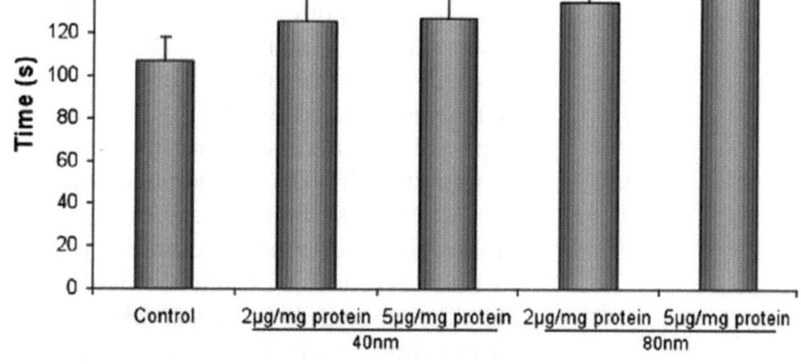

Fig. 3 Lag phase in succinate-energized liver mitochondria, upon incubation with silver nanoparticles. $\Delta\Psi$ was measured with a TPP^+-selective electrode. Reactions were carried out in 1 mL of reaction medium, supplemented with 3 μM Rotenone and 1 mg of freshly isolated mitochondria, as described in materials and methods. Data are means ± SEM of experiments performed with ten different mitochondrial preparations. *Indicates statistically significant difference versus control ($P < 0.05$)

IL-8, and vascular endothelial growth factor (VEGF) release indicates hMSC activation, and they demonstrated cyto- and genotoxic potentials of Ag NPs in hMSCs, especially at high concentrations (10 μg/mL for all test exposure periods), and that migration ability was not impaired at subtoxic concentrations [85]. One step further, the toxicity of four AgNPs representing various surface charging scenarios ranging from highly negative to highly positive was investigated by El Badawy et al. [86]. These Ag NPs were (1) uncoated H(2)–Ag NPs, (2) citrate-coated Ag NPs (citrate–Ag NPs), (3) polyvinylpyrrolidone-coated Ag NPs (PVP–Ag NPs), and (4) branched polyethyleneimine-coated Ag NPs (BPEI–Ag NPs). Their results clearly demonstrate that the Ag NPs exhibited surface charge-dependent toxicity which could help to suggest certain strategies to manipulate the toxicity of Ag NPs.

Recently, the identification of cytotoxicity of NPs toward mammalian germ line stem cells has aroused great concern over the biosafety of nanomaterials [81]. In their study, they used a cell line with spermatogonial stem cell characteristics to test in vitro toxicity of several types of NPs. The results showed that of all the tested materials (Ag, MoO_3, and Al), silver NPs were the most toxic with manifestations like drastic reduction of mitochondrial function, increased membrane leakage, necrosis, and induction of apoptosis. The findings are of significant practical implications because silver NPs are now able to access human sperms via a variety of commercialized products like contraceptive devices and maternal hygiene items. Based on this, fertility problems may occur.

Liver appears to be a major accumulation site of circulatory silver NPs [87]. Like germ line stem cells, similar patterns of cytotoxicity of silver NPs (decrease of mitochondrial function, LDH leakage, and abnormal cell morphologies) were observed with in vitro BRL 3A rat liver cells, but to a lesser extent [88]. In another study by the same researchers, a neuroendocrine cell line (PC-12 cells) was exposed to silver NPs as a control against Mn NPs and Mn^{2-} [89]. Experimental results showed that silver NPs were toxic to mitochondria than to Mn and Mn^2.

All of these findings are of importance because considerable amount of silver could be detected in rat brain following inhalation of silver NPs [87]. The neurological toxicity of silver is not clinically ascertained; however, several seizures cases have been related to exposure to silver or silver compounds [90]. Thus, mitochondria seem to be sensitive targets of cytotoxicity of silver NPs. However, the mechanism of silver NPs action on mitochondria is yet to be elucidated. In the study, with BRL 3A liver cell line, depletion of glutathione (GSH) level and increased ROS were found in association with mitochondrial perturbation, suggesting that oxidative stress might mediate the cytotoxicity of silver NPs. Based on these findings, a preliminary impression can be formed that silver NPs may interact with proteins and enzymes with thiol groups within mammalian cells. These proteins and enzymes like GSH, thioredoxin, superoxide dismutase, and thioredoxin peroxidase are key components of the cell's antioxidant defense mechanism which is responsible to neutralize the oxidative stress of ROS largely generated by mitochondrial energy metabolism. As these effects of Ag^- could be completely blocked by sulfhydryl reagents as GSH, the surface modification of silver NPs by phosphoryl disulfides was effective in improving silver biocompatibility and intracellular uptake [91]. They prepared the phospholipid derivatives containing disulfide groups to modify silver NP surfaces. By adding sodium borohydride to reduce both disulfide bonds of the derivatives and silver ion simultaneously, the generated thiol group can be reacted with newborn silver atoms immediately to generate nanoclusters. The assemblies consisted of either phosphorylcholine (PC) or phosphorylethanolamine (PE) head groups, which made the silver clusters biocompatible. In cell culture tests, the surface-modified NPs were internalized into platelet and fibroblast cells in a short period of incubation without harming the cells.

Recently, Pujalte et al. [92] reported the toxic effects of metallic NPs on glomerular and tubular human renal cells; ZnO and CdS NPs exerted cytotoxic effects that were correlated with metal composition, particle scale, and metal solubility. ROS production and oxidative stress induction clearly indicated their nephrotoxic potential. Cytotoxic effects were correlated with the physicochemical properties of NPs tested and the cell type used. Analysis of ROS and intracellular levels of reduced and oxidized GSH revealed that particles induced stress according to their composition, size, and solubility. Protein involved in oxidative stress such as NF-kB was activated with ZnO and CdS NPs. Such effects were not observed with TiO_2 NPs.

Recently, Naqvi et al. [58] studied in vitro the interaction of superparamagnetic iron oxide NPs of mean diameter 30 nm coated with Tween 80 and murine macrophage (J774) cells to evaluate the dose- and time-dependent toxic potentials as well as to investigate the role of oxidative stress in the toxicity. A 15–30-nm size range of spherical NPs was characterized by transmission electron microscopy and Zetasizer. MTT assay showed >95 % viability of cells in lower concentrations (25–200 μg/mL) and up to 3 h of exposure, whereas at higher concentrations (300–500 μg/mL) and prolonged (6 h) exposure viability reduced to 55–65 %. Necrosis–apoptosis assay by propidium iodide and Hoechst-33342 staining revealed loss of the majority of the cells by apoptosis. H_2DCFDDA assay to quantify generation of intracellular ROS indicated that exposure to a higher concentration of NPs resulted in enhanced ROS generation, leading to cell injury and death. The cell membrane injury induced by NPs was studied using the lactate dehydrogenase assay, showed both concentration- and time-dependent damage (Fig. 4). Thus, this study concluded that use of

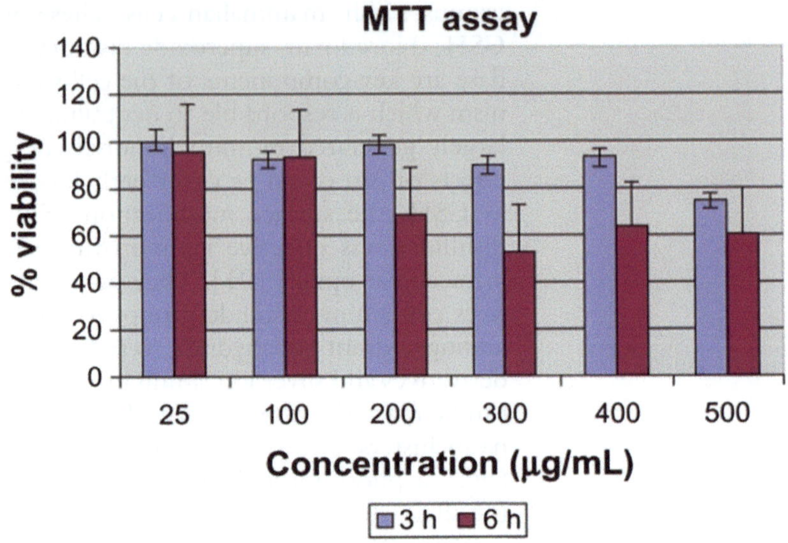

Fig. 4 The effects of superparamagnetic iron oxide nanoparticles on cell proliferation and viability of J774 cells as determined by MTT assay. Concentration-dependent cytotoxic effects of nanoparticles evaluated after 3 and 6 h of incubation. Results are represented as means ± standard error of the mean. Significant difference from control ($P < 0.05$)

a low optimum concentration of superparamagnetic iron oxide NPs is important for avoidance of oxidative stress-induced cell injury and death.

2 Ultrahigh Reactivity Provokes Nanotoxicity

Because the nanosize/surface area of the nanosubstance is directly correlated to many essential characteristics like surface properties, chemical reactivity, and physical absorption ability, all these factors strongly dominate nanotoxicological behavior in vivo [93]. The study of Chen et al. [78] on the acute oral toxicity of copper particles has shown a significant correlation with its size distribution. With the particle size reducing from 17 µm (particle number, 44 per µg; surface area 3.99×10^2 cm^2/g) to 23.5 nm (particle number, 1.7×10^{10} per µg; surface area 2.95×10^2 cm^2/g), LD_{50} of copper particle sharply increases from >5,000 (nontoxic) to 413 mg/kg (moderately toxic) based on the Hodge and Sterner scale. For identical chemical composition, why do the nanocopper particles possess unique biobehavior (nanotoxicity) in vivo comparing to those in bigger size (in microscale)? Looking for answer for this question may provide an insight into nanotoxicity reducing or elimination. They explained this on the basis that nanocopper particles may not compromise the mice directly. The nanocopper retained in gastric lumen can continuously react with the secreted acid juice. The

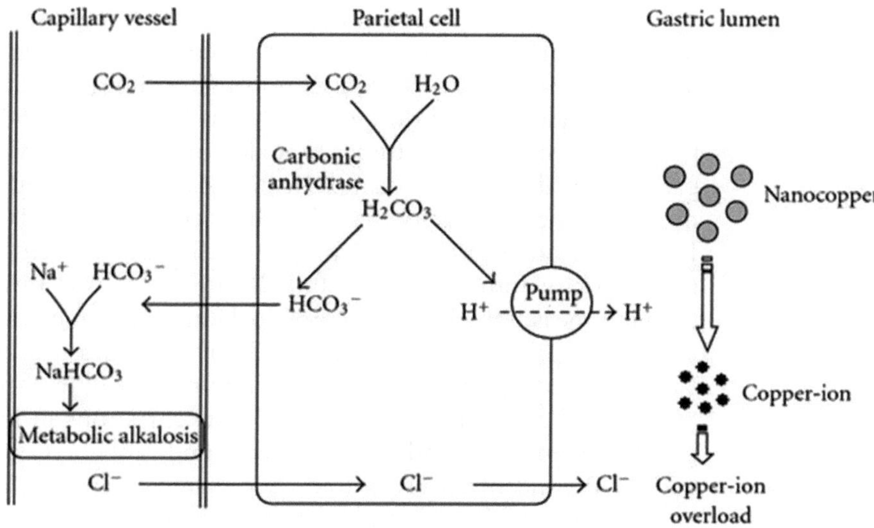

Fig. 5 Mechanism of metabolic alkalosis induced by nanocopper ion

depletion of H^+ results in metabolic alkalosis because the HCO_3 generated during production of gastric acid will return to the circulation resulting in the formation of large quantities of sodium bicarbonate which increase the arterial blood pH (Fig. 5).

The elevation of blood pH motivates a set of compensatory effects: (a) a respiratory compensation which is limited [94] and (b) renal compensation which starts relatively later but can sustain for a long time (several days) [94, 95]. However, a series of abnormalities were found in the pathological examination such as swollen glomerulus, dwindling in lumen Bowman's capsules, and being signs of glomerulonephritis. The renal dysfunction may largely weaken renal compensation in nanocopper group and deteriorate the metabolic alkalosis.

It is reported that copper ions ingested are metabolized in liver and excreted via urine [96]. If the intake of copper exceeds the range of the tolerance, it would cause toxic effects to hepatic and renal tissues, which is consistent with the finding of Meng et al. [97] that nanocopper possesses extremely high bioavailability; hence, the original safety limit may be modified to much lower level. Based on these findings, we can suggest that nanocopper and microcopper exhibit different biological behaviors in vivo via oral exposure routine. In terms of nanocopper particle, both copper overload and metabolic alkalosis contribute to their grave toxicity.

High chemical reactivity of Ag NPs was observed in the reaction with hydrochloric acid: $Ag\ (NPs) + HCl \rightarrow AgCl + H_2$; the reaction product silver chloride was characterized by X-ray powder diffraction to give a direct evidence for the reaction which has been proved impossible for the bulk Ag [98].

The microscale titanium dioxide (TiO_2), widely used in pharmaceutical and cosmetics industries, is considered as biologically inert [99]. Such that, there was no obvious lung toxicity in rats when a single instilled dose of Ti O_2 was 5 mg/rat or 50 mg/kg [100]. However, many studies have demonstrated that when TiO_2 particle size decreased to nanoscale dimension, they could produce more pulmonary toxicity than their bulk counterparts [101–104]. In a comparative study done by Li et al. [105], the acute pulmonary toxicity induced by 3- and 20-nm TiO_2 was investigated through measurement of selected biochemical parameters in bronchoalveolar lavage fluid (BALF). At 3-day postexposure, the 3-nm TiO_2 induced significant increase of albumin, alkaline phosphatase (ALP), and acid phosphatase (ACP) concentrations in high-dose group (40 mg/kg) and also induced significant increase of ALP and ACP concentrations in mid-dose group (4 mg/kg), but did not induce significant increase of total protein and LDH concentrations in any dose group. On the other hand, 20-nm TiO_2 induced significant increase of all biochemical parameters in high- and mid-dose groups. At 3-day postexposure, both TiO_2 particles did not induce obvious pulmonary toxicity in their low-dose (0.4 mg/kg) groups as evidence of no significant increase of all biochemical parameters. The pH values of 3-nm TiO_2 particles colloid were 5.38 ± 0.12, 4.55 ± 0.07, and 4.42 ± 0.13 at concentrations of 0.1 mg/mL, 1 mg/mL, and 10 mg/mL, respectively, while at the corresponding concentration, the pH values of 20-nm TiO_2 particles suspension were 5.5 ± 0.19, 4.64 ± 0.11, and 3.75 ± 0.04, respectively. They reported that pH value of TiO_2 particles in medium, other than particle size, surface area, and aggregation, plays an important role in affecting TiO_2 NPs pulmonary toxicity.

The potentiated toxicity of nanoscale vanadium oxide (V_2O_3) compared to bulk material is demonstrated in human endo- and epithelial lung cells and might be due to the higher catalytic surface of the particles [97]. Reduction in cell viability is almost ten times stronger and starts with the lowest concentrations of "nanoscaled" material (10 μg/mL). Vanadium oxide leads to an induction of heme oxygenase 1 (HO^{-1}) in a dose-dependent manner in ECV304 cells, whereas a reduction in protein levels can be observed for the epithelial cells (A549). Lipid peroxidation can be observed also for "nanoscaled" vanadium oxide to a much stronger extent in macrophages (RAW cells) than for bulk material. The observed effects cannot only be explained by oxidation from V_2O_3 to V_2O_5 as there are significant differences between the novel nano-vanadium and all used bulk materials (V_2O_3 and V_2O_5). It appears rather to be a nanoeffect of a high surface reactivity, here coupled with a yet unknown toxicity potentiating effect of a technically important catalyst.

3 Strategies to Reduce Nanoparticle Toxicity

Recently, many studies focus on the safety issue of manufactured nanomaterials to minimize or eliminate their nanotoxicity even before they are widely used. Many trials were done for modification of these products aiming to improve their biocompatibility and minimize their toxicity.

Equal attention is being paid to realize the potential of antioxidants that are thought to play a major role to prevent/treat/cure the diseases that are mediated by free radical pathways by scavenging them [106]. Though there were several ongoing clinical trials attempting to explore the potential of these antioxidants, the success is far from the reach, considering their association with poor biopharmaceutical properties. Table 1 enlists the antioxidants that were successfully attempted for NP delivery, namely, coenzyme Q10 (CoQ10) [107], curcumin [108], ellagic acid [109–111], and epigallocatechin gallate (EGCG) [112]. These molecules are delivered in very high doses due to bioavailability problems and no proper preformulation or formulation strategies have been applied, and therefore, therapeutic doses of these molecules are not known.

Chomouckaa et al. [113] was able to conjugate a nanostructure composed of Fe_2O_3 NPs and CdSe/CdS quantum dots (QDs) with core/shell structure. In the first step, they prepared streptavidin (STP)-conjugated Fe2O3 magnetic particles. The next stage was the preparation of QDs using method based on surface modification of hydrophobic core/shell CdSe/CdS with biotinylated GSH. GSH belongs to the most abundant peptides in organisms maintaining redox status which makes the QDs "water-soluble" and stabilized in aqueous solution. Biotin molecule is known for its excellent affinity to streptavidin and hence allows the conjugation of both nanostructures (magnetic and light emitting). They tested the influence of both prepared nanostructures on growth and viability of fibroblasts and BY-2 tobacco cells, respectively. NPs-STP and QDs-GSH may not pose a threat to cells, but nonmodified NPs and QDs without GSH had a very adverse effect on cells.

The relative impermeability of the blood–brain barrier BBB results from tight junctions and efflux transport systems limits drug delivery to the central nervous system (CNS) and thus severely restricts the therapy of many central nervous system diseases. In order to enhance the brain-specific drug delivery, Li et al. [114] was able to isolate peptides that could target the drug delivery system to the brain. A 12-amino acid peptide (denoted as Pep TGN) displayed by bacteriophage Clone 12-2 was selected by rounds of in vivo screening. Pep TGN was covalently conjugated onto the surface of poly(ethyleneglycol)-poly(lactic-co-glycolic acid) (PEG-PLGA)-based NPs. They reported that cellular uptake of Pep TGN-decorated NPs was significantly higher than that of

Table 1
Nanoparticulate approach for antioxidants

Alternative medicines		CoQ10	Curcumin	Ellagic acid	EGCG
Biopharmaceutic/pharmacokinetics	Metabolism	• Hepatic metabolism • Elimination through biliary and fecal excretion	• Intestinal and hepatic metabolism	• Intestinal and hepatic metabolism	• Extensive metabolism via phase II
	$t_{1/2}$ of drug (h)	34	1.45	8	2
	Hurdles in oral delivery	• Low solubility • Low permeability	• Low solubility • Low permeability Intestinal degradation	• Low solubility • Low permeability • Intestinal degradation and binding	• Intestinal degradation • Low permeability
Conventional means of delivery	Dosage form	• Tablets • Capsules • Suspension • Liposomal	• Tablet • Capsules	• Tablet • Capsule	• Tablet • Capsule
	Routes	• Oral	• Oral	• Oral	• Oral
	Scope for improvement	• Improving oral bioavailability • New indications	• Improving oral bioavailability • New indications	• Improving oral bioavailability • New indications	• Improving oral bioavailability • New indications
Nanoparticulate delivery (data from our lab)	Nanoparticles (nm)	112	242	125	130
	Polymer	PLGA	PLGA	PLGA	PLGA
	Solvent	Ethyl acetate	Ethyl acetate	DMSO/ethyl acetate	Water/ethyl acetate
	Surfactant	DMAB	PVA	DMAB	DMAB
	Entrapment efficiency (%)	82	84	52	70
	In vivo release (h)	–	48	–	–
	Pharmacological evaluation	• Improved efficacy in treatment of hypertension, hyperlipidemia and diabetes	• Improved oral bioavailability	• Improved prevention of CsA-mediated nephrotoxicty at three times lower dose • Improved efficacy in treatment of hyperlipidemia and diabetes	• Improved prevention of CsA-mediated nephrotoxicity at three times lower dose compared to oral and intraperitoneal administrations

unmodified NPs when incubated with bEnd.3 cells. Enhanced brain accumulation efficiency together with lower accumulation in liver and spleen as nontargeted organs was observed in the nude mice intravenously injected with Pep TGN-conjugated NPs compared with those injected with plain NPs, showing powerful brain selectivity of Pep TGN. In vitro cytotoxicities of plain NP, TGN-NP (1:3), and TGN-NP (1:1), were investigated with bEnd.3 cells after 4 h, 24 h incubation at 37 C by the CCK-8 assay. There was no significant difference in the toxicity of different formulations at any of the given concentrations ($P > 0.05$). Cell viability was slightly reduced at higher concentrations. The PLGA polymers are generally accepted as being of low cytotoxicity with good biocompatibility and biodegradability. Thus, PLGA-NP was regarded as the safety control group. No notable difference was observed between NP and TGN-NP and this confirmed the relative safety of TGN-NP compared with NP. Accordingly, TGN-NP is a promising drug carrier without any significant cytotoxic effects on cells tested.

4 Protein–Nanoparticle Interaction

Within the medical device community, it is now well accepted that material surfaces are modified by the adsorption of biomolecules such as proteins in a biological environment [115, 116], and there is some consensus that cellular responses to materials in a biological medium reflect the adsorbed biomolecule layer, rather than the material itself. However, the importance of the adsorbed protein layer in mediating interactions with living systems has been slower to emerge in the case of NP–protein interactions. The key role of protein–NP interactions in nanomedicine and nanotoxicity has begun to emerge recently with the development of the idea of the NP–protein "corona." This dynamic layer of proteins (and other biomolecules) adsorbs to NP surfaces immediately upon contact with living systems. The composition of the protein corona at any given time will be determined by the concentrations of the over 3,700 proteins in plasma [117] and the kinetic on and off rates (or equilibrium binding constants) of each protein for the particular NP. This corona may not immediately reach equilibrium when exposed to a biological fluid. Proteins with high concentrations and high association rate constants will initially occupy the NP surface but may also dissociate quickly to be replaced by proteins of lower concentration, slower exchange, and higher affinity. Thus, the protein corona is the biological identity of a NP, as it is what the cell "sees" and interacts with. Functional changes of proteins of such complexes may be another mechanism by which particularly small NP, with their large surface area as a binding interface, may induce protein malfunctioning, which may lead to the pathogenesis and adverse health effects [118]. Survey of the literature on

NP–protein binding shows that the vast majority of NP types studied, so far, bind apolipoproteins [116]. At first sight, this is a surprising result and quite distinct from that for a flat surface. However, the fact that apolipoproteins are known to be involved in lipoprotein complexes, which themselves have sizes on the nanoscale ranging from 100 nm (chylomicron) to 10 nm (high-density lipoprotein), may mean that there are specific size-dependent interactions that drive the binding of apolipoprotein to NPs. This is interesting from the point of view of NP interaction with cells, as lipoprotein complexes are involved in the general cellular processes of cholesterol metabolism [2]. Thus, there are multiple receptors for apolipoprotein complexes at cell surfaces that NPs with surface-adsorbed apolipoproteins can be potentially exploit to enter cells [119]. If we consider the issues of NP transport and fate in animals and humans, then it is also relevant that apolipoprotein E has been found to associate to some NPs [115]. This has potentially significant consequences for neurotoxicity and the development of neurotherapies, as apolipoprotein E is known to be involved in trafficking to the brain [119].

The first reports of the direct biological influence of proteins adsorbed to NPs are now emerging. SWNTs and 10-nm amorphous silica coated with albumin have been shown to induce anti-inflammatory responses in macrophages, measured as inhibited induction of cyclooxygenase-2 (Cox-2) by lipopolysaccharide under serum-free conditions [120]. Blocking the adsorption of albumin by precoating, the NPs with nonionic surfactant (Pluronic F127) also inhibits the anti-inflammatory properties of the NPs. These observations suggest an important role for the adsorbed proteins in modulating the uptake and toxicity of SWNTs and nanosized amorphous silica [120]. However, as these studies were conducted under serum-free conditions, it is unclear whether the albumin would remain bound to NPs under competitive binding conditions, such as those occur in plasma or in a cellular milieu. The interaction between human adult hemoglobin (Hb) and bare CdS QDs dramatically alters the conformation of Hb and decreasing the α-helix content of the secondary structure from 72.5 to 60.8 %. Raman spectra results indicate that the sulfur atoms of the cysteine residues form direct bonds on the surface of the CdS QDs [121]. Functionalization of NP surfaces with peptides is increasingly being used to control the interaction of NPs with proteins [117].

Although NP protein binding and the cytotoxicity of NPs have been separately reported, there has been no study linking the nature of NP protein clusters to cell uptake and the dynamic cellular responses. Mu et al. [122] reported that water-soluble iron oxide-based MNPs with different sizes and surface chemistry bind different serum proteins in terms of protein identity and quantity, without changing the protein secondary structures. Carboxylated MNPs (and aminated one in smaller MNPs) result in higher cytotoxicity, and PEG coating reduces both cell uptake and the cytotoxicity. Smaller MNPs (especially the carboxylated one) bind more serum proteins, are much less taken up by cells as compared to larger particles, and yet elicit more dynamic cytotoxic responses. Besides the intrinsic effects of size and surface charge of the water-soluble MNPs, the cellular effects of MNPs/protein clusters were also attributed to the identity and quantity of the adsorbed proteins rather than the binding-induced new epitopes on the proteins [122, 123].

Interestingly, several systems exist where binding to certain proteins makes it easier for the specific nanoparticles to be transported to the site of action. For example, the adsorption of apolipoproteins has been shown to be important for the transportation of drugs across the BBB and into the brain [124]. Researchers have shown an enhanced binding of ApoE and ApoB-100 to PEGylated NPs compared to the non-PEGylated NPs [125]. Even a difference in a coating of the same NP can make a difference. Gessner et al. [125] coated solid lipid NPs with either Tween 80 or poloxamer 188 and showed that particles coated with Tween 80 showed an abundance of binding of various apolipoproteins, especially ApoE. Additionally, the poloxamer 188-coated NPs bound various apolipoproteins, but did not bind ApoE. It did, however, bind albumin, which was not abundant on the Tween 80-coated particles. Goppert and Muller [124] also showed varying binding of ApoE to solid lipid NPs based on the different types of polysorbate coatings. One can expect that the direct attachment of apolipoproteins to a particle surface would aid in drug delivery to brain tissue. Indeed a study showed that the covalent attachment of either ApoE3, ApoA-I, or ApoB-100 to albumin NPs significantly facilitated transport of a bound study drug, loperamide, across the BBB as opposed to controls of loperamide alone or NPs with no apolipoproteins attached [125].

5 Discussion

Nanotoxicology refers to the biokinetic evaluation of engineered nanostructures and nanodevices. The need for this area of investigation became apparent after the intensive expansion of nanotechnology, which in the last two decades has been widely used in the pharmaceutical industry, medicine, and engineering technology [126, 127]. Particle toxicology and the consequent adverse health effects of asbestos fibers and coal dust serve as a historical reference points to the development of nanotoxicological concepts.

In the area of medicine, nanomedicine has been defined as the monitoring, repair, construction, and control of human biological systems at the molecular level, using engineered nanodevices and nanostructures [128]. Macrophages as specialized host defense cells, endothelium as thin specialized epithelial cells that line the inner surface of lymph vessels and blood vessels serve as gate keeper

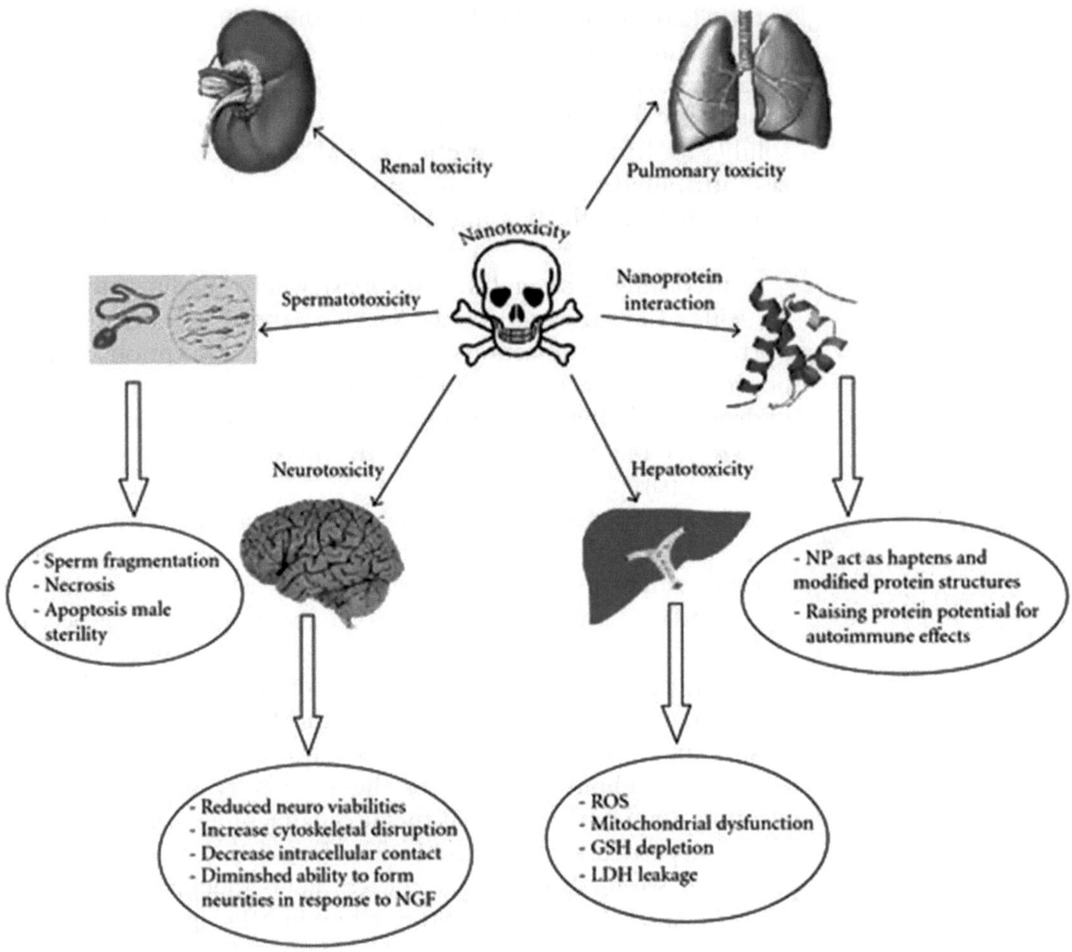

Fig. 6 A summary of the most important recorded toxic effects of therapeutically used nanoparticles reviewed in this chapter

to control passage of materials together, and tumors are the most common targets of NPs. Within these biological targets, NPs favor the formation of prooxidants, especially under exposure to light, ultraviolet light, or transition metals, thereby destabilizing the balance between the production of ROS and the biological system's ability to detoxify or repair the system [126, 127]. ROS can also be produced by the NADPH oxidase in phagocytic cells as target of NP devices. NPs can modify mitochondrial function as well as cellular redox signaling. Oxidative stress induced by NPs is reported to enhance inflammation through upregulation of redox-sensitive transcription factors including nuclear factor kappa B (NF-kB), activating protein (AP-1), and extracellular signal regulator kinases (ERK) C-Jun, N-terminal kinases JNK, and p38 mitogen-activated protein kinases pathways. Figure 6 summarizes the most important recorded toxic effects of therapeutically used NPs reviewed in this chapter.

6 Concluding Remarks

1. Nanotechnology is growing at an exponential rate and will undoubtedly have both beneficial and toxicological impact and consequences on health and environment.
2. As a result of their properties, nanomaterials differ substantially from those bulk materials of the same composition, allowing them to perform exceptional feats of conductivity, reactivity, and optical sensitivity. Possible undesirable results of these capabilities are harmful interactions with biological systems with the potential to generate toxicity.
3. Development of new techniques to show accurate correlations between in vitro and in vivo studies is imperative to accurately portray NP effects. Moreover, toxicity studies are critical to establish the full in vivo potential of nanomedicine. Understanding the physiochemical, molecular, and physiological processes of NPs is important for nanomedicine to become a reliable and sustainable treatment modality.
4. In the future, NPs could be classified in terms of their biomolecule corona which mediates their interaction with cellular machinery. This would represent a truly new paradigm in the field of nanoscale toxicology and in the design of safe nanocarriers for nanomedicine.
5. With this new opportunity to utilize the unique properties of NPs for research, industry, and medicine, there is a responsibility to test and optimize these new nanomaterials early during the development process to eliminate or ameliorate identified toxic characteristics.
6. The rapid commercialization of NPs requires focused environmental, health, and safety research; meaningful and open discussion of broader societal impacts; and urgent oversight.

Acknowledgments

The authors extend their appreciation to the Deanship of Scientific Research at King Saud University for funding the work through the research group project No. (RGP-VPP-005).

References

1. Shvedova AA, Kisin ER, Porter D, Schulte P, Kagan VE, Fadeel B et al (2009) Mechanisms of pulmonary toxicity and medical applications of carbon nanotubes: two faces of Janus? Pharmacol Ther 121:192–204
2. Zhang S, Bian Z, Gu C, Zhang Y, He S, Gu N et al (2007) Preparation of anti-human cardiac troponin I immunomagnetic nanoparticles and biological activity assays. Colloids Surf B Biointerfaces 55:143–148

3. Oberdörster G, Oberdörster E, Oberdörster J (2005) Nanotoxicology: an emerging discipline evolving from studies of ultrafine particles. Environ Health Perspect 113: 823–839
4. Oberdörster G, Stone V, Donaldson K (2007) Toxicology of nanoparticles: a historical perspective. Nanotoxicology 1:2–25
5. Dreher KL (2004) Health and environmental impact of nanotechnology: toxicological assessment of manufactured nanoparticles. Toxicol Sci 77:3–5
6. Allen TM, Cullis PM (2004) Drug delivery systems: entering the main stream. Science 303:1818–1822
7. Oberdörster G (2000) Toxicology of ultrafine particles; in vivo studies. Phil Trans R Soc Lond A 358:2719–2740
8. Borm PJA, Schins RPF, Albrecht A (2004) Inhaled particles and lung cancer, part B: paradigms and risk assessment. Int J Cancer 110:3–14
9. Chalupa DC, Morrow PE, Oberdörster G, Utell MJ, Frampton MW (2004) Ultrafine particle deposition in subjects with asthma. Environ Health Perspect 112:679–682
10. Johanston CJ, Finkelstein N, Gelein R, Bagos R, Oberdörster G (1996) Characterization of the early pulmonary inflammatory response associated with PTFE fume exposure. Toxicol Appl Pharmacol 140:154–163
11. Donaldson K, Brown D, Clouter A, Duffin R, MacNee W, Renwick L et al (2002) The pulmonary toxicology of ultrafine particles. J Aerosol Med 15:213–220
12. Boswell C (2002) Nanocrystalline materials for personal care. Chemical Market Reporter November 11:6
13. Dunford R, Salinaro A, Cai L, Serpone N, Horikoshi S, Hidaka H et al (1997) Chemical oxidation and DNA damage catalysed by inorganic sunscreen ingredients. FEBS Lett 418:87–90
14. Hidaka H, Horikoshi S, Serpone N, Knowland J (1997) In vitro photochemical damage to DNA, RNA and bases by an inorganic sunscreen agent on exposure to UVA and UVB radiation. J Photochem Photobiol 111:203–213
15. Li X, Quan X, Kutal C (2004) Synthesis and photocatalytic properties of quantum confined titanium dioxide nanoparticle. Scr Mater 50:499–505
16. Menzel F, Reinert T, Vogt J, Butz T (2004) Investigations of percutaneous uptake of ultrafine TiO_2 particles at the high energy ion nanoprobe LIPSION. Nucl Inst Meth Phys Res B Beam Interact Mater Atoms 219–220: 82–86
17. Powell JJ, Harvey RSJ, Ashwood P, Wolstencroft R, Gershwin ME, Thompson RPH (2000) Immune potentiation of ultrafine dietary particles in normal subjects and patients with inflammatory bowel disease. J Autoimmun 14:99–105
18. Hugot J (2004) Inflammatory bowel disease: a complex group of genetic disorders. Best Pract Res Clin Gastroenterol 18:451–462
19. Urban RM, Jacobs JJ, Tomlinson MJ, Gavrilovic J, Black J, Peoch M (2000) Dissemination of wear particles to the liver, spleen and abdominal lymph nodes of patients with hip and knee replacement. J Bone Joint Surg Am 82A:457–477
20. Gatti AM, Rivasi F (2002) Biocompatibility of micro- and nanoparticles. Part I: in liver and kidney. Biomaterials 23:2381–2387
21. Gatti AM (2004) Biocompatibility of micro and nano-particles in the colon. Part II. Biomaterials 25:385–392
22. Szebeni J, Baranyi L, Savay S, Milisevits J, Bunger R, Laverman P et al (2002) Role of complement activation in hypersensitivity reactions to Doxil and HYNIC PEG liposomes: experimental and clinical studies. J Liposome Res 12:165–172
23. Derfus AM, Chan WCW, Bhatia SN (2004) Probing the cytotoxicity of semiconductor quantum dots. Nano Lett 4:11–18
24. Webster TJ (2006) Nanomedicine-what's in a definition. Int J Nanomedicine 1:115–116
25. Chen H, Langer R (1998) Oral particulate delivery: status and future trends. Adv Drug Deliv Rev 34:339–350
26. Panyam J, Labhasetwar V (2003) Biodegradable nanoparticles for drug and gene delivery to cells and tissue. Adv Drug Deliv Rev 55:329–347
27. Kreuter J, Ramge P, Petrov V, Hamm S, Gelperina SE, Engelhardt B et al (2003) Direct evidence that polysorbate-80-coated poly(butylcyanoacrylate) nanoparticles deliver drugs to the CNS via specific mechanisms requiring prior binding of drug to the nanoparticles. Pharm Res 20:409–416
28. Bhardwaj V, Hariharan S, Bala I, Lamprecht A, Kumar N, Panchagnula R et al (2005) Pharmaceutical aspects of polymeric nanoparticles for oral delivery. J Biomed Nanotech 1:235–258
29. Peer D, Karp JM, Hong S, Farokhzad OC, Margalit R, Langer R (2007) Nanocarriers as an emerging platform for cancer. Nat Nanotechnol 2:751–760
30. Alivisatos AP (2002) Less is more in medicine. In: Understanding nanotechnology, Warner Book, New York, NY, pp 56–69

31. Ratner M, Ratner D (2003) Nanotechnology: a gentle introduction to the next big idea. Prentice-Hall, Upper Saddle River, NJ
32. Kipen HM, Laskin DL (2005) Smaller is not always better: nanotechnology yields nanotoxicology. Am J Physiol 289:L696–L697
33. Hagens WI, Oomen AG, de Jong WH, Cassee FR, Sips AJAM (2007) What do we (need to) know about the kinetic properties of nanoparticles in the body? Regul Toxicol Pharmacol 49:217–229
34. Moghimi SM, Hunter AC, Murray JC (2005) Nanomedicine: current status and future prospects. FASEB J 19:311–330
35. Garnett MC, Kallinteri P (2006) Nanomedicines and nanotoxicology: some physiological principles, Occup Med 56:307–311
36. Vega-Villa KR, Takemoto JK, Yáñez JA, Remsberg CM, Forrest ML, Davies NM (2008) Clinical toxicities of nanocarrier systems. Adv Drug Deliv Rev 60:929–938
37. Wiesenthal A, Hunter L, Wang S, Wickliffe J, Wilkerson M (2011) Nanoparticles: small and mighty. Int J Dermatol 5:247–254
38. Hardman R (2006) A toxicological review of quantum dots: toxicity depends on physicochemical and environmental factors. Environ Health Perspect 114:165–172
39. Denison RA (2005) Environmental and safety impacts on nanotechnology: what research is needed? http://www.environmentaldefense.org/documents/5136_DenisonHousetestimonyOnNano.pdf
40. Liang HD, Blomley MJK (2003) The role of ultrasound in molecular imaging. Br J Radiol 76:S140–S150
41. Stern JM, Cadeddu JA (2008) Emerging use of nanoparticles for the therapeutic ablation of urologic malignancies. Urol Oncol 26: 93–96
42. Müller RH, Keck CM (2004) Drug delivery to the brain—realization by novel drug carriers. J Nanosci Nanotechnol 4:471–483
43. Moghimi SM, Hunter AC, Murray J (2001) Long-circulating and target-specific nanoparticles: theory to practice. Pharmacol Rev 53:283–318
44. Ito A, Shinkai M, Honda H, Kobayashi T (2005) Medical application of functionalized magnetic nanoparticles. J Biosci Bioeng 100: 1–11
45. Pankhurst QA, Connolly J, Jones SK, Dobson J (2003) Applications of magnetic nanoparticles in biomedicine. J Phys D 36: R167–R181
46. Vasir JK, Labhasetwar V (2005) Targeted drug delivery in cancer therapy. Technol Cancer Res Treat 4:363–374
47. Moroz P, Jones SK, Gray BN (2002) Magnetically mediated hyperthermia: current status and future directions. Int J Hyperthermia 18:267–284
48. Sun L, Chow LC (2008) Preparation and properties of nano-sized calcium fluoride for dental applications. Dent Mater 24:111–116
49. Saad M, Garbuzenko OB, Ber E, Chandna P, Khandare JJ, Pozharov VP et al (2008) Receptor targeted polymers, dendrimers, liposomes: which nanocarrier is the most efficient for tumor-specific treatment and imaging? J Control Release 130:107–114
50. Ferrara K, Pollard R, Borden M (2007) Ultrasound microbubble contrast agents: fundamentals and application to gene and drug delivery. Annu Rev Biomed Eng 9: 415–447
51. Bull JL (2007) The application of microbubbles for targeted drug delivery. Expert Opin Drug Deliv 4:475–493
52. Chen S, Shohet RV, Bekeredjian R, Frenkel P, Grayburn PA (2003) Optimization of ultrasound parameters for cardiac gene delivery of adenoviral or plasmid deoxyribonucleic acid by ultrasound-targeted microbubble destruction. J Am Coll Cardiol 42:301–308
53. Lum AF, Borden MA, Dayton PA, Kruse DE, Simon SI, Ferrara KW (2006) Ultrasound radiation force enables targeted deposition of model drug carriers loaded on microbubbles. J Control Release 111:128–134
54. Souza GR, Christianson DR, Staquicini FI, Ozawa MG, Snyder EY, Sidman RL et al (2006) Networks of gold nanoparticles and bacteriophage as biological sensors and cell-targeting agents. Proc Natl Acad Sci USA 103:1215–1220
55. Souza GR, Staquicini FI, Christianson DR, Ozawa MG, Miller JH, Pasqualini R et al (2010) Combinatorial targeting and nanotechnology applications. Biomed Microdevices 12:597–606
56. Sengupta S, Eavarone D, Capila I, Zhao G, Watson N, Kiziltepe T et al (2005) Temporal targeting of tumour cells and neovasculature with a nanoscale deliverysystem. Nature 436: 568–572
57. Garcia-Bennett A, Nees M, Fadeel B (2011) In search of the holy grail: folate-targeted nanoparticles for cancer therapy. Biochem Pharmacol 181:976–984
58. Naqvi S, Samim M, Abdin M, Ahmed FJ, Maitra A, Prashant C et al (2010) Concentration-dependent toxicity of iron oxide nanoparticles mediated by increased oxidative stress. Int J Nanomedicine 16:983–989
59. Behlke MA (2006) Progress towards in vivo use of siRNAs. Mol Ther 13:644–670

60. Xie FY, Woodle MC, Lu PY (2006) Harnessing in vivo siRNA delivery for drug discovery and therapeutic development. Drug Discov Today 11:67–73
61. De Fougerolles A, Vornlocher HP, Maraganore J, Lieberman J (2007) Interfering with disease: a progress report on siRNA-based therapeutics. Nat Rev Drug Discov 6:443–453
62. Pouton CW, Seymour LW (2001) Key issues in non-viral gene delivery. Adv Drug Deliv Rev 46:187–203
63. Elouahabi A, Ruysschaert JM (2005) Formation and intracellular trafficking of lipoplexes and polyplexes. Mol Ther 11:336–347
64. Torchilin VP (2000) Drug targeting. Eur J Pharm Sci 11:S81–S91
65. Nighswonger G (1999) A Medical Device Link MD& DI column: new polymers and nanotubes add muscle to prothetic limbs, http://www.devicelink.com/mddi/archive/99/08/004.html
66. Oberdörster G (2006) Toxicology of airborn environment and occupational particles. http://www2.envmed.rochester.edu/envmed/tox/faculty/oberdoerster.html
67. Lam CW, James JT, McCluskey R, Hunter RL (2004) Pulmonary toxicity of single-wall carbon nanotubes in mice 7 and 90 days after intractracheal instillation. Toxicol Sci 77: 126–134
68. Warheit DB, Laurence BR, Reed KL, Roach DH, Reynolds GA, Webb TR (2004) Comparative pulmonary toxicity assessment of single-wall carbon nanotubes in rats. Toxicol Sci 77:117–125
69. Poland CA, Duffin R, Kinloch I, Maynard A, Wallace WAH, Seaton A et al (2008) Carbon nanotubes introduced into the abdominal cavity of mice show asbestos-like pathogenicity in a pilot study. Nat Nanotechnol 3:423–428
70. Chou CC, Hsiao HY, Hong QS, Chen CH, Peng YW, Chen HW et al (2008) Single-walled carbon nanotubes can induce pulmonary injury in mouse model. Nano Lett 3: 437–445
71. Mitchell LA, Gao J, Vander WR, Gigliotti A, Burchiel SW, McDonald JD (2007) Pulmonary and systemic immune response to inhaled multiwalled carbon nanotubes. Toxicol Sci 100:203–214
72. Muller J, Huaux F, Moreau N, Misson P, Heilier JF, Delos M et al (2005) Respiratory toxicity of multi-wall carbon nanotubes. Toxicol Appl Pharmacol 207:221–231
73. Di Giorgio ML, Di Bucchianico S, Ragnelli AM, Aimola P, Santucci S, Poma A (2011) Effects of single and multi walled carbon nanotubes on macrophages: cyto and genotoxicity and electron microscopy. Mutat Res 722:20–31
74. Donaldson K, Stone V, Tran CL, Kreyling W, Borm PJA (2004) Nanotoxicology. Occup Environ Med 61:727–728
75. Limbach LK, Wick P, Manser P, Grass RN, Bruinink A, Stark WJ (2007) Exposure of engineered nanoparticles to human lung epithelial cells: influence of chemical composition and catalytic activity on oxidative stress. Environ Sci Technol 41:4158–4163
76. Wright JB, Lam K, Buret AG, Olson ME, Burrell RE (2002) Early healing events in a porcine model of contaminated wounds: effects of nanocrystalline silver on matrix metalloproteinases, cell apoptosis, and healing. Wound Repair Regen 10:141–151
77. Supp AP, Neely AN, Supp DM, Warden GD, Boyce ST (2005) Evaluation of cytotoxicity and antimicrobial activity of Acticoat burn dressing for management of microbial contamination in cultured skin substitutes grafted to athymic mice. J Burn Care Rehabil 26: 238–246
78. Chen HW, Su SF, Chien CT, Lin WH, Yu SL, Chou CC et al (2006) Titanium dioxide nanoparticles induce emphysema-like lung injury in mice. FASEB J 20:2393–2395
79. Muangman P, Chuntrasakul C, Silthram S, Suvanchote S, Benjathanung R, Kittidacha S et al (2006) Comparison of efficacy of 1 % silver sulfadiazine and Acticoat™ for treatment of partial-thickness burn wounds. J Med Assoc Thai 89:953–958
80. Paddle-Ledinek JE, Nasa Z, Cleland HJ (2006) Effect of different wound dressings on cell viability and proliferation. Plast Reconstr Surg 117:110S–118S
81. Braydich-Stolle L, Hussain S, Schlager JJ, Hofmann MC (2005) In vitro cytotoxicity of nanoparticles in mammalian germline stem cells. Toxicol Sci 88:412–419
82. Ahamed M, AlSalhi MSM, Siddiqui KJ (2010) Silver nanoparticle applications and human health. Clin Chim Acta 411:1841–1848
83. Ahamed M, Karns M, Goodson M, Rowe J, Hussain SM, Schlager JJ et al (2008) DNA damage response to different surface chemistry of silver nanoparticles in mammalian cells. Toxicol Appl Pharmacol 233:404–410
84. Teodoro JS, Simões AM, Duarte FV, Rolo AP, Murdoch RC, Hussain SM et al (2011) Assessment of the toxicity of silver nanoparticles in vitro: a mitochondrial perspective. Toxicol In Vitro 25:664–670

85. Hackenberg S, Scherzed A, Kessler M, Hummel S, Technau A, Froelich K et al (2011) Silver nanoparticles: evaluation of DNA damage, toxicity and functional impairment in human mesenchymal stem cell. Toxicol Lett 201:27–33
86. El Badawy AM, Silva RG, Morris B, Scheckel KG, Suidan MT, Tolaymat TM (2011) Surface charge-dependent toxicity of silver nanoparticles. Environ Sci Technol 45: 283–287
87. Takenaka S, Karg E, Roth C, Schulz H, Ziesenis A, Heinzmann U et al (2001) Pulmonary and systemic distribution of inhaled ultrafine silver particles in rats. Environ Health Perspect 109:547–551
88. Hussain SM, Hess KL, Gearhart JM, Geiss KT, Schlager JJ (2005) In vitro toxicity of nanoparticles in BRL 3A rat liver cells. Toxicol in Vitro 19:975–983
89. Hussain SM, Javorina AK, Schrand AM, Duhart HM, Ali SF, Schlager JJ (2006) The interaction of manganese nanoparticles with PC-12 cells induces dopamine depletion. Toxicol Sci 92:456–463
90. Mirsattari SM, Hammond RR, Sharpe MD, Leung FY, Young GB (2004) Myoclonic status epilepticus following repeated oral ingestion of colloidal silver. Neurology 62: 1408–1410
91. Chung YC, Chen IH, Chen CJ (2008) The surface modification of silver nanoparticles by phosphoryl disulfides for improved biocompatibility and intracellular uptake. Biomaterials 29:1807–1816
92. Pujalte I, Passagne I, Brouillaud B, Treguer M, Durand E, Ohayon-Courtes C et al (2011) Cytotoxicity and oxidative Stress induced by different metallic nanoparticles on human kidney cells. Part Fibre Toxicol 8:10
93. Zhao Y, Meng H, Chen Z, Zhao F, Chai ZF (2007) Dependence of nanotoxicity on nanoscale characteristics and strategies for reducing and eliminating nanotoxicity. In: Zhao YL, Nalwa HS (eds) Nanotoxicology. American Scientific, Stevenson Ranch, CA, pp 265–280
94. Galla JH (2000) Metabolic alkalosis. J Am Soc Nephrol 11:369–375
95. Williams AJ (1998) ABC of oxygen. Assessing and interpreting arterial blood gases and acid–base balance. Br Med J 317:1213–1216
96. Tao TY, Liu F, Klomp L, Wijmenga C, Gitlin JD (2003) The copper toxicosis gene product Murr 1 directly interacts with the Wilson disease protein. J Biol Chem 278:41593–41596
97. Meng H, Chen Z, Xing G, Yuan H, Chen C, Zhao F et al (2007) Ultrahigh reactivity provokes nanotoxicity: explanation of oral toxicity of nano-copper particles. Toxicol Lett 175:102–110
98. Li L, Zhu YJ (2006) High chemical reactivity of silver nanoparticles toward hydrochloric acid. J Colloid Interface Sci 303:415–418
99. Driscoll KE, Lindenschmidt RC, Maurer JK, Higgins JM, Ridder G (1990) Pulmonary response to silica or titanium dioxide: inflammatory cells, alveolar macrophage-derived cytokines, and histopathology. Am J Respir Cell Mol Biol 2:381–390
100. Ferin J, Oberdorster G (1985) Biological effects and toxicity assessment of titanium dioxides: anatase and rutile. Am Ind Hyg Assoc J 46:69–72
101. Bermudez E, Mangum JB, Asgharian B, Wong BA, Reverdy EE, Janszen DB et al (2002) Long-term pulmonary responses of three laboratory rodent species to subchronic inhalation of pigmentary titanium dioxide particles. Toxicol Sci 70:86–97
102. Bermudez E, Mangum JB, Wong BA, Asgharian B, Hext PM, Warheit DB et al (2004) Pulmonary responses of mice, rats, and hamsters to subchronic inhalation of ultrafine titanium dioxide particles. Toxicol Sci 77:347–357
103. Warheit DB, Webb TR, Reed KL (2003) Pulmonary toxicity studies with TiO_2 particles containing various commercial coatings. Toxicologist 72:298A
104. Warheit DB, Webb TR, Reed KL, Frerichs S, Sayes CM (2007) Pulmonoary toxicity study in rats with three forms of ultrafine-TiO_2 particles: differential responses related to surface properties. Toxicology 230:90–104
105. Li J, Li Q, Xu J, Li J, Cai X, Liu R et al (2007) Comparative study on the acute pulmonary toxicity induced by 3 and 20 nm TiO_2 primary particles in mice. Environ Toxicol Pharmacol 24:239–244
106. Ratnam DV, Ankola DD, Bhardwaj V, Sahana DK, Kumar MNVR (2006) Role of antioxidants in prophylaxis and therapy: a pharmaceutical perspective. J Control Release 113: 189–207
107. Ankola DD, Viswanand B, Bhardwaj V, Ramarao P, Kumar MNVR (2007) Development of potent oral nanoparticulate formulation of coenzyme Q 10 for treatment of hypertension: can the simple nutritional supplements be used as first line therapeutic agents for prophylaxis/therapy? Eur J Pharm Biopharm 67:361–369
108. Shaikh J, Ankola DD, Beniwal V, Singh D, Kumar MNVR (2009) Nanoparticle encapsulation improves oral bioavailability of curcumin

by at least 9-fold when compared to curcuminn administered with piperine as absorption enhancer. Eur J Pharm Sci 37:223–230

109. Bala I, Bhardwaj V, Hariharan S, Kharade SV, Roy N, Kumar MNVR (2006) Sustained release nanoparticulate formulation containing antioxidant-ellagic acid as potential prophylaxix system for oral administration. J Drug Target 14:27–34
110. Bala I, Bhardwaj V, Hariharan S, Sitterberg S, Bakowsky U, Kumar MNVR (2005) Design of biodegradable nanoparticles: a novel approach to encapsulating poorly soluble phytochemical ellagic acid. Nanotechnology 16:2819–2822
111. Sonaje K, Italia JL, Sharma G, Bhardwaj V, Tikoo K, Kumar MNVR (2007) Development of biodegradable nanoparticles for oral delivery of ellagic acid and evaluation of their antioxidant efficacy against cyclosporine a-induced nehrotoxicity in rats. Pharm Res 24:899–908
112. Italia JL, Datta P, Ankola DD, Kumar MNVR (2008) Nanoparticles enhance per oral bioavailability of poorly available molecules: epigallocatechin gallate nanoparticles ameliorates cyclosporine induced nephrotoxicity in rats at three times lower dose than oral solution. J Biomed Nanotechnol 4:304–312
113. Chomouckaa J, Drbohlavovaa J, Babulab P, Adamc V, Hubaleka J, Provaznikd I et al (2010) Cell toxicity and preparation of streptavidin-modified iron nanoparticles and glutathione-modified cadmium-based quantum dots. Procedia Eng 5:922–925
114. Li J, Feng L, Fan L, Zha Y, Guo L, Zhang Q et al (2011) Targeting the brain with PEG-PLGA nanoparticles modified with phage-displayed peptides. Biomaterials 32: 4943–4950
115. Lynch I, Cedervall T, Lundqvist M, Cabaleiro-Lago C, Linse S, Dawson KA (2007) The nanoparticle—protein complex as a biological entity; a complex fluids and surface science challenge for the 21st century. Adv Colloid Interface Sci 134–135:167–174
116. Cedervall T, Lynch I, Foy M, Berggård T, Donnelly SC, Cagney G et al (2007) Detailed identification of plasma proteins adsorbed on copolymer nanoparticles. Angew Chem Int Ed 46:5754–5756
117. Lynch I, Dawson KA (2008) Protein-nanoparticle interactions. Nanotoday 3:40–47
118. Borm PJA, Kreyling W (2004) Toxicological hazards of inhaled nanoparticles—potential implications for drug delivery. J Nanosci Nanotechnol 4:521–531
119. Kim HR, Andrieux K, Gil S, Taverna M, Chacun H, Desmaële D et al (2007) Translocation of poly(ethylene glycol-*co*-hexadecyl)cyanoacrylate nanoparticles into rat brain endothelial cells: role of apolipoproteins on receptor-mediated endocytosis. Biomacromolecules 8:793–799
120. Dutta D, Sundaram SK, Teeguarden JG, Riley BJ, Fifield LS, Jacobs JM et al (2007) Adsorbed proteins influence the biological activity and molecular targeting of nanomaterials. Toxicol Sci 100:303–315
121. Shen XC, Liou XY, Ye LP, Liang H, Wang ZY (2007) Spectroscopic studies on the interaction between human hemoglobin and CdS quantum dots. J Colloid Interface Sci 311:400–406
122. Mu Q, Li Z, Li X, Mishra SR, Zhang B et al (2009) Characterization of protein clusters of diverse magnetic nanoparticles and their dynamic interactions with human cells. J Phys Chem C 113:5390–5395
123. Karmali PP, Simberg D (2011) Interactions of nanoparticles with plasma proteins. Implications on clearance and toxicity of drug delivery system. Expert Opin Drug Deliv 8:343–357
124. Goppert TM, Muller RH (2005) Adsorption kinetics of plasma proteins on solid lipid nanoparticles for drug targeting. Int J Pharm 302:172–186
125. Gessner A, Lieske A, Paulke BR, Müller RH (2003) Functional groups on polystyrene model nanoparticles: influence on protein adsorption. J Biomed Mater Res 65:319–326
126. Kurath M, Maasen S (2006) Toxicology as a nanoscience?—disciplinary identities reconsidered. Part Fibre Toxicol 3:1–13
127. Kabanov AV (2006) Polymer genomics: an insight into pharmacology and toxicology of nanomedicines. Adv Drug Deliv Rev 58: 1597–1621
128. Miller J (2003) Beyond biotechnology: FDA regulations of nanomedicine. Columbia Sci Technol Law Rev 4:E5

Chapter 5

Fullerenol Nanoparticles: Toxicity and Antioxidant Activity

Rade Injac, Matevz Prijatelj, and Borut Strukelj

Abstract

Fullerenes are a relatively new group of compounds and represent a class of sphere-shaped molecules made exclusively of carbon atoms. Since their discovery in 1985, many aspects of both fullerene and its analogues have been intensively studied to reveal their physical and chemical reactivity, as well as potential use in biological systems. Both in vitro and in vivo studies have shown that polyhydroxylated fullerene derivatives, fullerenol nanoform ($C_{60}(OH)_n$, $n = 2–72$), can be potential antioxidative agents in biological systems. This chapter represents a review of published studies of fullerenes' biological activities with special accent on the most tested fullerenol nanoform $C_{60}(OH)_{24}$.

Key words Fullerene, Fullerenol, Nanoparticles, Antioxidant, Toxicity, Bioapplication

1 Introduction

Fullerenes are a type of carbon molecule first discovered in the mid-1980s. The most well known of these is the buckyball, which is a closed spherical molecule, shaped like a soccer ball. The fullerenes discovered by Peter Harris and colleagues were completely closed and comparatively large, containing around 6,000–10,000 carbon atoms each, and with a generally disordered structure. Since the Allende meteorite has been shown to have formed at the same time as our solar system's planets, the results, which show fullerenes are capable of surviving for billions of years in space, present a further step toward accepting where life came from. The findings also support previous claims by astrobiologist Luann Becker at the University of Washington in Seattle that fullerenes were present in carbonaceous material from meteorites [1].

Studies on biological properties of fullerene C_{60} and its derivatives started a decade ago as curiosity-driven studies and are now flourishing as an area of transdisciplinary research. Fullerenes are the third pure crystal form of carbon in addition to diamond and graphite. Molecules consist of a spherical (C_{60}), ellipsoid (C_{70}), or cylindrical (carbon nanotubes) carbon atoms. They were named

Donald Armstrong and Dhruba J. Bharali (eds.), *Oxidative Stress and Nanotechnology: Methods and Protocols*, Methods in Molecular Biology, vol. 1028, DOI 10.1007/978-1-62703-475-3_5,

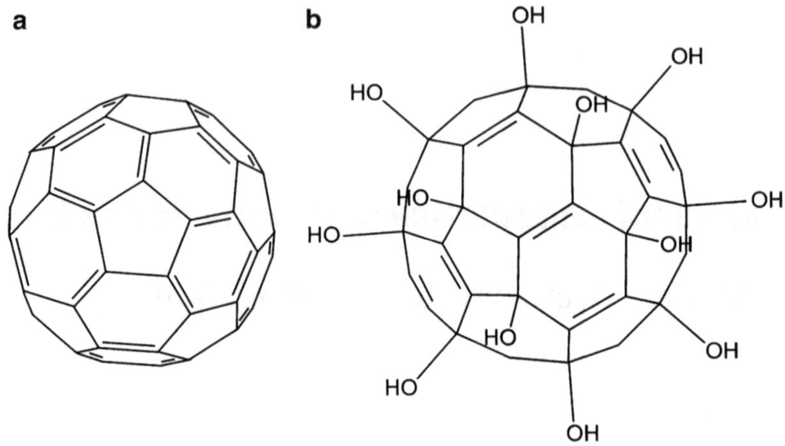

Fig. 1 Structure of (**a**) fullerene C_{60} and (**b**) fullerenol $C_{60}(OH)_{24}$

after Richard Buckminster Fuller. A spherical fullerene C_{60} looks like a soccer ball, and therefore, it is often called "buckyball" or "footballene" (Fig. 1). Robert F. Curl, Jr., Richard E. Smalley, and Sir Harold W. Kroto discovered and defined fullerenes' structure as an unexpected surprise during laser spectroscopy experiments at Rice University in September 1985. The Nobel Prize in Chemistry in 1996 was awarded to this American-English research team for discovery of this molecule with the perfect symmetry existing in nature [2].

2 Fullerene

Fullerenes C_{60} are polyhedral closed cages made up entirely of n three-coordinate carbon atoms and having 12 pentagon and 20 hexagonal faces. Fullerene C_{60} molecule possesses geometry of truncated icosahedrons fullerene-60-Ih. Diameter of the molecule C_{60} is 0.710 ± 0.007 nm. The outer and inner diameters of π electron cloud can be estimated as 0.340 nm and 0.350 nm, respectively, and bond energy per atom is 7.40 eV. Pure fullerenes are black crystals and powder insoluble in water and proton-acceptor solvents. It is soluble in halogen and alkyl-substituted benzene (1,2-dichlorbenzen 27 mg/ml). Using combinations of nucleophilic and electrophilic additions, cycloadditions, and radical additions, it is possible to covalently bond any class of organic compounds to a fullerene core. These structural and electronic characteristics of fullerene C_{60} provide possibility for conducting various types of chemical transformations that resulted in a wide variety of biologically active water-soluble fullerenes. Fullerene derivatives take important place in the development of nanobiotechnology- and nanomedicine-related products. The chemical modification of fullerene C_{60} molecule by attachment of hydroxyl groups was an easy and straightforward method to synthesize water-soluble fullerenes, named fullerols or fullerenols ($C_{60}(OH)_n$ $n=2$–72). Structure of fullerenol $C_{60}(OH)_{24}$ is shown in Fig. 1 [1–4].

Fullerenes are hydrophobic molecules; therefore, potential biomedical applications are restricted by their extremely poor solubility. Different methods can be used for improving poor water solubility of fullerene, such as surfactants [5], and incorporation into water-soluble supramolecular structures such as γ-cyclodextrin [6]. Additionally, liposomes with incorporated fullerene (C_{60}) and photoactive C_{60} were also used in order to facilitate poor solubility of this molecule [7]. Early experiments demonstrated that modifications of fullerenes resulting from interactions with solutes present in environmental and biological systems may have a significant influence on the metabolism, transport, and reactivity of fullerene [8]. One applied solution for improving negligible solubility of fullerene is its derivatization. Derived fullerenes usually form stable molecular dispersed suspensions as a result of hydrophilization or encapsulation of C_{60} in order to prevent contact between the fullerene cores [9]. However, derivatization as itself does not totally avoid cluster formation [9–11]. On the other hand, the unique electronic π-system of fullerenes makes them potential photosensitizers upon absorption of UV or visible light. The ground state adsorption spectrum of C_{60} is characterized by intense absorption in the UV region with weaker bands extending throughout the visible spectrum up to 700 nm. The short-lived singlet state is formed by irradiation of C_{60} and converts almost quantitatively into a longer-lived triplet state. Furthermore, singlet oxygen is often generated in the process of transferring energy to molecular oxygen. The triplet lifetime is essential for photoactivity in cells. Actually, it was suggested that only triplet states with lifetimes longer than 100 μs have enough cytotoxic potential. Singlet oxygen possesses intensive oxidative potential. Zhao et al. [12] have shown that a variety of water-soluble fullerenes can efficiently generate singlet oxygen ($^1O_2$) upon irradiation via energy transfer from the excited triplet of fullerene to oxygen [13, 14]. There are also reports that photo-irradiation of fullerenes in aqueous systems results in the production of the radical anions ($C_{60}^{\bullet}$) followed by generation of superoxide anion radical ($O_2^{\bullet -}$) and hydroxyl radical ($^{\bullet}OH$) via electron transfer, particularly in the presence of electron donors (such as NADH or amines) [15–17]. These photochemical mechanisms, occurring during photodynamic therapy, demonstrate the main pathways for the photoinduced toxicity of fullerenes [18]. This dual nature of C_{60} to either reduce or generate cell-damaging ROS could therefore be used for its development as a cytoprotector or as a cytotoxic anticancer/antimicrobial agent [19].

3 Fullerenol

Fullerenol $C_{60}(OH)_{24}$ has diameter of around 1 nm with symmetrically arranged hydroxyl groups on the C_{60} sphere. Fullerenol is a dark brown amorphous substance, soluble in water and in DMSO.It can be dissolved in water, in concentration ranging from 10^{-3} mol/l to 5×10^{-3} mol/l, and in pH interval from 3.5 to 9.8; it forms nano-aggregates, with up to 40 presences of aggregates larger than 400 nm (Fig. 2) [20]. Fullerenol may react rapidly and irreversibly with a variety of metal salts under ambient aqueous conditions to produce insoluble metal–hydroxyfullerene cross-linked polymers (M-fullerenol). Interaction of fullerenols with bio-metal is important with regard to fullerene-based pharmaceutical agents and waste treatment as well. Several chemical properties of fullerenols may significantly influence their biological activity. Similar to

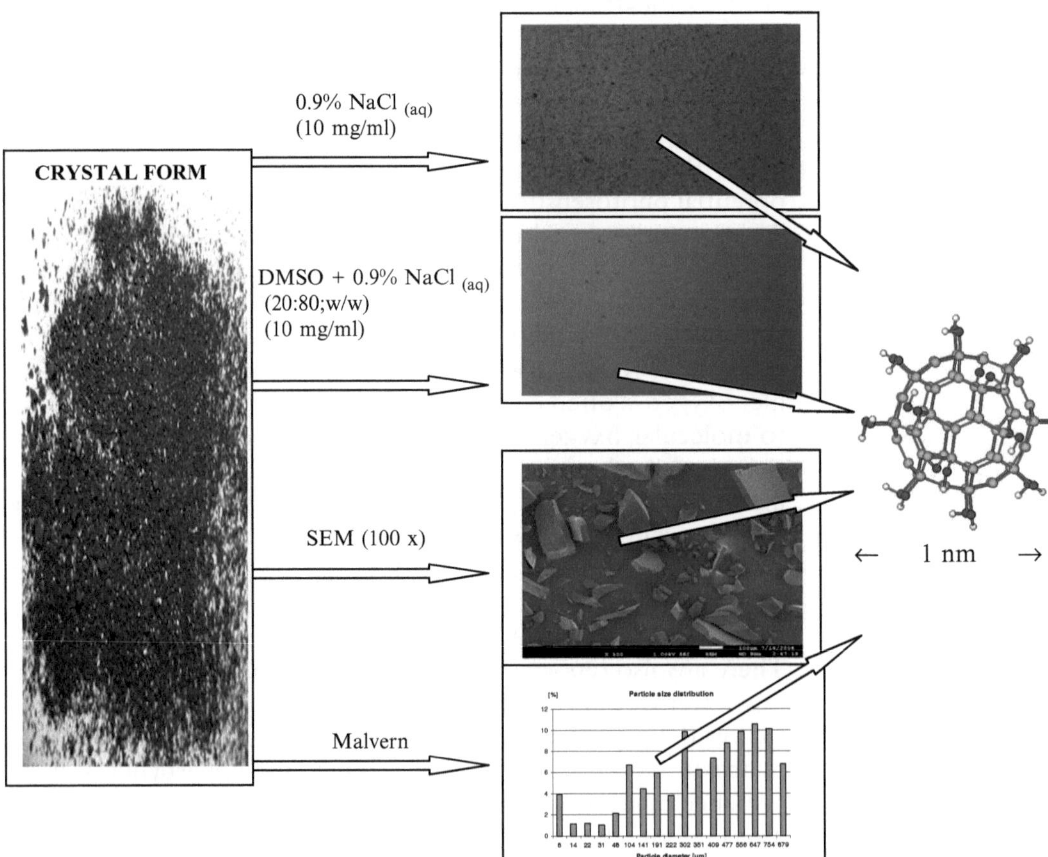

Fig. 2 Fullerenol $C_{60}(OH)_{24}$: macroscopic view (crystal form); dissolution in physiological fluid (optical microscope; a lot of aggregates of fullerenol with a size of 50–100 μm); dissolution in DMSO/physiological fluid (optical microscope; some aggregates of fullerenol; <20 μm); SEM view; particle size distribution (Malvern method); structure scheme (With permission) [20]

nonhydroxylated water-soluble fullerene derivatives, the most important characteristics of fullerenols regarding mechanisms of their biological activity are photosensitizing property and free radical-scavenging activity [20–30].

The electronic properties and reactivity of fullerenols strongly depend on the number of hydroxyl groups as well as their precise position on the carbon cage. The specific behavior of such molecules is a result of their structural flexibility, the rotation of the –OH groups around the axes going through the C–O bonds, and the distribution of these groups across different carbon sites of the fullerene surface [31].

Fullerenols at the same time have both attractive (C–OH) and repulsive (C–O–) sites. The acidic protons may be involved in attractive hydrogen bonding interactions with other fullerenol molecules, driving cluster formation. Experiments related to physicochemical properties of fullerenols confirmed that hydroxylated fullerenes C_{60} and C_{70} form colloids and are not likely to form a molecular solution in water, which is in conflict with the description of this material as water soluble [10]. These clusters have a mean size of 100 nm or more and are often of different shapes, which depend on the characteristics of the hydroxyl groups.

In relation to biological activity of polyhydroxylated fullerenes $C_{60}(OH)_x$, it is of value to mention its protective effect against oxidative stress in RAW 264.7 macrophage cell line, ischemia-reperfused rat lungs [32], and its ability to significantly reduce doxorubicin toxicity against human breast cancer cell lines [33]. On the other hand, possible mechanism of its cytotoxic and cytostatic action is in inhibition of protein tyrosine kinase activity [34].

Antioxidative and free radical-scavenging activities of fullerenols are their most exploited properties in biomedical studies. Antiproliferative and cytotoxic effects of various water-soluble fullerenols $C_{60}(OH)_n$ have been observed in different experimental models (in vivo as well as in in vitro).

4 Fullerenol In Vitro Antioxidative Activity

Antioxidative potency of fullerenol is being extensively evaluated in both in vitro and in vivo models. In relation to biological antioxidative activity of polyhydroxylated fullerenes $C_{60}(OH)_x$ in vitro, it is of value to mention its protective effect against oxidative stress in RAW 264.7 macrophage cell line, ischemia-reperfused rat lungs [32], and its ability to significantly reduce doxorubicin toxicity against human breast cancer cell lines and HepG2 cells [21, 33].

One of the most studied antioxidative effects of fullerenol is its protective potency against doxorubicin-induced cell death. Fullerenol modulated doxorubicin-induced cytotoxicity on human breast cancer cell lines and human hepatocellular carcinoma cells (HepG2) [21, 35]. Fullerenols have protective effects on doxorubicin-induced cytotoxicity in three human breast cancer cell lines (T47D, MDA-MB-231, and MCF). Fullerenol in combination with doxorubicin strongly inhibited doxorubicin cytotoxicity by more than 50 % for most concentrations at each time point. The best protection of cells was obtained when fullerenol was added to culture one hour prior to doxorubicin. The inhibition of doxorubicin cytotoxicity with fullerenol in estrogen receptor-positive cell lines was the most prominent after 48 h at concentration 0.44 μmol/l. The rates of inhibition of doxorubicin–fullerenol combination were similar to or lower than inhibition induced by the fullerenol alone. Assuming that doxorubicin cytotoxicity can be explained by induction of oxygen free radical formation and proposed antioxidative activity of the fullerenols, the effect of fullerenol was compared with the effect of natural antioxidant proanthocyanidin in the same system [33]. The results show that both proanthocyanidin (20 μg/ml) and fullerenol (1.9 μg/ml = 1.7 μmol/l) added as single agents to MCF-7 cell culture for 48 h induce low growth inhibition (5 % and 17 % for proanthocyanidin and fullerenol, respectively). In combination with doxorubicin (1 μmol/ml; CI = 41.05 %), both agents strongly decrease doxorubicin cytotoxicity. Doxorubicin cytotoxicity was decreased by 80 % (fullerenol) and 87 % (proanthocyanidin) if the agents were added to the culture one hour before doxorubicin. Similarly fullerenol-mediated protection against doxorubicin-induced cell death was also observed in human hepatocellular carcinoma cells (HepG2). Doxorubicin-induced cytotoxicity in the HepG2 cell line is time and dose dependent. When co-added with doxorubicin (5 μmol/l), fullerenol (8.8 μmol/l and 39 μmol/l), with its antioxidant activity, significantly increased the cell survival rate for almost twofold within the first 24 h. After 24 h of treatment, the survival fractions of the HepG2 cells were as follows: 28.68 %, 61.47 %, and 71.86 % (doxorubicin, doxorubicin/fullerenol 8.8 μmol/l, doxorubicin/fullerenol 39 μmol/l, respectively). After 96 h, the survival rate of the HepG2 cells was almost seven times higher among the doxorubicin-/fullerenol-treated cells compared to doxorubicin alone. The level of cytotoxicity of the doxorubicin/fullerenol combination was similar to or lower than the cytotoxicity induced by the fullerenol alone [21] (Fig. 3). Localization of fullerenol close to the mitochondria [36], its free radical scavenger activity, and the ability to act as an artificial electron acceptor leading to inhibition of monooxygenase activity could explain protective effect of fullerenol in the model of doxorubicin-induced cytotoxicity on human breast cancer cell lines and human hepatocellular carcinoma cells [33].

Fullerenol effects on antioxidative enzyme activity in irradiated human erythroleukemia cell line were also investigated. Experiments have provided evidence of reactive oxygen and

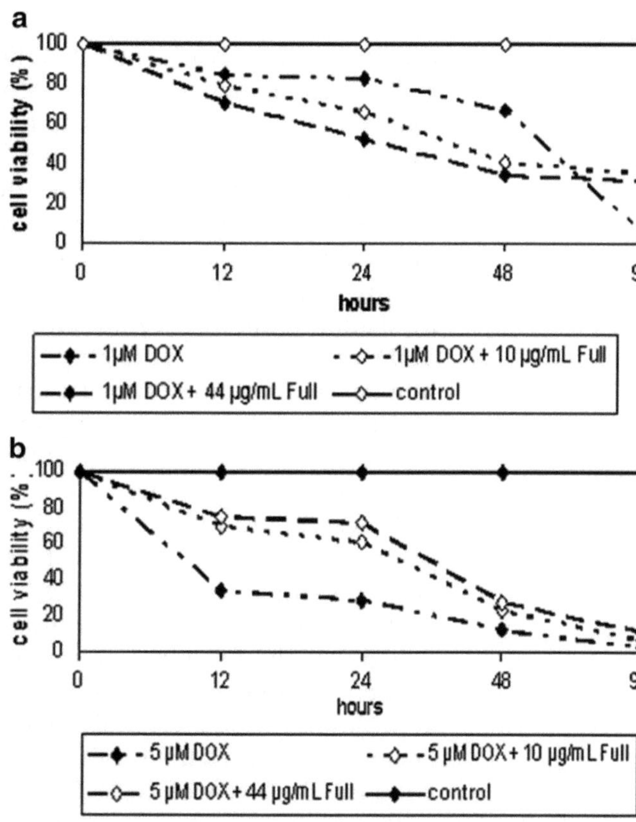

Fig. 3 Viability of HepG2 cells treated with (**a**) 1 μM of Dox—doxorubicin (alone or in combination with 10 or 44 μg/ml of Full—fullerenol) or (**b**) 5 μM of Dox (alone or in combination with 10 μg/ml or 44 μg/ml of Full) over a 96 h time period (With permission) [21]

nitrogen species (ROS/RNS) scavenger properties in vitro and radioprotective efficiency in vivo. Pretreatment of irradiated cells by fullerenol exerted the statistically significant effects on cell number and response of antioxidative enzymes to x-ray irradiation-induced oxidative stress in human erythroleukemia K562 cells. The cellular area of 100 cells in each experimental set was measured with optical resolution of 6.57 pixels, using Scion image software. The enzyme activities of γ-glutamyltransferase (GGT), total superoxide dismutase (SOD), and glutathione peroxidase (GSH-Px) were determined by kinetic method [37–39]. At 48 h after irradiation, the number of K562 cells was increased, indicating that the survived fraction of cells continued to proliferate. This effect was more pronounced in fullerenol pretreated cells. Activity of GGT in irradiated K562 cells was significantly increased, in time-related manner, being fourfold increased at 24 h postirradiation time and 2.8-fold at 48 h postirradiation time. The fullerenol pretreatment significantly reduced the levels of GGT in irradiated cells, indicating the important role of fullerenol for maintenance of redox homeostasis [40]. Published data are showing the significant increase of total SOD levels in irradiated cells, at 24 and 48 h postirradiation time. More interestingly, we observed fourfold elevated SOD levels at 1 h time point in fullerenol pretreated nonirradiated cells and 5.5-fold elevated level in fullerenol pretreated irradiated cells. This effect was attenuated at 24 and 48 h postirradiation time. These results are in accordance with fullerenol ROS-scavenging activity, observed in both in vivo and in vitro systems [41, 42]. The administration of fullerenol before irradiation induced a significant increase in the activities of both SOD and GSH-Px. This effect was followed by preserved cell number and colony-forming capacity in the same experimental set. The hypothesis concerning the possible mechanisms responsible for the increased activity of SOD in K562 cells pretreated by fullerenol may be viewed in light of the possibility that the •NO-scavenging fullerenol activity may prevent superoxide consummation in the reaction of formation of peroxynitrite anion, concomitantly increasing the $O_2^{\bullet-}$ concentration, and increasing SOD activity as a consequence of superoxide excess. It was hypothesized that the fullerenol pretreatment prevented the deleterious effects of ROS, directly by increasing the antioxidant enzyme activities of irradiated cells. Fullerenol pretreatment did not change the control levels of GSH-Px activity (1 h time point). In contrast, the levels of GSH-Px at 1 h time point were higher in both irradiated cellular sets. Following fullerenol pretreatment and 24 Gy irradiation, the GSH-Px level was significantly elevated at 1 h postirradiation time in comparison with other experimental sets. This effect of fullerenol pretreatment in irradiated cells may be partially explained by consecutive activation of GSH-Px due to SOD increased activity, as well as by substrate accumulation in cell cytosol, which inhibits the GSH-Px enzyme active center [43]. The fullerenol capacity to form clusters in water, with a mean size of approximately 100 nm or more, at physiological pH and temperature, may be partially responsible for the biological effects that were observed [44].

5 Fullerenol In Vitro Toxicity

The assessment of fullerenol toxicity is an absolute and obvious prerequisite for its potential use in biomedicine and consumer products. Recent studies describe lower toxicity of fullerenol and other water-soluble fullerenes against dermal fibroblasts and liver carcinoma HepG2 cells in comparison to pure C_{60}. It was demonstrated that the cytotoxic activity of C_{60} colloid was caused by ROS-mediated cell membrane lipid peroxidation [45]. Furthermore, Isakovic et al. demonstrated that pure C_{60} and fullerenols $C_{60}(OH)_x$ apparently employ different cytotoxic mechanisms resulting in preferential induction of caspase-independent necrosis and caspase-dependent apoptosis, respectively [46]. From molecular point of view, C_{60} induces production of oxygen radicals, which are involved in lipid peroxidation, and consequently stimulates necrotic cell death (vacuolation of the cytoplasm, breakdown of the plasma membrane, release of cellular components and pro-inflammatory molecules). On the other hand, $C_{60}(OH)$ x-triggered apoptosis in some cell types (chromatin condensation, activation of caspases, fragmentation of DNA without membrane breakdown, absence of inflammation) seems to be ROS independent [46]. Pro-oxidant activity of pure C_{60} is related to its chemical structure. It was established that derivatization of C_{60} decreases ROS generation and cytotoxicity [47].

Although most of the scientific literature supports a protective role of fullerenol in biological systems, there is a growing literature detailing the cytotoxic effects of this nanoparticle [47–50]. Fullerenol's mechanism of cell death appears to be cell type specific, and both apoptotic and non-apoptotic mechanisms have been reported in the literature [48, 51]. Water-soluble fullerene derivatives have been reported to cause cell cycle arrest at the G1 phase in Chinese hamster lung and ovary cells [50]. Fullerenol alone has also suppressed proliferation of human breast cancer cells in a cell line-dependent manner [33]. Derivatized fullerenes have also been reported to exhibit differential cytotoxicity in human dermal fibroblasts, human colon adenocarcinoma cells, and liver carcinoma cell lines, with the more water-soluble derivatives demonstrating lesser adverse effects in culture [21, 47].

To exemplify, fullerenol selectively inhibits proliferation of the human colon adenocarcinoma cell line (Caco-2) and human hepatocellular carcinoma cells (HepG2) in a concentration- and time-dependent manner, with greater influence on Caco-2 cells. Knowing that low levels of free oxygen species promote cell proliferation and that redox alterations play a significant role in a signal transduction pathway, important for cell growth regulation [52, 53], it is sound to propose that fullerenol, as antioxidant and free radical scavenger, might influence the HepG2 and Caco-2 cell redox state, leading to decreased cell proliferation [21]. Such observation could explain the selectivity of fullerenol on growth inhibition for the HepG2 cells in comparison to Caco-2 cells (Fig. 4). For the cells to protect themselves against oxidative stress and to regenerate after injury, there is a high requirement for enzymes to regulate cellular processes [54]. Fullerenols are good free radical trappers in biological systems [55–57], and Ueng et al. [49] reported that they can suppress the activities of mitochondrial oxidative phosphorylation in vitro. Our research group demonstrated that fullerenol does disturb the mitochondrial activity since a significant dose-depended drop of mitochondrial potential incubation with fullerenol was seen (Figs. 5 and 6). The effect was more pronounced

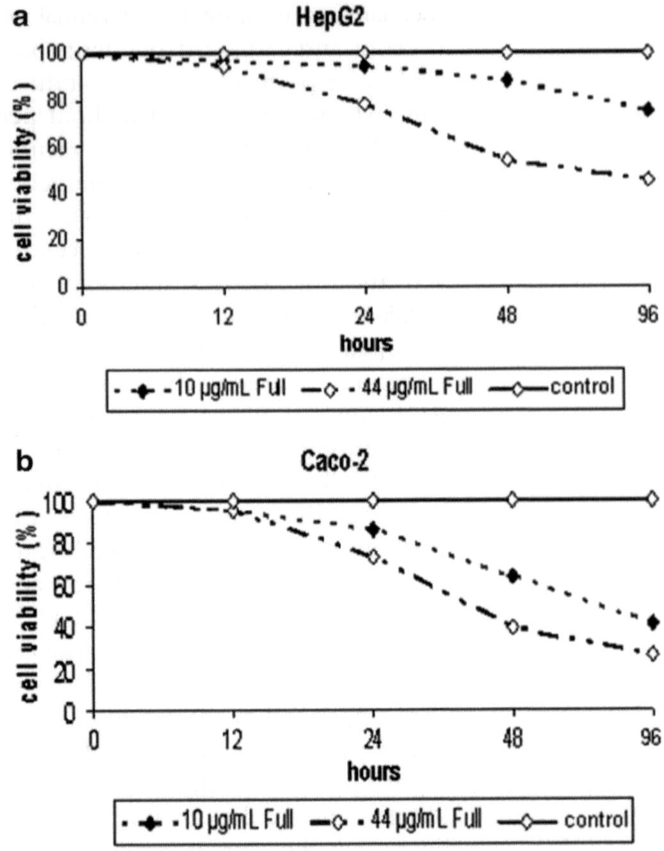

Fig. 4 Viability (%) of (**a**) HepG2 and (**b**) Caco-2 cells treated with 10 and 44 μg/ml of Full—fullerenol—over a 96 h time period (With permission) [21]

in Caco-2 cells than in HepG2 cells, which can result from higher mitochondrial capacity and regeneration properties of the HepG2 cells compared to Caco-2 cells [21].

Loss of mitochondrial membrane potential after addition of fullerenol to cell culture was also observed in kidney cells [58]. Fullerenol was found to be cytotoxic in the millimolar range. Fullerenol-induced cell death was associated with cytoskeleton disruption and autophagic vacuole accumulation. Addition of fullerenol to cell culture also resulted in concurrent loss of cellular mitochondrial membrane potential and depletion of ATP. Fullerenol-induced ATP depletion and loss of mitochondrial potential were partially inhibited by co-treatment with the autophagy inhibitor, 3-methyladenine. Based on these data, it is hypothesized that in kidney cells, cytoskeleton disruption may be an initiating event in fullerenol cytotoxicity, leading to subsequent autophagy dysfunction and loss of mitochondrial capacity. The proposed mechanism may be relevant for a variety of nanomaterials, since nanoparticle-induced cytoskeleton disruption, autophagic vacuole

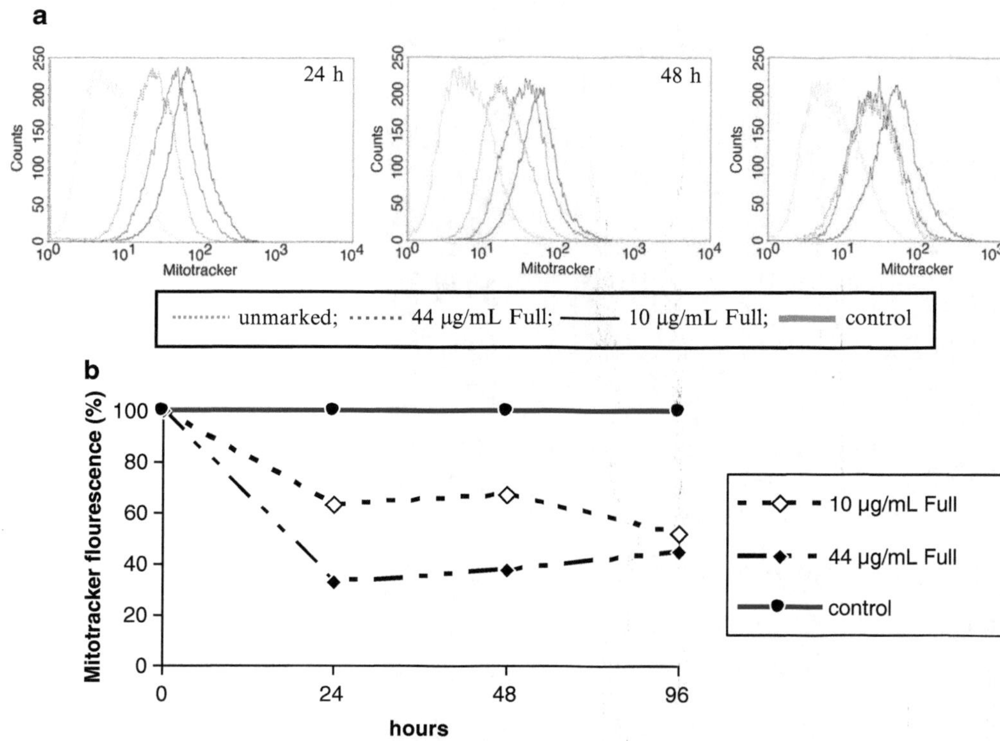

Fig. 5 Concentration-dependent effect of Full—fullerenol—on MitoTracker fluorescence in human hepatocellular carcinoma cells, evaluated by FACS (With permission) [21]

accumulation, and mitochondrial dysfunction are frequently reported in the literature [58].

Since many of the fullerene applications are based on intravascular administration of these compounds, it is of great significance to know about the effects of fullerenes on endothelial cells. Yamawaki et al., for example, reported that addition of fullerenol to endothelial cells decreases cell density, decreases cell proliferation and cell attachment, promotes LDH release, and increases accumulation of polyubiquitinated proteins [48]. These authors concluded that fullerenol did not cause apoptosis based on the failure to demonstrate cleavage of caspase 3 and PARP but rather, similarly as in kidney cells [58], induced autophagy. Further study by Gelderman et al. [51] on human epithelial HUVEC cells revealed that the fullerenol had both pro-inflammatory and pro-apoptotic effects on cells, exemplified by fullerenol-induced DNA fragmentation. In addition, cell cycle arrest and increase of intracellular Ca^{2+} were observed in cells treated with hydroxyfullerene $C_{60}(OH)_{24}$. It is therefore reasonable to speculate that both apoptotic and autophagic cell death may occur in fullerenol-treated endothelium.

With the wide potential applications of fullerenes in mind, the safety hazards of these materials need to be thoroughly evaluated.

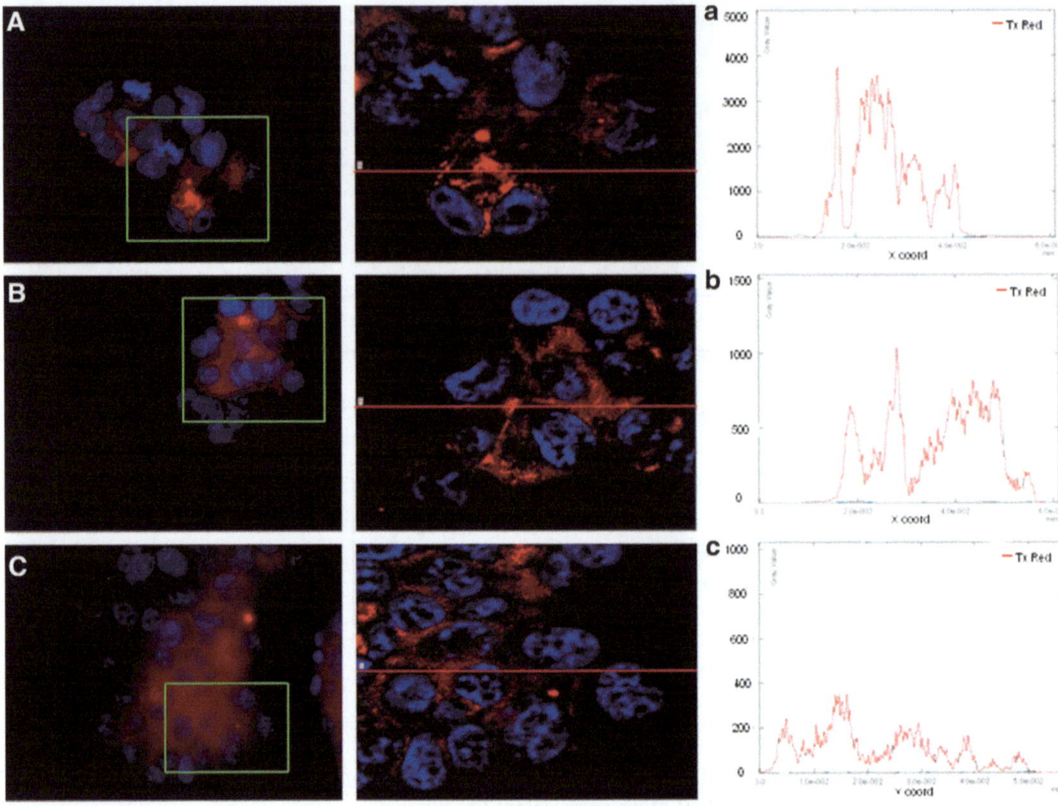

Fig. 6 Fluorescent pictures (**A**, **B**, **C**) and fluorescence intensity profiles (**a**, **b**, **c**) of human hepatocellular carcinoma cells: (**A/a**) control, (**B/b**) treated with 10 μg/ml of Full, and (**C/c**) 44 μg/ml Full. Nuclei were stained with Hoechst (*blue*) and mitochondria with MitoTracker (*red*) (With permission) [21]

6 Fullerenol In Vivo Antioxidative Activity

The antioxidant properties of fullerenol were tested by measuring their ability to scavenge stable 2,2-diphenyl-1-picrylhydrazyl (DPPH) free radical and reactive hydroxyl radical (•OH) during the Fenton reaction using electron spin resonance (ESR) spectroscopy. The addition of fullerenol to the reaction system (from 0.18 to 0.88 mmol/l) resulted in a dose-dependent inhibition of the ESR signal intensity of DPPH radical. As in the case of inhibition of DPPH radical, higher concentrations of fullerenol were more effective in reducing the ESR signal of DMPO-OH spin adducts. The concentration of 0.88 mmol/l of fullerenol was more effective in inhibition of hydroxyl radical (inhibition ratio 83 %) in comparison with the percentage of inhibition of DPPH radical. On the basis of these results, the possible mechanism of antioxidative activity of fullerenol is radical-addition reaction of $2n$ •OH radicals to remaining olefinic double bonds of fullerenol core to yield $C_{60}(OH)_{24} + 2n^{\bullet}OH$ ($n = 1$–12). The other proposed mechanism is the possibility of hydroxyl radical to interact with a hydrogen from fullerenol, including the formation of relatively stable fullerenol radical $C_{60}(OH)_{23}O^{\bullet}$ [29].

Prevention of lipid peroxidation by fullerenol in liposome model system was also demonstrated. As a model system, liposome was used as commercial preparation of biological membrane. Briefly, tested substances and liposome were added to $FeSO_4$/ ascorbic acid system and incubated 1 h at room temperature [59]. Antioxidant potential of fullerenol toward lipid peroxides was compared with commercial antioxidants butylated hydroxytoluene (BHT). Treatment of liposome with $FeSO_4$ and ascorbic acid leads to oxidation of polyunsaturated fatty acid in liposome and formation of TBA-RS. Results showed that fullerenol induced dose-dependent inhibition of $FeSO_4$-/ascorbic acid-stimulated formation of TBA-RS. Parallel with this, we examined the effect of BHT on lipid peroxidation, and obtained results demonstrated that fullerenol possesses similar efficiency in blockade of lipid peroxidation as BHT (calculated ED_{50} for fullerenol was 2.5 mmol/l and for BHT 1.2 mmol/l) [30].

In addition, superoxide radical-scavenging activity of fullerenol in xanthine/xanthine oxidase system was tested. To estimate possible superoxide radical-quenching activity of fullerenol, xanthine/xanthine oxidase system was used. Superoxide radicals, generated through enzymatic activity of xanthine oxidase, reduce cytochrome C. Removing superoxide radicals from the system leads to decreased rate of cytochrome C reduction. Application of fullerenol into xanthine/xanthine oxidase system caused decrease in reduction rate of cytochrome C compared to control. Obtained result demonstrated that fullerenol in the range of nanomolar and micromolar concentrations decreased reduction of cytochrome C between 5 and 20 %, while concentration of 1 mmol/l decreased reduction of cytochrome C for 40 % [30].

Finally, nitric oxide-scavenging activity in chemical model system was used to confirm previous statement regarding antioxidative potential of fullerenol. Investigation of the possible nitric oxide-scavenging activity of fullerenol demonstrated that it expressed direct scavenging activity toward nitric oxide radical (NO) liberated within solution of sodium nitroprusside (SNP), a NO donor. To estimate possible NO-scavenging activity of fullerenol, different concentrations of SNP (1, 2, or 5 mmol/l) were incubated alone or in combination with different concentrations of fullerenol (0.05 or 0.1 mmol/l). A release of NO from SNP dissolved in phosphate buffer and held on light and room temperature was time and dose dependent. Co-incubation of SNP with fullerenol resulted in the dose-dependent decrease of the level of nitrite in comparison to the nitrite levels obtained when SNP was dissolved alone. However, the percentage of the decrease was the same disregarding to the concentration of SNP present in the solution. Average ED_{50} for fullerenol, calculated for all 6 dose-dependent curves, was 10^{-4} mol/l [30].

Radioprotective effect of fullerenol is connected to its capability to react and has influence on the level of free radical in in vivo systems. Ionizing radiation produces harmful effects on living organisms by inducing enhanced production of free radical species. Exposure of cells to ionizing radiation leads to the formation of ROS that are related to radiation-induced cytotoxicity. The antioxidative enzyme SOD catalyzes the dismutation of superoxide anions into hydrogen peroxide. SOD is a very important parameter, which contributes to increased protection from lethal irradiation to hematopoietic cells. Trajkovic et al. investigated potential radioprotective effects of fullerenol in dose of 100 mg/kg i.p. which was given 30 min before x-irradiation [60]. Previous reports about radioprotective effect of fullerenol established that i.p. administration of 100 mg/kg of fullerenol 30 min before irradiation is suitable in comparison to a smaller dose of 10 mg/kg [33]. In relation to fullerenol radioprotectivity, emphasis must be put on its hemato- and tissue-protective effects. The tissue-protective outcome was particularly confirmed on the spleen, small intestine, and lung tissue [61].

Fullerenols $C_{60}(OH)_n$ are being expansively studied due to its great prospective as an antioxidant. It is suggested that fullerenols may act as a free radical scavenger in biological systems, in xenobiotics, as well as radioactive irradiation-s-induced oxidative stress. It has repeatedly demonstrated protective effects against cytotoxicity of doxorubicin in animal models, especially by fullerenol $C_{60}(OH)_{24}$ [1–4, 20–30].

Sprague–Dawley outbred rats with chemically induced mammary carcinomas were used for to study a potential protective role of fullerenol against cardio-, hepato-, nephro-, and pulmo-toxicity (Fig. 7) induced by doxorubicin [21–24]. According to preliminary studies on healthy adult Wistar rats [3, 25], it was confirmed that 100 mg/kg (i.p.) of fullerenol administered 30 min before anticancer agent has a protective influence on heart and liver tissue. Therefore, in further investigations [20–30], that dose was chosen as effective one.

Doxorubicin caused nephrotoxicity, but pretreatment with fullerenol prevented oxidative stress, lipid peroxidation, and a disbalance of GSH/GSSG level in kidney tissue caused by doxorubicin. Latest results approve the satisfactory nephroprotective efficacy of fullerenol in the acute phase of toxicity and encourage further studies of it as a possible nephroprotector [23]. An in vivo study was examined to confirm the potential protective role of fullerenol $C_{60}(OH)_{24}$ (25, 50, and 100 mg/kg/week) on doxorubicin-induced (1.5 mg/kg/week) liver and heart toxicity within 3 weeks of treatment, using female Sprague–Dawley rats with colorectal carcinoma [21]. The outcome for the treatment with doxorubicin

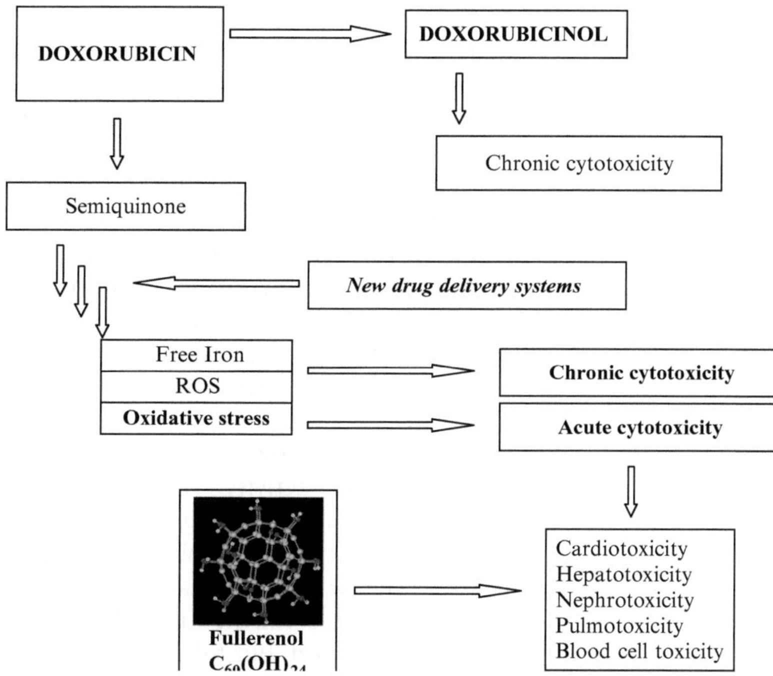

Fig. 7 Doxorubicin toxicity and fullerenol nanoform protection scheme

alone was significant variations in the serum levels of ALT, AST, LDH, and α-HBDH within 3 weeks of treatment as well as in the levels of MDA, GSH, GSH-Px, TAS, GR, CAT, and SOD in the liver as well as in the heart tissue at the end of treatment period. These toxicity effects were drastically reduced for all considered parameters by pretreatment with fullerenol. Chronic cardio- and hepatotoxicity was confirmed, and fullerenol antioxidative–organoprotective effect was compared with a well-known antioxidant, vitamin C (100 mg/kg/week). According to macroscopic, microscopic, hematological, biochemical, physiological, pharmacological, and pharmacokinetic results, it was confirmed that, at all examined doses, fullerenol exhibits a protective influence on the heart and liver tissue against chronic toxicity induced by doxorubicin. On the other hand, vitamin C in the investigated dose was not a sufficient organoprotector [21].

Tissue injury to the heart muscle after doxorubicin administration was also confirmed by changes in the ultrastructural pathology results and SOD, MDA, CAT, GSSG, GR, and TAS levels, as was a potential cardioprotective influence of fullerenol as a pretreatment agent for doxorubicin therapy in the acute phase. Fullerenol itself did not affect heart injury in rats with breast cancer in a dose of 100 mg/kg. The presented results suggested that fullerenol might be a potential cardioprotector in doxorubicin-treated subjects, but only after a progress of the preferred physical properties of fullerenol, especially the ones regarding solubility of the solid-state nanoform in water or cell acceptable solvent [22].

Oxidative stress has a significant role in the pathogenesis of doxorubicin-induced nephrotoxicity. In one of the recently published studies, the aim was to investigate the nephroprotective effects of fullerenol $C_{60}(OH)_{24}$ on doxorubicin-induced nephrotoxicity. Study was carried out on adult female Sprague–Dawley outbred rats with chemically induced breast cancer (1-methyl-1-nitrosourea; 50 mg/kg; i.p.). Animals were divided into control healthy, control cancer, doxorubicin alone (8 mg/kg; i.p.; cancer), doxorubicin plus fullerenol as pretreatment agent (8 mg/kg and 100 mg/kg, respectively; i.p.; cancer), and fullerenol alone (100 mg/kg; i.p.; cancer) groups. Forty-eight hours after drug administration, samples (blood and kidney tissues) were taken for analysis. The activity of lactate dehydrogenase and alpha-hydroxybutyrate dehydrogenase as serum enzymes and the level of malondialdehyde, glutathione, glutathione peroxidase, glutathione reductase, catalase, and superoxide dismutase were determined. It was confirmed that doxorubicin caused nephrotoxicity on the acute level, but fullerenol pretreatment prevented oxidative stress, lipid peroxidation, and disbalance of GSH/GSSG level in kidney tissue caused by doxorubicin. Published results confirm acceptable nephroprotective efficacy of fullerenol in acute phase of toxicity and inspire further studies as a potential nephroprotector [23]. The same model and approach were used for pulmoprotective investigation fullerenol $C_{60}(OH)_{24}$ on doxorubicin-induced lung toxicity using biochemical and histopathological approaches. The levels of malondialdehyde and oxidized glutathione in the lung tissue were higher in the group treated with doxorubicin alone than in the both control groups. The activities of catalase (CAT), glutathione reductase (GR), superoxide dismutase (SOD), and lactate dehydrogenase (LDH) were found to be increased in the lung tissue of the animals in the cytostatic group over the all other groups, while GSH-Px significantly decreased in the activity compared with the control and fullerenol nanoform groups. There was no significant difference in MDA and GSSG levels and enzyme activities in either control or fullerenol groups. The acute macroscopic and microscopic change found in doxorubicin group was subpleural edema. On the contrary, these histopathological changes appeared almost normal in the groups treated with fullerenol. Finally, this study evidently indicated that doxorubicin treatment obviously impaired pulmonary function and pretreatment with fullerenol nanoform might avoid toxicity in rats through inhibition of oxidative stress [24].

Recently, it was published by Icevic et al. [62] that fullerenol nanoform might have also protective effect after oral administration in rat's animal healthy model. The goal of that work was to investigate the potential protective effects of orally applied fullerenol nanoform on rats after a single dose of doxorubicin (8 mg/kg (i.p.)) 6 h after the last application of fullerenol. After the last drug administration, the rats were sacrificed, and the blood and tissues were taken for the analysis. Biochemical and pathological results taken in this study showed that fullerenol in solution form H_2O/DMSO (80:20, w/w) given orally in final doses of 10, 14.4, and 21.2 mg/kg 3 days sequentially has the protective (hepatoprotective and nephroprotective) effect against doxorubicin-induced cytotoxicity through its antioxidant properties.

All previously presented studies and the biological behavior of fullerene nanoform derivatives show their considerable potential for medical application. Polyhydroxylated, water-soluble fullerenol $C_{60}(OH)_{22}$, used in Maksim et al. [63] studies, was synthesized by procedure complete substitution of bromine atoms from polybromine derivatives $C_{60}Br_{24}$ in alkaline media. In order to investigate the pharmacological behavior and metabolism of the fullerenol, they presented the way and possibilities for its labeling with 99mTc ($T_{1/2}$ = 6.02 h, E = 141 keV), which bind to the outside, leaving the cage intact. Two different labeling approaches with technetium-99 m were performed: direct labeling by tin (II) chloride method and labeling with $^{99m}Tc(I)$ using $[^{99m}Tc(CO)_3(H_2O)_3]^+$ precursor. The radiochemical study results have shown that content of free $^{99}mTcO_4$ in the samples was depending on fullerenol–stannous chloride ratio and increased if ratio was higher. HPLC quality control results of $^{99m}Tc(CO)_3$–$[C_{60}(OH)_{22}]$ have shown that the labeling yield, in alkaline media, was 95.35 %. Protein binding to HA was species-dependent process. The biological behavior of labeled fullerenol was different in dependence of labeling method and the time of sacrifice of animals. From recent investigations, it can be concluded that oxidation state in technetium coordination complexes has a great impact on in vitro and in vivo behavior of the these complexes [63].

Fullerenol $C_{60}(OH)_{24}$ nanoparticles are a promising candidate for many biomedical applications due to, besides other properties, strong free radical-scavenging and antioxidative potential. Using the carrageenan-induced rat footpad edema test, the anti-inflammatory effect of fullerenol nanoparticles has been estimated in comparison with those of amifostine and indomethacin. Fullerenol and indomethacin have been dissolved in dimethyl sulfoxide, and amifostine was prepared in saline, and were intraperitoneally injected to rats in a dose range 12.5–75 mg/kg, 3–10 mg/kg, and 50–300 mg/kg, respectively. The control groups were given matching vehicles. The drugs or vehicles were given 30 min before carrageenan injection. Footpad swelling was measured 3 h after carrageenan application, followed by calculation of a percentage of inhibition derived through comparison with the control groups. Histopathological examination of the inflamed foot skin biopsies was also performed. Fullerenol dose dependently and significantly reduced the extent of footpad edema, comparable to that of indometacin and significantly better than amifostine; their ED_{50} values (in μmol/kg) were 35.36, 32.12, and 871.0, respectively. Results were also confirmed on histopathological level. The largest therapeutic index of fullerenol suggests their safety for prospective use in humans. Presented results support the hypothesis that fullerenol nanoparticles produce a strong acute anti-inflammatory activity [28].

There are many ongoing studies with fullerenols as potential strong antioxidative agent and organoprotector in different medical and pathological cases. Anyhow, the way is still very long from present to future in vivo human studies.

7 Fullerenol In Vivo Toxicity

During past decade, there are many published papers that included different in vivo and in vitro topics with fullerenes, especially C_{60} and C_{70}, but very limited number of fullerenol investigations including $C_{60}(OH)_{18}$, $C_{60}(OH)_{22}$, and $C_{60}(OH)_{24}$. The situation with in vivo toxicity regarding fullerenols is even worst in comparison to antioxidative studies and fullerenes toxicity testing. In most of the cases, toxicity was presented as potential cytostatic application in different tumor cell lines [4].

A recently published paper by Zhu et al. [64] demonstrated the favorable tumor-inhibitory effect and immunomodulatory activity of $C_{60}(OH)_x$, which significantly inhibits the growth of murine H22 hepatocarcinoma through a 17-day treatment after tumor inoculation and reduces the damnification of liver function in tumor-bearing mice. $C_{60}(OH)_x$ thus exhibiting effective tumor therapeutic activity by improving the innate immunity. Macrophages, which act as regulatory and effector immunocytes, are efficiently activated by $C_{60}(OH)_x$. It contributed to building and consolidation of the immunological defense systems against malignancies during the cancer cell development. Interestingly, the accumulation and tumor-inhibitory effect (Fig. 8) of $C_{60}(OH)_{24}$ were also observed in one of our investigations, which is of great importance, because this opens new possibilities for the further therapeutic usage of cytostatic properties of fullerenol and adds certain advantages to its side [21].

Wu et al. in their study systematically investigated the interaction of fullerenes/fullerenols with model proteins using a widely used computational docking program AutoDock 4.0. They found that pi–pi interaction existed in all the proteins–fullerene/fullerenol systems investigated and contributed greatly to the overall interaction energy. It was also confirmed that with the increase of the carbon cage size, the binding strength between proteins and fullerenes/fullerenols increased constantly. In addition, presented

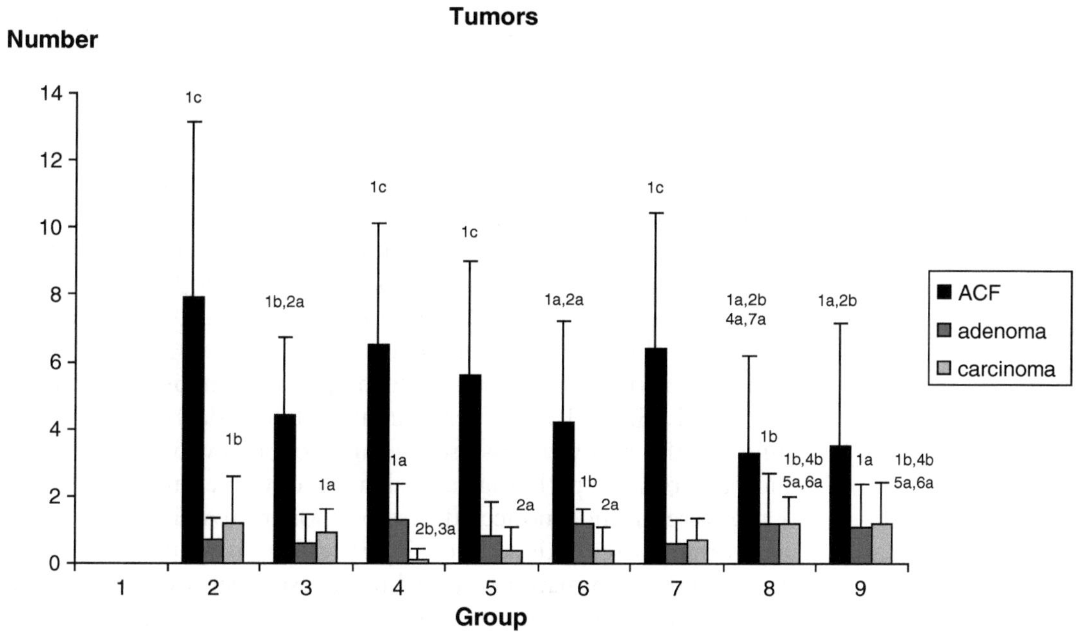

Fig. 8 Differences in colorectal tumor number between investigated groups. Numbers show significant difference from the corresponding group (a: $p<0.05$; b: $p<0.01$; c: $p<0.001$). The histological evaluation of tissue samples was carried out blindly without knowledge of the treatment status of each animal. ACFs were assessed by criteria used routinely in human pathology. Microscopic lesions composed of two to five dysplastic crypts were classified as ACF and those composed of more than five as adenomas. The histological criteria for the determination of adenomas and carcinomas were based on the International Classification of Rodent Tumors (With permission) [20]

results show that functionalization of fullerenes with hydroxyl groups decreases the binding between proteins and fullerene derivatives [65].

The purpose of Jiao et al. [66] study with $C_{60}(OH)_{20}$ was to test the potential antitumor and antimetastatic activities of fullerenol and their linked mechanisms. Thirty EMT-6 tumor-bearing mice were injected intraperitoneally with 0.1 ml saline or 0.1 ml saline containing fullerenol $C_{60}(OH)_{20}$ (0.08 and 0.4 mg/ml) daily for 16 days. Using tumor tissues, imbalances in the oxidative defense system have been investigated as well as the expression of several angiogenesis factors. Fullerenol exhibits antitumor and antimetastatic activities in EMT-6 breast cancer metastasis model, and it was also found to modulate oxidative stress significantly. On the other hand, the expression of some angiogenesis factors was reduced in tumor tissues after treatment with fullerenol. Outstandingly, CD31 (also known as platelet endothelial cell adhesion molecule) expression and vessel density were obviously reduced in tumors from fullerenol-treated mice compared with controls. Modulation of oxidative stress in tumor tissues, inhibition of the formation of angiogenesis factors, and subsequent reduction in tumor vessel density and the nutrient supply to tumor cells could be important mechanisms by which fullerenol aggregates inhibit tumor growth and suppress carcinoma metastasis in vivo [66].

Even if many scientists have been fascinated by polyhydroxylated fullerenes for their radical-scavenging and antioxidant capacity in vivo and in vitro and their prospective use as a drug or for drug delivery, there is very few information on their pulmonary toxicological properties. The aim of this study performed by Xu et al. was to investigate the effect of fullerenols on Sprague–Dawley rats after intratracheal instillation. Fullerenols [$C_{60}(OH)_x$, $x = 22, 24$; combination of two polyhydroxylated forms] were administered intratracheally (1, 5, or 10 mg per animal). After 3-day exposures, the lungs of the rats were assessed using bronchoalveolar lavage fluid biomarkers and a pathological evaluation of tissue. Introductions to 1 mg dose did not induce adverse pulmonary toxicity, whereas the two other doses induced cell injury effect, oxidative/nitrosative stress, and inflammation. Accordingly, results showed that fullerenols produced reactions in a dose-dependent manner. The dosage of $C_{60}(OH)_x$ retained in lung and the ensuing aggregation might be the main factor in the process, which is very similar to aggregation in the fat liver tissue after i.p. administration [20–22]. Presented results might be in conflict on the possible use of fullerene derivates as an inhaled drugs or drug delivery systems [67].

In the following investigation, the authors presented the novel application of polyhydroxylated fullerenes in cancer drug delivery. Doxorubicin, a first-line cancer therapeutic agent, was conjugated to fullerenols through a carbamate linker, achieving ultrahigh loading efficiency. The fullerenol-drug conjugate was found to be quite stable in phosphate-buffered saline but temporally released the active drug when incubated with tumor cell lysate. It was found that fullerenol–doxorubicin conjugate suppressed the proliferation of cancer cell lines in vitro through a G2–M cell cycle block, resulting in apoptosis. Additionally, in an in vivo murine tumor model, conjugate exhibited similar antitumor efficacy as free drug without the systemic toxicity of free doxorubicin. Moreover, it was shown that the fullerenol platform can be extended to other chemotherapeutic agents, such as the slightly water-soluble cisplatin [68]. Before final conclusions as anticancer agent, fullerenol has to be tested in patients with different kinds of tumors as well as in combination with well-known chemotherapeutics. Even before that, some investigations on bigger animal models are strongly recommended.

Bring about nanomaterials are at the primary advantage of the rapidly developing nanosciences and are establishing an important class of new materials with specific physicochemical properties different from bulk materials with the same configurations. The distinctive size-dependent properties of nanomaterials make them very attractive for pharmaceutical applications and new drug delivery systems.

Cytotoxic effects of certain engineered nanomaterials toward malignant cells form the basis for one aspect of nanomedicine as well as its potential use as carrier for high toxic drugs with goal to reduce side effects [69].

Injection of gadolinium endohedral metallofullerenol [Gd@ $C_{82}(OH)_{22}]_n$ nanoparticles could express reduction of enzyme activities connected with the metabolism of reactive oxygen species (ROS) in tumor-bearing mice [70]. Its application can powerfully repair the functions of injured liver as well as kidney tissue. It has presented that investigated metallofullerenol nanoparticles were delivered to almost all types of tissues in mice after i.p. administration which is in very good correlation with fullerenol $C_{60}(OH)_{24}$ distribution in rats [20]. The fullerene cage cannot be destroyed during metabolism in organisms. The presence of Gd in vivo represents the exact distribution of $[Gd@C_{82}(OH)_{22}]_n$ nanoparticles. In one of the published papers, it was found that the accumulation of Gd in liver was higher than that in tumor when they examined the accumulation of Gd-incorporating nanoemulsion in mice tumor for neutron-capture study [71]. Fullerenols could be entrapped in the reticuloendothelial system and distributed to all tissues by blood circulation in vivo [72]. On the other hand, after intravenous administration of $Gd@C_{82}(OH)_{40}$ to mice caused essentially in delivery to lung, liver, spleen, and kidney [72]. The thrombin- and fibrinogen-related parameters in plasma were measured after $[Gd@C_{82}(OH)_{22}]_n$ administration. The increased fibrinogen was observed due to $[Gd@C_{82}(OH)_{22}]_n$ treatment, and the prothrombin time was shortened.

Unexpectedly, in the $[Gd@C_{82}(OH)_{22}]_n$-treated nude mice, hepatomegaly and splenomegaly were less. This shows that the liver injury was inhibited. It is in agreement with the serum enzyme levels reported before, in which the serum AST and ALT activities were significantly decreased by i.p. injection of $[Gd@C_{82}(OH)_{22}]_n$. Presented results are not in good correlation with ALT and AST level in rats treated with fullerenol $C_{60}(OH)_{24}$ by i.p. way of administration [20, 21]. Indicators of kidney injury were repaired to the standard level by nanoparticle treatments established on renal function tests. The activities of well-known oxidative stress parameters like hepatic superoxide dismutase (SOD), glutathione peroxidase (GSH-Px), glutathione S-transferase (GST), and catalase (CAT) were measured in mice administered with $[Gd@C_{82}(OH)_{22}]_n$ nanoparticles, to confirm the molecular level of fullerenol–cell interaction. The GSH-Px, CAT, and SOD activities were depressed after treatment with $[Gd@C_{82}(OH)_{22}]_n$ nanoparticles. Furthermore, the levels of glutathione, protein-bound thiols, and malondialdehyde were also significantly lower. With all these tests and investigations, it was once again confirmed that hydroxylated fullerene derivatives might be novel antioxidants in vivo, with very high potential in different therapeutic areas [73].

Recently, water-soluble fullerene derivatives (–OH and –COOH) of different kinds have been synthesized, and it has been found that they have potent and selective pharmacological effects on organs, cells, enzymes, and nucleic acids. In the presence of fullerene C_{60} derivative (10^{-5} mol/l), endothelium-dependent relaxations induced by agonists in the vascular system were eliminated and acetylcholine-induced contractile response of smooth muscle was observed. Some investigators have reported free radical-scavenging activity and direct nitric oxide-quenching activity of fullerene derivatives, but mostly with carboxyl and hydroxyl configuration. Awareness of the chemical modifications, biological importance, and material applications of functionalized fullerenes is growing rapidly. The antineoplastic efficacy of metallofullerenol particles (22 nm) in H22 hepatoma-implanted animals has been shown to be much higher than that of known chemotherapeutic agents such as cisplatin and cyclophosphamide. The very specific in vivo behavior has been observed after administration of nanofullerenol forms. Efficacy was different in the different levels of distribution. Namely, direct toxic effects on tumor cells were not the result of high antitumor efficacy by fullerenol molecule. It was a combination of low-dose cytostatic influence (less than 0.05 % of the given dose) as well as improvement of immunity and interferes with tumor invasion in normal muscle cells. In addition, it was shown that such small particles of fullerenol metalloform can regulate the oxidative defense system in vivo [74].

The acute toxicity of fullerenol-1 was published using mice pretreated i.p. with polyhydroxylated C_{60} derivatives. The LD_{50} value of fullerenol-1 was estimated to be 1.2 g/kg. Pretreatments with 0.5 and 1.0 g/kg fullerenol reduced significantly cytochrome P450 and b [5] level as well as NADPH–cytochrome P450 reductase, benzo[a]pyrene hydroxylase, 7-ethoxycoumarin O-deethylase, aniline hydroxylase, and erythromycin N-demethylase activities in liver microsomes. Pretreatments with 0.01 and 0.1 g/kg fullerenol had no influence on these monooxygenases. Added extras of fullerenol to mouse liver microsomes inhibited monooxygenases activities toward benzo[a]pyrene, 7-ethoxycoumarin, aniline, and erythromycin with IC50 values of 42 μmol/l, 94 μmol/l, 102 μmol/l, and 349 μmol/l, respectively. Fullerenol presented noncompetitive and mixed-type of inhibition in benzo[a]pyrene hydroxylation and 7-ethoxycoumarin O-deethylation. Additions of fullerenol to rat liver mitochondria caused in a dose-dependent inhibition of ADP-induced uncoupling and markedly repressed mitochondrial Mg^{2+}-ATPase activity with an IC_{50} value of 7.1 μmol/l. These results prove that fullerenol-1 can decrease the activities of P450-dependent monooxygenase and mitochondrial oxidative phosphorylation in vitro and suppress the levels of the microsomal enzymes in vivo [49].

8 Potential Bioapplication of Fullerenol Nanoform

The potential environmental and health effects of fullerenes have attracted increasing attention in recent years, especially with the development of water-soluble forms that alleviate their use in biological systems [12]. These compounds have potential for suppression of metastases, treatment of cerebral conditions such as Alzheimer's and Parkinson's diseases, type-C hepatitis therapy, and HIV treatment. C_{60} derivatives with better water solubility are synthesized by attaching various functional groups (-OH, –COOH, $-NH_2$) to the C_{60} cage. Some of them are or were candidates for many biomedical applications, including cytoprotection, DNA photocleavage, enzyme inhibition, diagnostic imaging, and antimicrobial and anticancer therapy [75].

The chemical modification of fullerenes by adding the –OH groups to their carbon surface yields a variety of polyhydroxylated structures $C_{60}(OH)_x$ exhibiting different degrees of solubility and antioxidant activity in the aqueous environment [76, 77]. The crucial advantage of fullerenol compared to other branded antioxidants or cytoprotectors is its dual function as radioprotector and organoprotector (heart, liver, kidney, and lung) during the anticancer therapy (radio- and chemotherapy) [1–4, 20–29]. However, before a categorical statement may be made about the potential efficacy of fullerenol as an adjunct to doxorubicin therapy, there is a need to carry out further studies, including investigation in bigger animal models (pigs and/or monkeys) and human trials.

Acknowledgment

This work received partial financial support from the Slovenian Research Agency, Ljubljana, Slovenia, grant No.: P4-0127.

References

1. Djordjevic A, Bogdanovic G, Dobric S (2006) Fullerenes in biomedicine. J BUON 11:391–404
2. Injac R, Kocevar N, Strukelj B (2008) Fullerenol $C_{60}(OH)_{24}$ as potential drug. Farm Vest 59:257–262
3. Injac R, Strukelj B (2008) Recent advances in protection against doxorubicin-induced toxicity. Technol Cancer Res Treat 7:497–516
4. Injac R, Radic N, Govedarica B et al (2008) Bioapplication and activity of fullerenol $C_{60}(OH)_{24}$. Afr J Biotechnol 25:4940–4950
5. Deguchi S, Alargova RG, Tsujii K (2001) Stable dispersions of fullerenes, C-60 and C-70, in water. Preparation and characterization Langmuir 17:6013–6017
6. Makha M, Purich A, Raston CL et al (2006) Structural diversity of host-guest and intercalation complexes of fullerene C_{60}. Eur J Inorg Chem 37:507–515
7. Husebo LO, Sitharaman B, Furukawa T et al (2004) Fullerenols revisited as stable radical anions. J Am Chem Soc 126:12055–12064
8. Brant JA, Labille J, Bottero J et al (2006) Characterizing the impact of preparation method on fullerene cluster structure and chemistry. Langmuir 22:3878–3885
9. Gao Y, Tang Z, Watkins E et al (2005) Synthesis and characterization of amphiphilic fullerenes and their Langmuir–Blodgett films. Langmuir 21:1416–1423

10. Brant JA, Labille J, Robichaud CO et al (2007) Fullerol cluster formation in aqueous solutions: implications for environmental release. J Colloid Interface Sci 314:281–288
11. Vileno B, Sienkiewicz A, Lekka M et al (2004) In vitro assay of singlet oxygen generation in presence of water-soluble derivatives of C_{60}. Carbon 42:1195–1198
12. Zhao B, He Z, Bilsk PJ et al (2008) Pristine (C_{60}) and hydroxylated [$C_{60}(OH)_{24}$] fullerene phototoxicity towards HaCaT keratinocytes: Type I vs Type II mechanisms. Chem Res Toxicol 21:1056–1063
13. Nagano T, Arakane K, Ryu A et al (1994) Comparison of singlet oxygen production efficiency of C-60 with other photosensitizers, based on 1268-Nm emission. Chem Pharm Bull 42:2291–2294
14. Sera N, Tokiwa H, Miyata N (1996) Mutagenicity of the fullerene C_{60} -generated singlet oxygen dependent formation of lipid peroxides. Carcinogenesis 17:2163–2169
15. Nakanishi I, Ohkubo K, Fujita S et al (2002) Direct detection of superoxide anion generated in C-60-photosensitized oxidation of NADH and an analogue by molecular oxygen. J Chem Soc Perkin Trans 2:1829–1833
16. Nakanishi I, Fukuzumi S, Konishi T et al (2002) DNA cleavage via superoxide anion formed in photoinduced electron transfer from NADH to gamma-cyclodextrin-bicapped C-60 in an oxygen-saturated aqueous solution. J Phys Chem B 106:2372–2380
17. Yamakoshi Y, Umezawa N, Ryu A et al (2003) Active oxygen species generated from photoexcited fullerene (C_{60}) as potential medicines:$O_2^{-\bullet}$ versus $^1O_2$. J Am Chem Soc 125:12803–12809
18. Tokuyama H, Yamago S, Nakamura E et al (1993) Photoinduced biochemical-activity of fullerene carboxylic-acid. J Am Chem Soc 115:7918–7919
19. Markovic Z, Trajkovic V (2008) Biomedical potential of the reactive oxygen species generation and quenching by fullerenes (C_{60}). Biomaterials 29:3561–3573
20. Injac R, Perse M, Cerne M et al (2009) Protective effects of fullerenol $C_{60}(OH)_{24}$ against doxorubicin-induced cardiotoxicity and hepatotoxicity in rats with colorectal cancer. Biomaterials 30:1184–1196
21. Injac R, Perse M, Obermajer N et al (2008) Potential hepatoprotective effects of fullerenol $C_{60}(OH)_{24}$ in doxorubicin-induced hepatotoxicity in rats with mammary carcinomas. Biomaterials 29:3451–3460
22. Injac R, Perse M, Boskovic M et al (2008) Cardioprotective effects of fullerenol $C_{60}(OH)_{24}$ on a single dose doxorubicin-induced cardiotoxicity in rats with malignant neoplasm. Technol Cancer Res Treat 7:15–25
23. Injac R, Boskovic M, Perse M et al (2008) Acute doxorubicin nephrotoxicity in rats with malignant neoplasm can be successfully treated with fullerenol $C_{60}(OH)_{24}$ via suppression of oxidative stress. Pharmacol Rep 5:742–749
24. Injac R, Radic N, Govedarica B et al (2009) Acute doxorubicin pulmotoxicity in rats with malignant neoplasm is effectively treated with fullerenol $C_{60}(OH)_{24}$ through inhibition of oxidative stress. Pharmacol Rep 2:335–342
25. Djordjevic-Milic V, Djordjevic A, Dobric S et al (2006) Influence of fullerenol $C_{60}(OH)_{24}$ on doxorubicin induced cardiotoxicity in rats. Mater Sci Forum 518:525–529
26. Djordjevic-Milic V, Stankov K, Injac R et al (2009) Activity of antioxidative enzymes in erythrocytes after a single dose administration of doxorubicin in rats pretreated with fullerenol $C_{60}(OH)_{24}$. Toxicol Mech Methods 19:24–28
27. Injac R, Djordjevic A, Strukelj B (2008) Doxorubicin-induced myocardial failure in rats with malignant neoplasm: protective role of fullerenol $C_{60}(OH)_{24}$. Hem Ind 62:197–204
28. Dragojevic-Simic V, Jacevic V, Dobric S et al (2011) Anti-inflammatory activity of fullerenol $C_{60}(OH)_{24}$ nano-particles in a model of acute inflammation in rats. Dig J Nanomater Bios 6:819–827
29. Djordjevic A, Canadanovic-Brunet J, Vojinovic-Miloradov M et al (2005) Antioxidant properties and hypothetical radical mechanism of fullerenol $C_{60}(OH)_{24}$. Oxid Commun 4: 806–812
30. Mirkov S, Djordjevic A, Andric N et al (2004) Nitric oxide scavenging activity of polyhydroxylated fullerenol, $C_{60}(OH)_{24}$. Nitric Oxide 11:201–207
31. Fileti EE, Rivelino R, de Brito Mota F et al (2008) Effects of hydroxyl group distribution on the reactivity, stability and optical properties of fullerenols. Nanotechnology 19:365703
32. Chen YW, Hwang KC, Yen CC et al (2004) Fullerene derivatives protect against oxidative stress in RAW 264.7 cells and ischemia-reperfused lungs. Am J Physiol Regul Integr Comp Physiol 287:R21–R26
33. Bogdanovic G, Kojic V, Djordjevic A et al (2004) Modulating activity of fullerol $C_{60}(OH)_{22}$ on doxorubicin-induced cytotoxicity. Toxicol In Vitro 18:629–637
34. Lu LH, Lee YT, Chen HW et al (1998) The possible mechanism of the antiproliferative effects of fullerenol, polyhydroxylated C_{60} on vascular smooth muscle cells. Br J Pharmacol 123:1097–1102

35. Kojic V, Jakimov D, Bogdanovic G et al (2005) Effects of fullerenol $C_{60}(OH)_{24}$ on cytotoxicity induced by antitumor drugs on human breast carcinoma cell lines. Mater Sci Forum 492: 543–548
36. Foley S, Crowley C, Smaihi M et al (2002) Cellular localization of a water soluble fullerene derivative. Biochem Biophys Res Commun 294:116–119
37. Bergmeyer HU (1983) Methods of enzymatic analysis. Weinheim, Basel
38. Misra HP, Fridovich I (1972) The generation of superoxide radical during the autoxidation of hemoglobin. J Biol Chem 247:6960–6962
39. Beutler E (1984) Red cell metabolism. A Manual of Biochemical Methods, Grune & Stratton, New York
40. Niva Y, Iwai N (2006) Genotoxicity in cell lines induced by chronic exposure to water-soluble fullerenes using micronucleus test. Environ Health Prev Med 11:292–297
41. Zhao Q, Li Y, Xu J et al (2005) Radioprotection by fullerenols of Stylonychia mytilus exposed to gamma-rays. Int J Radiat Biol 81: 169–175
42. Daroczi B, Kari G, McAleer MF et al (2006) In vivo radioprotection by the fullerene nanoparticle DF-1 as assessed in a zebrafish model. Clin Cancer Res 12:7086–7091
43. Bogdanovic V, Stankov K, Icevic I et al (2008) Fullerenol $C_{60}(OH)_{24}$ effects on antioxidative enzymes activity in irradiated human erythroleukemia cell line. J Radiat Res 49:321–327
44. Brant JA, Labile J, Robichaund CO et al (2007) Fullerol cluster formation in aqueous solutions: implications for environmental release. J Colloid Interface Sci 314:281–288
45. Sayes CM, Gobin AM, Ausman KD et al (2005) Nano-C-60 cytotoxicity is due to lipid peroxidation. Biomaterials 26:7587–7595
46. Isakovic A, Markovic Z, Todorovic-Markovic B et al (2006) Distinct cytotoxic mechanisms of pristine versus hydroxylated fullerene. Toxicol Sci 91:173–183
47. Sayes C, Fortner J, Lyon D et al (2004) The differential cytotoxicity of water-soluble fullerenes. Nano Lett 4:1881–1887
48. Yamawaki H, Iwai N (2006) Cytotoxicity of water-soluble fullerene in vascular endothelial cells. Am J Physiol Cell Physiol 290: C1495–C1502
49. Ueng T-H, Kang J-J, Wang H-W et al (1997) Suppression of microsomal cytochrome P450-dependent monooxygenases and mitochondrial oxidative phosphorylation by fullerenol, a polyhydroxylated fullerene C_{60}. Toxicol Lett 93:29–37
50. Su YY, Xu JY, Shen PP et al (2010) Cellular uptake and cytotoxic evaluation of fullerenol in different cell lines. Toxicology 269:155–159
51. Gelderman MP, Simakova O, Clogston JD et al (2008) Adverse effects of fullerenes on endothelial cells: fullerenol $C_{60}(OH)_{24}$ induced tissue factor and ICAM-1 membrane expression and apoptosis in vitro. Int J Nanomedicine 3:59–68
52. Burdon RH, Gill V (1993) Cellularly generated active oxygen species and HeLa cell proliferation. Free Radic Res Commun 19:203–213
53. Wei YH, Lee HC (2002) Oxidative stress, mitochondrial DNA mutation, and impairment of antioxidant enzymes in aging. Exp Biol Med 227:671–682
54. Tagmatarchis N, Shinohara H (2001) Fullerenes in medicinal chemistry and their biological applications. Mini Rev Med Chem 1:339–348
55. Lai HS, Chen Y et al (2000) Free radical scavenging activity of fullerenol on grafts after small bowel transplantation in dogs. Transplant Proc 32:1272–1274
56. Lai HS, Chen WJ et al (2000) Free radical scavenging activity of fullerenol on the ischemia-reperfusion intestine in dogs. World J Surg 24:450–454
57. Zimmermann C, Winnefeld K et al (2004) Antioxidant status in acute stroke patients and patients at stroke risk. Eur Neurol 51: 157–161
58. Johnson-Lyles DN, Peifley K, Lockett S et al (2010) Fullerenol cytotoxicity in kidney cells is associated with cytoskeleton disruption, autophagic vacuole accumulation, and mitochondrial dysfunction. Toxicol Appl Pharmacol 248:249–258
59. Buege AL, Aust DS (1978) Methods in enzymology. Academic, New York
60. Trajkovic S, Dobric S, Djordjevic A et al (2005) Radioprotective efficiency of fullerenol in irradiated mice. Mater Sci Forum 494: 549–554
61. Trajkovic S, Dobric S, Jacevic V et al (2007) Tissue-protective effects of fullerenol $C_{60}(OH)_{24}$ and amifostine in irradiated rats. Colloids Surf B Biointerfaces 58:39–43
62. Icevic I, Vukmirovic S, Srdjenovic B et al (2011) Protective effects of orally applied fullerenol nanoparticles in rats after a single dose of doxorubicin. Hem Ind 65:329–337
63. Maksim T, Djokic D, Jankovic D et al (2007) Comparison of some physico-chemical parameters and biological behaviour of fullerenol labelled with technetium-99 m. J Optoelectron Adv Mater 9:2571–2577

64. Zhu J, Ji Z, Wang J et al (2008) Tumor-inhibitory effect and immunomodulatory activity of fullerenol $C_{60}(OH)_x$. Small 4:1168–1175
65. Wu XA, Yang ST, Wang HF et al (2010) Influences of the size and hydroxyl number of fullerenes/fullerenols on their interactions with proteins. J Nanosci Nanotechnol 10: 6298–6304
66. Jiao F, Liu Y, Qu Y et al (2010) Studies on anti-tumor and antimetastatic activities of fullerenol in a mouse breast cancer model. Carbon 48:2231–2243
67. Xu J-Y, Han K, Li S-X et al (2009) Pulmonary responses to polyhydroxylated fullerenols, $C_{60}(OH)_x$. J Appl Toxicol 29:578–584
68. Chaudhuri P, Paraskar A, Soni S et al (2009) Fullerenol cytotoxic conjugates for cancer chemotherapy. ACS Nano 3:2505–2514
69. Liang X-J, Chen C, Zhao Y et al (2008) Biopharmaceutics and therapeutic potential of engineered nanomaterials. Curr Drug Metab 9:697–709
70. Wang J, Chen C, Li B et al (2006) One-dimensional metallofullerene crystal generated inside single-walled carbon nanotubes. Biochem Pharmacol 71:872–881
71. Watanabe T, Ichikawa H, Fukumori Y (2002) Tumor accumulation of gadolinium in lipid-nanoparticles intravenously injected for neutron-capture therapy of cancer. Eur J Pharm Biopharm 54:119–124
72. Mikawa M, Kato H, Okumura M et al (2001) Paramagnetic water-soluble metallofullerenes having the highest relaxivity for MRI contrast agents. Bioconjug Chem 12:510–514
73. Sies H (1999) Glutathione and its role in cellular functions. Free Radic Biol Med 27: 916–921
74. Satoh M, Takayanagi I (2006) Pharmacological Studies on Fullerene (C_{60}), a novel carbon allotrope, and Its derivatives. J Pharmacol Sci 100:513–518
75. Bossi S, Da Ros T, Spalluto G et al (2003) Fullerene derivatives: an attractive tool for biological applications. Eur J Med Chem 38: 913–923
76. Deguchi S, Mukai SA, Tsudome M et al (2006) Facile generation of fullerene nanoparticles by hand-grinding. Adv Mater 18:729–732
77. Brettreich M, Hirsch A (1998) A highly water-soluble dendro[60]fullerene. Tetrahedron Lett 39:2731–2734

Chapter 6

Hydrogen Peroxide (H_2O_2) Detection with Nanoprobes for Biological Applications: A Mini-review

Gwangseong Kim, Yong-Eun Koo Lee, and Raoul Kopelman

Abstract

Hydrogen peroxide (H_2O_2) is an important member of the reactive oxygen species, playing various roles in biology and medicine. The conventional detection methods for H_2O_2 are often restricted by their limited sensitivity, poor selectivity towards H_2O_2, inappropriate physicochemical properties for detection in biological environments, long response time, etc. We briefly review here some recent nanotechnology-based approaches for H_2O_2 detection, which present an effective improvement, overcoming some of the limitations of the conventional H_2O_2 sensing techniques.

Key words Hydrogen peroxide, Nanoparticles, Biosensing, Optical detection

1 Introduction

Hydrogen peroxide (H_2O_2) is a member of the reactive oxygen species (ROS) that play many important roles in biology and medicine, including immune response [1–3], ageing [4–7], cell signaling [8–10], and wound healing [11, 12]. The major source of H_2O_2 in mammalian cytosols is the leakage from the mitochondrial electron transportation chain. This leakage generates superoxide anion radical ($O_2^{-\bullet}$). The generated superoxide anion can be rapidly transformed into H_2O_2 by superoxide dismutase activity [4, 13–15]. Also, H_2O_2 can be generated by activity of oxidases, after which it participates in various physiological events [16]. Under normal ("healthy") conditions, the endogenous H_2O_2 can be effectively neutralized by the natural antioxidant machinery, including activities of catalases, peroxidases, and peroxiredoxins [9]. However, when the delicate redox balance between ROS and antioxidants is disrupted, causing accumulation of H_2O_2 and other ROS, various pathological consequences occur, including cancer [17, 18], atherosclerosis [19], neurodegenerative diseases [13, 20], autoimmune diseases [21, 22], and diabetes mellitus [23, 24].

Donald Armstrong and Dhruba J. Bharali (eds.), *Oxidative Stress and Nanotechnology: Methods and Protocols*, Methods in Molecular Biology, vol. 1028, DOI 10.1007/978-1-62703-475-3_6, © Springer Science+Business Media New York 2013

Like other ROS, H_2O_2 has the capability to cause oxidative damage to DNAs, lipids, and proteins, disrupting those biomolecules' natural function through a conformation change induced by an undesired oxidation. However, regarding reactivity, H_2O_2 is relatively mild, compared to other ROS, and its role for redox-based cytotoxicity is often achieved by being just a precursor to further species with higher reactivity and toxicity, for example, hydroxyl radical ($^{\bullet}OH$) [25], hypochlorous acid (HOCl) [26], and peroxynitrite ($ONOO^-$) [27], rather than by its own action.

Growing evidence is being discovered about H_2O_2's other functions, not directly related with oxidative toxicity. H_2O_2 has the longest lifetime (about 2 s in water) among the ROS, which enables the longest free diffusion distance; it also is known to be lipophilic, so that it can readily permeate cellular membranes [28]. Due to these characteristics, H_2O_2 is capable of diffusing from one cell to another. These unique properties make H_2O_2 suitable for functioning as a cell-signaling molecule [29]. This H_2O_2 function of cell signaling has been only recently uncovered and is raising great interest. H_2O_2 can regulate various cellular cascades (such as proliferation, angiogenesis, cell migration, apoptosis, and cell cycle arrest) initiated by external stimuli, e.g., growth factors [30, 31], hormones [32], enzymes [33], neurotransmitters [34], and physical/chemical stresses [11, 35], by modifying the conformation of proteins/enzymes (protein kinases and phosphatases) [36]. In addition, H_2O_2 was recently discovered to be involved in the wound healing process [11, 12].

In spite of the significant roles of H_2O_2 in the aforementioned physiological activities, H_2O_2's mechanisms of action are still poorly understood. Thus, effective means to detect or monitor H_2O_2's actions would be highly desirable. Traditionally, the detection methods of H_2O_2 have been established by titration [37], chemiluminescence [38], fluorescence [39], and electrochemistry [40, 41]. Because titration-based methods and chemiluminescence-based method are difficult to apply in biological measurements, only fluorescence-based methods and electrochemical methods have been widely utilized for H_2O_2 detection in biomedical studies. Even those applicable techniques are often limited by various drawbacks, including poor H_2O_2 specificity, low sensitivity, difficulty of application in biological environments, and invasiveness of measurement (e.g., electrode-based method).

Nanotechnology can provide possible working solutions to overcome these drawbacks and promise more accurate and precise means to investigate H_2O_2's action in various biological events. Various advantages obtained by employing nanoprobes/sensors are as follows: (1) In general, nanoprobes are in the size range of 10–500 nm, which is much smaller than the size of biological cells. This small size minimizes physical perturbations to cells or tissues while performing measurements. Also, it favors the detection of unstable analytes with limited diffusion distances, like H_2O_2. (2) Compared to free molecular probes, nanoprobes can have stronger functional effects, because a single nanoprobe can be loaded with a large number of molecular probes. (3) Nano-sized structures have a high surface-area-over-volume ratio, enabling a higher probability for analyte detection. (4) H_2O_2 (or ROS)-sensitive molecular probes are chemically unstable, and encapsulating them into nanoparticles can improve their stability. (5) The use of molecular probes exhibiting inappropriate properties for biological measurements (e.g., poor aqueous solubility, uneven distribution due to sequestration, chemical interference, or toxicity) can be enabled by encapsulating them in a nanoparticle matrix with appropriate properties. (6) The nanoparticle often serves as a protecting device for the sensory contents, protection from external interference present in biological environments (e.g., undesirable enzymatic reactions and nonspecific uptake by proteins). (7) Nanoprobes can possess more than one functionality (multifunctionality), and (8) the nanoprobes can be targeted to specific cells or tissues of interests by conjugating target-specific ligand moieties onto the nanoparticle surface. A number of nanoprobes and nanosensor techniques have been developed utilizing the above-described advantages to address difficulties in analyte detection in vitro/ in vivo. How nanoprobes can serve for H_2O_2 detection and analysis in biological environments, resolving shortfalls in traditional techniques, will be given in this brief review.

2 Categories of Nanoprobes for H_2O_2 Detections in Physiological Systems

H_2O_2 detection approaches using nanoprobes have been categorized in many ways, including detection principles, materials of nanoparticle matrix, and dependence on enzymes. Here, we organized the recent reports about H_2O_2-detecting nanoprobes based on detection principles: (1) chemiluminescence, (2) fluorescence, (3) localized surface plasmon resonance, (4) near-infrared absorption, and (5) nonoptical (electrochemical) methods, as summarized in Table 1. Nanoprobes based on optical principles have emerged only very recently and are the main focus of this review. Although electrochemical methods (amperometric sensors) have been widely used for H_2O_2 sensing, we only briefly overview their principle in this chapter, because amperometric H_2O_2 sensors using nanoparticles have already been well documented and the roles of nanoparticles in this techniques are rather secondary (e.g., as stabilizer, electrode surface modifier, substrate for enzyme immobilization).

Table 1
Summary of nanoprobe-based H_2O_2 detection methods

Nanoprobes category	Detection principle	Nanoparticle matrix	Role of nanoparticle	Detection range (μM)
Chemiluminescence based	Peroxalate chemiluminescence	Polymers [42, 43] micelles [44, 45]	Confine CL donor and receptor closely	0.25–10 [42]
Fluorescence based (with enzyme)	HRP-catalyzed oxidation of fluorophores	Polyethylene glycol [49] Polyacrylamide [50] Fe_3O_4/SiO_2 [51]	Encapsulating HRP Catalyst	0.5–250 [49] 0.5–10 [50] 0.005–10 [51]
Fluorescence based (enzymeless)	H_2O_2-specific fluorescence	Organically modified silica (ORMOSIL) [52]	Blocking interferences Targeted carrier	0.01–100 [52]
Localizes surface plasmon resonance based	H_2O_2-induced aggregation	Polymer-coated silver (Ag)	Optical reporter Radical initiator	N/A
Luminescence quenching based	Removal of surface functional group by oxidation	Gold (Au) nanodot [61]	Surface-sensitive optical reporter	0.1–1,000 [61]
Near-infrared absorption with SWNT	Absorption change by surface oxidation	Single-walled carbon nanotube [62, 63]	Electron donor	10s [62, 63] (lower limit)
Electrochemistry based	Current generation by redox reaction on electrode	Au Metal oxides Silica	Conductor Surface modifier Surface immobilizer	Various

3 Chemiluminescence-Based Nanoprobes

It has been reported that H_2O_2 can be detected using the peroxalate (or peroxyoxalate) chemiluminescence (POCL) technique. The mechanism was well documented by Rauhut et al. [38]. The POCL technique involves three components: peroxalate ester, a fluorophore, and H_2O_2. In principle, peroxalate reacts with H_2O_2, forming a high-energy cyclic peroxide (1,2-dioxetandione). This high-energy dioxetanedione can chemically excite the adjacent fluorophore through an electron exchange luminescence mechanism, resulting in H_2O_2-dependent light emission, as shown in Fig. 1. POCL is known to be specific to H_2O_2 and has been widely used as a chemical assay for H_2O_2.

The first nanoprobes based on the H_2O_2/POCL method were developed recently as an imaging contrast for H_2O_2 detection by Lee et al. [42]. A polymer that contains peroxalate esters in its

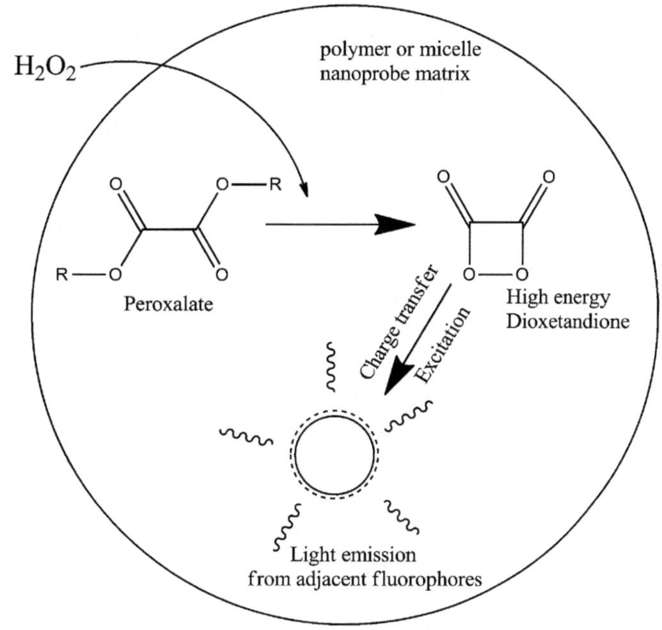

Fig. 1 Schematic description of the principle for POCL nanoprobes

backbone was synthesized by using hydroxybenzyl alcohol, octanediol, and oxalyl chloride as monomers. The nanoparticles with 550 nm in size were prepared using the synthesized peroxalate polymer, followed by the solvent extraction method. A fluorophore, e.g., 6,13-bis (4-pentylphenylethynyl) pentacene, was co-loaded into the peroxalate polymer particle matrix. Upon introduction of H_2O_2, POCL generated 630 nm light emission.

The presence of H_2O_2 is detected by monitoring the chemiluminescence emission, and the concentration of H_2O_2 is linearly proportional to the chemiluminescence intensity. In order to supply sufficient energy, the particles were formulated with excess peroxalate polymer compared to the pentacene (molar ratio 60:1). This formulation exhibited a 250 nM to 10 μM detection range for H_2O_2 in the presence of 10 μM H_2O_2 as well as negligible interferences against other ROS: hydroxyl radical ($^{\bullet}OH$), butoxyl radical superoxide anion ($O2^{-\bullet}$), and a superoxide–catalase mixture. The peroxalate polymer particles have been applied to an in vivo rat model and successfully imaged exogenous and endogenous H_2O_2.

Notably, in this approach, the nanoparticles serve as energy sources for chemiluminescence and also as carrier devices to keep the chemiluminescence donor and receptor together, in close proximity. More recently, the same research group and others have reported new designs of POCL nanoprobes, using polyethylene glycol (PEG) and surfactant micelles, featuring a smaller particle size, an improved stability, and a better aqueous solubility [43–45].

4 Fluorescence-Based Nanoprobes with Enzymatic Catalysis

Various fluorescence molecular probes for H_2O_2 detection are commercially available, including 2′,7′-dichlorofluorescein diacetate (DCFDA) [46, 47], dihydrorhodamine-123 [47], and 10-acetyl-3,7-dihydroxyphenoxazine (Amplex Red) [48]. Various synthetic approaches towards derivatizing those molecular probes, to improve their properties, have been attempted as well. The fluorescent molecular probes exhibited a change in their optical properties (increase/decrease of fluorescence emission intensity or peak shift) by reaction with H_2O_2. The H_2O_2 can be detected by monitoring the change in optical property and quantified by analyzing the kinetics of the change. In spite of significant advances in the quality of molecular probes and assay methods for H_2O_2 detection, fluorescence molecular probes share common problems, such as poor specificity (interference by other ROS), instability, enzyme dependence/interference, and difficulty in localization at the target site.

Recently, nanoparticles containing horse radish peroxidase (HRP) have been introduced as fluorescence-based H_2O_2-detecting nanoprobes by Kim et al. [49], Poulsen et al. [50], and Chang et al. [51]. Their approach depends on the catalytic activity of HRP with some fluorescent molecular probes, as shown in Fig. 2.

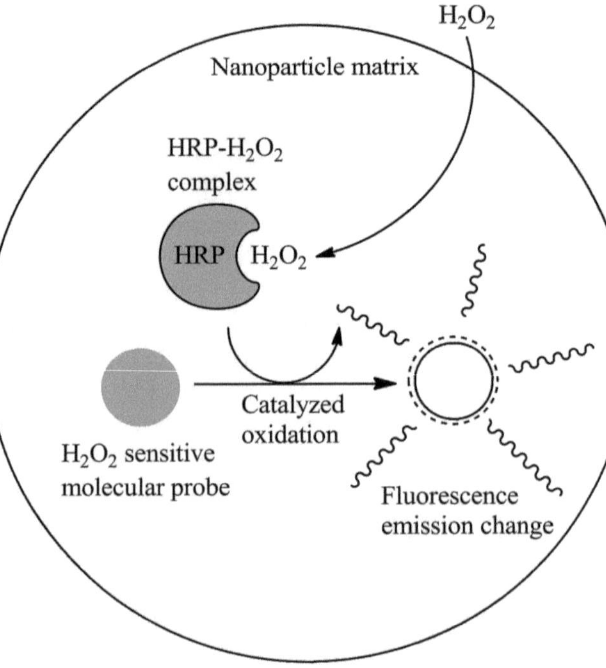

Fig. 2 Schematic description of the principle for HRP-encapsulated fluorescence nanoprobes

In principle, HRP can react with H_2O_2 forming an HRP–H_2O_2 complex. Then, this enzyme–H_2O_2 complex can rapidly oxidize the fluorescent molecular probes, resulting in optical property changes.

Kim et al. developed HRP-encapsulated nanoparticles with an organic polymer matrix, polyethylene glycol (PEG) [49]. The nanoparticles were synthesized by reverse microemulsion radical polymerization and gave an average particle size of 300 nm. The HRP-encapsulated nanoprobes were exposed to H_2O_2, allowing formation of the HRP–H_2O_2 complex within the nanoprobe matrix. Then, the nanoprobes were treated with a solution containing the fluorescent H_2O_2 molecular probe, Amplex Red. In the presence of HRP, Amplex Red can react with H_2O_2 in a 1:1 stoichiometry, generating red fluorescence at 585 nm. The presence of H_2O_2 was determined by observing the fluorescence generation from the Amplex Red. This enabled a linear detection range covering three decades, from 500 nM to 250 μM of H_2O_2.

Poulsen et al. reported on another method, based on a similar principle to the above [50]. They encapsulated HRP and a fluorescent molecular probe (fluorescein or fluorescein–dextran) together in polyacrylamide nanoparticles (PAA), at a size of 30 nm, which were also produced by the reverse microemulsion radical polymerization technique. The presence of H_2O_2 was estimated using the fluorescence emission intensity change at 525 nm. Their nanoprobes were able to quantify H_2O_2 in the concentration range from 0.5 to 10 μM. Although both above research groups have presented their results based on H_2O_2 detection, their methods still inherit the relatively poor H_2O_2 specificity from the "molecular probes" that they employed. Their H_2O_2 detection approach may be limited to complex biological events associated with multiple ROS activities.

Chang et al. have suggested a glucose/H_2O_2 sensor using HRP immobilized on Fe_3O_4/SiO_2 magnetic core–shell nanoparticles, produced by sol–gel reaction chemistry [51]. A nonfluorescent chemical, 3-(4-hydroxyphenyl)propionic acid (HPPH), was used as an H_2O_2-sensitive fluorogenic indicator. In the presence of HRP, H_2O_2 can oxidize the nonfluorescent HPPH molecule to a strongly fluorescent dimer form. The fluorescence emission at 409 nm, from the oxidized HPPH dimer, is directly proportional to the H_2O_2 concentration. The reported detection range was from 5 nM to 10 μM.

In the aforementioned approaches, the nanoparticles serve as catalysts to run reactions between H_2O_2 and fluorescent molecular probes within a fixed location, which can track and monitor the reaction progress.

5 Enzymeless, Fluorescence-Based Nanoprobes

Our group developed a new type of fluorescence-based nanoprobes for H_2O_2 detection, based on a different principle [52]. A conventional ROS-sensitive molecular probe, 2′,7′-dichlorofluorescein diacetate (DCFDA), was post-loaded into preformed organically modified silica (ORMOSIL) nanoparticles with a size of 100 nm in diameter by a solvent displacement method [52]. By carefully matching the physicochemical properties of the H_2O_2, ORMOSIL matrix, and DCFDA probe, H_2O_2 detection with extremely enhanced specificity could be enabled. DCFDA is the most widely used fluorescence probe for general ROS detection, and often used for H_2O_2 detection as well, due to its high sensitivity [46, 47]. It can be present in three different forms: (1) nonfluorescent/hydrophobic/cell membrane-permeable precursor form (2′,7′-dichlorofluorescein diacetate, DCFDA), (2) weakly fluorescent/hydrophilic/activated intermediate form (2′,7′-dichlorofluorescein, DCFH), and (3) strongly fluorescent/hydrophilic/oxidized final product (2′,7′-dichlorofluorescein, DCF) [53]. In spite of its widely accepted uses, this popular probe has many significant drawbacks. (1) DCFDA does not have specificity towards H_2O_2. (2) The intermediate DCFH is not stable and tends to be oxidized by autooxidation [52]. (3) Its reaction rate is relatively slow; thus, it is often used with HRP as a catalyst, but can be oxidized by HRP alone [46, 54]. Therefore, unlike in the previously described methods, HRP is defined as an interference in this approach. (4) DCFDA is highly pH sensitive (loss of activity at acidic pH and uncontrolled rapid transformation to the DCF form at basic pH) [53]. Due to these shortcomings, this dye has only been used for H_2O_2 detection in relatively simple situations in which only minor interference by other ROS's activity can be allowed. In our approach, the relatively hydrophobic ORMOSIL nanoparticle matrix can provide a favorable environment for the lipophilic H_2O_2, extending its lifetime and stabilizing DCFDA in its hydrophobic precursor form within the nanoparticle matrix. Upon exposure to H_2O_2, DCFDA rapidly turns into the DCF form by its reaction with H_2O_2, thereby generating a strong fluorescence at 525 nm, in a concentration dependent manner, with a dynamic range from 10 nM to higher than 100 μM, i.e., over four magnitudes.

An important characteristic of these ORMOSIL nanoprobes is that the nanoparticles function as active screening devices against other ROS interferences. The glass-like silica matrix can effectively block the entrance of large molecules, including peroxidases and esterases, which could interfere with the action of the DCFDA probe (effect called "size barrier"). Also, it can provide an energetically favorable environment for the hydrophobic H_2O_2 while destabilizing the reactive hydrophilic/ionic ROS, including the

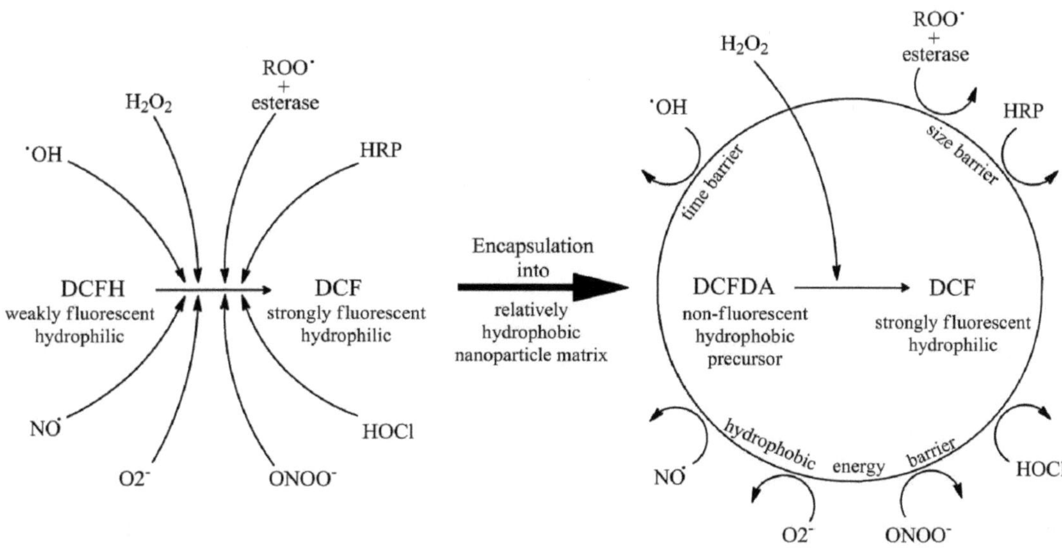

Fig. 3 Schematic description of DCFDA-encapsulated ORMOSIL nanoprobes and their principle for induced H_2O_2 specificity

superoxide anion ($O2^{-\bullet}$), nitric oxide ($NO^{\bullet}$), peroxynitrite ($ONOO^{-}$), and hypochloric acid (HOCl) ("hydrophobic energy barrier"). Furthermore, the hydroxyl radical ($^{\bullet}OH$) cannot penetrate the nanoparticle matrix because of its extremely short lifetime/diffusion distance ("lifetime barrier") [55]. These three "exclusion barriers" are summarized in Fig. 3. The combination of these effects enables a highly specific detection of H_2O_2, free from all common interferences, while using conventional molecular probes that, by themselves, have no H_2O_2 specificity. Using this novel nanoprobe led to successful detection of low nM endogenous H_2O_2, from stimulated macrophage cells in vitro.

6 Localized Surface Plasmon Resonance (LSPR)-Based Nanoprobes

Endo et al. recently reported on localized surface plasmon resonance (LSPR)-based nanoprobes for H_2O_2 detection [56, 57]. LSPR is defined as a class of surface modes that involves the collective excitation of conduction electrons in response to an incident electromagnetic field [56], and it is a unique optical characteristic for noble metal nanoparticles, such as gold and silver. This effect can be observed as a color or strong absorption in the visible region (380–750 nm). It is known to be extremely sensitive to various factors, including the immediate surrounding media. The Endo group applied this optical phenomenon to various biosensing purposes, including enzymatic reactions, DNA hybridization, and antigen–antibody interactions [58–60]. In their recent reports [56, 57], they suggested polyvinylpyrrolidone (PVP)-coated silver

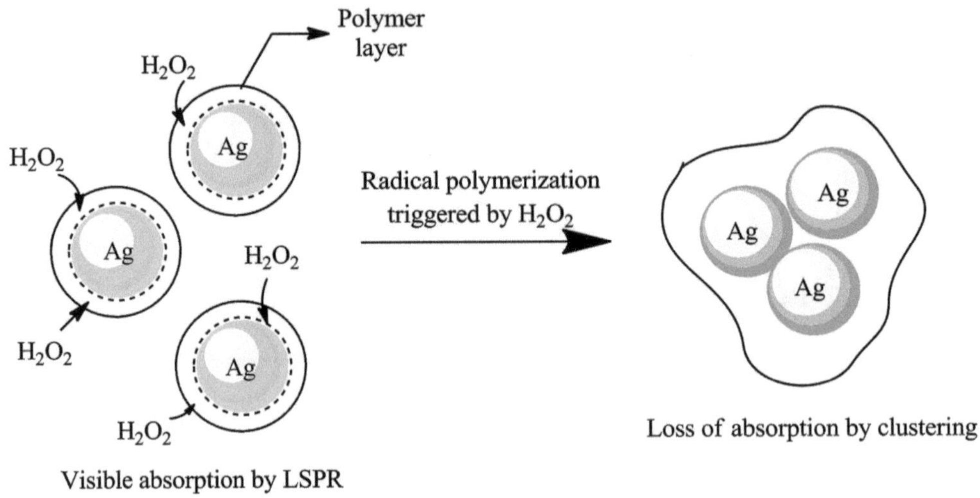

Fig. 4 Graphical explanation of the principle of LSPR-based H_2O_2 detecting PVP-coated Ag nanoprobes

nanoparticles as new nanoprobes for H_2O_2. PVP-coated silver nanoparticles with a size of 20 nm in diameter exhibit a strong visible absorption band at 405 nm. This visible absorption intensity can be significantly reduced by exposing the nanoprobes to H_2O_2. The detailed mechanism of this reaction is still not fully understood. However, they hypothesized that the silver nanoparticle core can effectively catalyze the decomposition of H_2O_2, resulting in the production of hydroxyl radicals (•OH), and then the hydroxyl radicals trigger a radical polymerization of the solubilizing PVP layer, so as to form aggregation among silver nanoparticles. Finally, this clustering effectively quenches the LSPR effect stemming from individual silver nanoparticles (Fig. 4). The detection range or limit and the interference by other ROS and relevant chemicals have not been investigated yet. We note that in this approach, the metal nanoparticle cores serve as a catalyst for H_2O_2 decomposition and as a reporter of optical signal as well, and the surface coating by PVP provides the H_2O_2-dependent machinery for H_2O_2 detection.

7 Luminescence Quenching-Based Nanoprobes

Shiang et al. have proposed another nanoprobe/sensor for H_2O_2 detection method using gold nanoparticles (called also Au nanodots) [61]. The surfaces of the Au nanodots were coated with 11-mercapto-undecanoic acid (11-MUA) by Au–sulfur adsorption. The optical property of the Au nanodots was found to be sensitive to their surface composition, though its mechanism is still poorly understood [61]. In Shiang et al.'s work, when the surface of the Au nanodots was well covered by 11-MUA units, the Au nanodots showed strong photoluminescence at 522 nm. On the

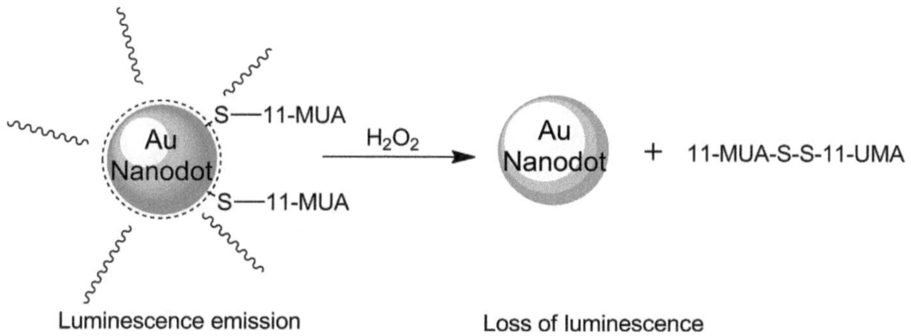

Fig. 5 Schematic description of the principle for luminescence quenching-based nanoprobes

other hand, as the surface coverage by the 11-MUA decreases, the Au nanodots lose their photoluminescence. H_2O_2 can readily cleave the Au–S bonding, forming an 11-MUA dimer with disulfide bonding (11-MUA-S-S-11-MUA). This oxidization by H_2O_2 results in the quenching of the Au nanodot's photoluminescence, as shown in Fig. 5. These 11-MUA–Au nanodots could detect H_2O_2 over a large concentration range, from 100 nM to 1 mM.

8 Near-Infrared Absorption/Fluorescence-Based Nanoprobes Using Carbon Nanotubes

Several approaches to utilize the optical properties of single-walled carbon nanotubes (SWNT) for H_2O_2 detection were recently reported by Song et al. [62] and Xu et al. [63]. Carbon nanotubes are known to have an intrinsic absorption in the near-infrared region (around 1,000–1,300 nm), and their optical property can be affected by a change in the surrounding environment and by surface functionalization. In this near-infrared region, the biological tissues are relatively transparent, allowing deeper penetration of light. Due to the difficulty in suspending carbon nanotubes in aqueous media (especially physiological media), most efforts have been made for finding a surface modification method to stabilize the aqueous suspension of carbon nanotubes. The H_2O_2 detection using SWNT was designed by solubilizing the SWNT with electroactive molecules, e.g., sodium dodecylsulfate (SDS) by the Song group, and with double-stranded DNA (dsDNA) by the Xu group. Once solubilized, H_2O_2 can diffuse into the surface coating layer to reach the SWNT surface, and then charge transfer from SWNT to H_2O_2 induces a reduction in the near-infrared absorption intensity, as well as fluorescence intensity. This optical property change was reported to be reversible. These SWNT systems showed detection limits for H_2O_2 concentration of 0.4 ppm (11 μM) at the 1,245 nm band and of 1 ppm (20 μM) at the 1,322 nm band. Interference by pH was reported by both groups, but interferences from other ROS or chemicals have not yet been studied.

9 Nanoparticles for Electrochemical Sensing of H_2O_2

H_2O_2 is an electroactive molecule. H_2O_2 can be oxidized or reduced on the surface of an electrode, generating an electric current. In order to improve current electrochemical sensing techniques, research interests in H_2O_2 detection have been focused on surface modification, so as to improve sensor performance. By modifying or architecting the electrode surface, using appropriate nanomaterials or nanostructures, various advantages have been obtained, such as large surface area for sensing, improved electric conductivity, ability to accumulate analyte, surface functionalization, and electrocatalytic activity. A very wide range of nanostructured materials, including gold nanoparticles, metal/inorganic oxides (nickel, zinc, titanium, zirconium, tin, silicon, cerium, etc.), and sol–gel silicas have been investigated for H_2O_2 sensing. Nanoparticles have been used in two major ways: (1) to catalyze, by themselves, the electrochemical reaction and improve sensor stability or (2) to serve as a matrix to immobilize electrocatalytic enzymes (HRP for H_2O_2 sensing) by facile physical adsorption/encapsulation or covalent conjugation. The electrochemical H_2O_2 sensing techniques have been well reviewed already [64, 65].

10 Conclusions

Above we showed that nanoprobes can provide promising tools for H_2O_2 detection at physiological conditions, with steadily improving current techniques, on many different aspects. Nanoprobes have an enormous potential for uncovering the ROS-related mechanisms occurring in various physiological/pathological events that still are poorly understood. How to deploy these advanced techniques for more accurate and fundamental understandings about complex physiological phenomena related to H_2O_2 and other ROS activities still remains an important future challenge. Also, the described advantages of nanoprobes could be readily applied to the biosensing of other analytes.

References

1. Babior BM (1978) Oxygen-dependent microbial killing by phagocytes. N Engl J Med 298(12): 659–668
2. Karnovsky ML, Lazdins J, Simmons SR (1975) Metabolism of activated mononuclear phagocytes at rest and during phagocytosis. Mononuclear Phagocytes in immunity, infection, and pathology. R. Van Furth, editor. BlackweU Scientific Publications, Oxford, pp 432–439
3. Oren R et al (1963) Metabolic patterns in three types of phagocytizing cells. J Cell Biol 17(3):487
4. Balaban RS, Nemoto S, Finkel T (2005) Mitochondria, oxidants, and aging. Cell 120(4):483–495
5. Cutler RG (2005) Oxidative stress and aging: catalase is a longevity determinant enzyme. Rejuvenation Res 8(3):138–140

6. Finkel T (2005) Radical medicine: treating ageing to cure disease. Nat Rev Mol Cell Biol 6(12):971–976
7. Finkel T, Holbrook NJ (2000) Oxidants, oxidative stress and the biology of ageing. Nature 408:239–247
8. Finkel T (2003) Oxidant signals and oxidative stress. Curr Opin Cell Biol 15(2):247–254
9. Rhee SG (2006) H_2O_2, a necessary evil for cell signaling. Science 312(5782):1882
10. Rhee SG et al (2003) Cellular regulation by hydrogen peroxide. J Am Soc Nephrol 14(suppl 3):S211–S215
11. Niethammer P et al (2009) A tissue-scale gradient of hydrogen peroxide mediates rapid wound detection in zebrafish. Nature 459(7249):996–999
12. Rojkind M et al (2002) Role of hydrogen peroxide and oxidative stress in healing responses. Cell Mol Life Sci 59(11):1872–1891
13. Eberhardt MK (2001) Reactive oxygen metabolites: chemistry and medical consequences. CRC press LLC, Boca Raton, FL
14. McBride AG, Borutaité V, Brown GC (1999) Superoxide dismutase and hydrogen peroxide cause rapid nitric oxide breakdown, peroxynitrite production and subsequent cell death. Biochim Biophys Acta 1454(3):275–288
15. McCord JM, Fridovich I (1969) Superoxide dismutase. J Biol Chem 244(22):6049
16. Lambeth JD (2004) NOX enzymes and the biology of reactive oxygen. Nat Rev Immunol 4(3):181–189
17. Emerit I (1994) Reactive oxygen species, chromosome mutation, and cancer: possible role of clastogenic factors in carcinogenesis. Free Radic Biol Med 16(1):99–109
18. Waris G, Ahsan H (2006) Reactive oxygen species: role in the development of cancer and various chronic conditions. J Carcinog 5(1):14
19. Harrison D et al (2003) Role of oxidative stress in atherosclerosis. Am J Cardiol 91(3):7–11
20. Simonian NA, Coyle JT (1996) Oxidative stress in neurodegenerative diseases. Annu Rev Pharmacol Toxicol 36(1):83–106
21. Griffiths H (2005) ROS as signalling molecules in T cells—evidence for abnormal redox signalling in the autoimmune disease, rheumatoid arthritis. Redox Rep 10(6):273–280
22. Namazi MR (2009) Cytochrome-P450 enzymes and autoimmunity: expansion of the relationship and introduction of free radicals as the link. J Autoimmune Dis 6(1):4
23. Houstis N, Rosen ED, Lander ES (2006) Reactive oxygen species have a causal role in multiple forms of insulin resistance. Nature 440(7086):944–948
24. Nishikawa T, Araki E (2007) Impact of mitochondrial ROS production in the pathogenesis of diabetes mellitus and its complications. Antioxid Redox Signal 9(3):343–353
25. Zepp RG, Faust BC, Hoigne J (1992) Hydroxyl radical formation in aqueous reactions (pH 3–8) of iron (II) with hydrogen peroxide: the photo-Fenton reaction. Environ Sci Technol 26(2):313–319
26. Weiss SJ, Slivka A (1982) Monocyte and granulocyte-mediated tumor cell destruction. A role for the hydrogen peroxide-myeloperoxidase-chloride system. J Clin Invest 69(2):255
27. Uppu RM, Pryor WA (1996) Synthesis of peroxynitrite in a two-phase system using isoamyl nitrite and hydrogen peroxide. Anal Biochem 236(2):242–249
28. Bienert GP, Schjoerring JK, Jahn TP (2006) Membrane transport of hydrogen peroxide. Biochim Biophys Acta 1758(8):994–1003
29. D'Autréaux B, Toledano MB (2007) ROS as signalling molecules: mechanisms that generate specificity in ROS homeostasis. Nat Rev Mol Cell Biol 8(10):813–824
30. Bae YS et al (1997) Epidermal growth factor (EGF)-induced generation of hydrogen peroxide. J Biol Chem 272(1):217
31. Sundaresan M et al (1995) Requirement for generation of H_2O_2 for platelet-derived growth factor signal transduction. Science 270(5234): 296
32. Goldstein BJ et al (2005) Role of insulin-induced reactive oxygen species in the insulin signaling pathway. Antioxid Redox Signal 7(7–8):1021–1031
33. Patterson C et al (1999) Stimulation of a vascular smooth muscle cell NAD (P) H oxidase by thrombin. J Biol Chem 274(28):19814
34. MukhinYVetal(2000)5-Hydroxytryptamine1A receptor/Gibetagamma stimulates mitogen-activated protein kinase via NAD (P) H oxidase and reactive oxygen species upstream of src in chinese hamster ovary fibroblasts. Biochem J 347(Pt 1):61
35. Hernández-García D et al (2010) Reactive oxygen species: a radical role in development Free Radic Biol Med 49(2):130–143
36. Paulsen CE, Carroll KS (2009) Orchestrating redox signaling networks through regulatory cysteine switches. ACS Chem Biol 5(1):47–62
37. Hurdis E, Romeyn H Jr (1954) Accuracy of determination of hydrogen peroxide by cerate oxidimetry. Anal Chem 26(2):320–325
38. Rauhut MM et al (1967) Chemiluminescence from reactions of electronegatively substituted aryl oxalates with hydrogen peroxide and fluorescent compounds. J Am Chem Soc 89(25):6515–6522

39. Soh N (2006) Recent advances in fluorescent probes for the detection of reactive oxygen species. Anal Bioanal Chem 386(3):532–543
40. Li J, Tan SN, Ge H (1996) Silica sol-gel immobilized amperometric biosensor for hydrogen peroxide. Anal Chim Acta 335(1–2):137–145
41. Razola SS et al (2002) Hydrogen peroxide sensitive amperometric biosensor based on horseradish peroxidase entrapped in a polypyrrole electrode. Biosens Bioelectron 17(11–12):921–928
42. Lee D et al (2007) In vivo imaging of hydrogen peroxide with chemiluminescent nanoparticles. Nat Mater 6(10):765–769
43. Dasari M et al (2009) Chemiluminescent PEG-PCL micelles for imaging hydrogen peroxide. J Biomed Mater Res A 89(3):561–566
44. Lee I et al (2009) Detection of hydrogen peroxide in vitro and in vivo using peroxalate chemiluminescent micelles. Bull Korean Chem Soc 32(7):2187
45. Rui C et al (2011) Chemiluminescent nanomicelles for imaging hydrogen peroxide and self-therapy in photodynamic therapy. BioMedical Research International 2011:1–9
46. Keston AS, Brandt R (1965) The fluorometric analysis of ultramicro quantities of hydrogen peroxide. Anal Biochem 11(1):1–5
47. Royall J, Ischiropoulos H (1993) Evaluation of 2′, 7′-dichlorofluorescin and dihydrorhodamine 123 as fluorescent probes for intracellular H_2O_2 in cultured endothelial cells. Arch Biochem Biophys 302(2):348–355
48. Votyakova TV, Reynolds IJ (2004) Detection of hydrogen peroxide with Amplex Red: interference by NADH and reduced glutathione auto-oxidation. Arch Biochem Biophys 431(1):138–144
49. Kim SH et al (2005) Encapsulation of enzymes within polymer spheres to create optical nanosensors for oxidative stress. Anal Chem 77(21):6828–6833
50. Poulsen AK, Scharff-Poulsen AM, Olsen LF (2007) Horseradish peroxidase embedded in polyacrylamide nanoparticles enables optical detection of reactive oxygen species. Anal Biochem 366(1):29–36
51. Chang Q et al (2009) Sensitive fluorescent probes for determination of hydrogen peroxide and glucose based on enzyme-immobilized magnetite/silica nanoparticles. Anal Bioanal Chem 395(7):2377–2385
52. Kim G et al (2010) Nanoencapsulation method for high selectivity sensing of hydrogen peroxide inside live cells. Anal Chem 82(6):2165–2169
53. Afri M, Frimer AA, Cohen Y (2004) Active oxygen chemistry within the liposomal bilayer: Part IV: locating 2′, 7′-dichlorofluorescein (DCF), 2′, 7′-dichlorodihydrofluorescein (DCFH) and 2′, 7′-dichlorodihydrofluorescein diacetate (DCFH-DA) in the lipid bilayer. Chem Phys Lipids 131(1):123–133
54. LeBel CP, Ischiropoulos H, Bondy SC (1992) Evaluation of the probe 2′, 7′-dichlorofluorescin as an indicator of reactive oxygen species formation and oxidative stress. Chem Res Toxicol 5(2):227–231
55. King M, Kopelman R (2003) Development of a hydroxyl radical ratiometric nanoprobe. Sens Actuator B Chem 90(1–3):76–81
56. Endo T et al (2010) Localized surface plasmon resonance optical characteristics for hydrogen peroxide using polyvinylpyrrolidone coated silver nanoparticles. Mater Lett 64(19): 2105–2108
57. Endo T, Yanagida Y, Hatsuzawa T (2008) Quantitative determination of hydrogen peroxide using polymer coated Ag nanoparticles. Measurement 41(9):1045–1053
58. Endo T et al (2008) Stimuli-responsive hydrogel-silver nanoparticles composite for development of localized surface plasmon resonance-based optical biosensor. Anal Chim Acta 611(2):205–211
59. Endo T et al (2006) Multiple label-free detection of antigen-antibody reaction using localized surface plasmon resonance-based core-shell structured nanoparticle layer nanochip. Anal Chem 78(18):6465–6475
60. Endo T et al (2005) Label-free detection of peptide nucleic acid-DNA hybridization using localized surface plasmon resonance based optical biosensor. Anal Chem 77(21):6976–6984
61. Shiang YC, Huang CC, Chang HT (2009) Gold nanodot-based luminescent sensor for the detection of hydrogen peroxide and glucose. Chem Commun 23:3437–3439
62. Song C, Pehrsson PE, Zhao W (2005) Recoverable solution reaction of HiPco carbon nanotubes with hydrogen peroxide. J Phys Chem B 109(46):21634–21639
63. Xu Y et al (2007) Double-stranded DNA single-walled carbon nanotube hybrids for optical hydrogen peroxide and glucose sensing. J Phys Chem C 111(24):8638–8643
64. Chen W et al (2012) Recent advances in electrochemical sensing for hydrogen peroxide: a review. Analyst 137(1):49–58
65. Rad AS et al (2011) A review on glucose and hydrogen peroxide biosensor based on modified electrode included silver nanoparticles. Int J Electrochem Sci 6:3671–3683

Chapter 7

Studying the Oxidative Stress Paradigm In Vitro: A Theoretical and Practical Perspective

Martin J.D. Clift and Barbara Rothen-Rutishauser

Abstract

Since the early 1990s, interest into the biological interaction of nanosized particles of various compositions has increased. Following the initial findings that nanoscaled particles can elicit an adverse biological response when compared to their larger (micron-scale) material counterparts, interest into how nanosized materials may elicit potentially adverse effects upon any biological system has been intensively investigated. Over the past 20 years, hundreds to thousands of research studies have been published highlighting the biological effects and interaction of the plethora of nanoparticles (NPs) that are being either accidentally or intentionally (engineered) produced. While a definitive knowledge of many aspects is required prior to investigating the biological interaction of NPs, such as the relevant exposure route to the biological system, the specific characteristics of the NPs being studied, and the realistic dose (concentration) that would interact with the biological system, understanding how the NPs affect the biological system is not based upon any defined theory. In fact, there is no specific understanding as to why particles show different effects when occurring within a certain nanosize range compared to their larger counterpart (micron size range). Despite this, certain paradigms and theories have been proposed and are studied, such as the fiber paradigm and theory of genotoxicity, in order to try and understand such nanoscale effects. The most studied and widely accepted paradigm, however, is the oxidative stress paradigm. This chapter will provide an insight into this paradigm, how it is perceived, how it is studied, why investigating this paradigm in vitro is advantageous, and how the findings associated with this paradigm can provide an insight into the (potentially adverse) biological interaction of nanoscale objects.

Key words Nano, Nanotoxicology, Nanoparticle, Nano-object, Oxidative stress, Oxidative stress paradigm, In vitro

1 Introduction

Throughout evolution, humans have been exposed to many different forms of environmental particles. An example of which is through man's inevitable interaction with fire, which results in exposure to combustion-derived (nano)particles [1]. Thus, from even the earliest stages in our evolutionary history, humans have been subjected to (accidentally produced) (nano)particles (NPs) of varying sizes, compositions, and shapes. In addition, due to the recent advent

Donald Armstrong and Dhruba J. Bharali (eds.), *Oxidative Stress and Nanotechnology: Methods and Protocols*, Methods in Molecular Biology, vol. 1028, DOI 10.1007/978-1-62703-475-3_7,

and growth of nanotechnology, a plethora of "new" NPs (or NPs formed through novel manipulation of fundamental chemistry) are constantly being manufactured from a diverse number of materials such as carbon, titanium, gold, polystyrene, cadmium, iron, and zinc for a wide range of industrial, consumer, and medical applications, thus interacting with the human body, or even the environment, through a variety of exposure routes and processes [2].

Despite this vast array of manufactured NPs already in use for a wide range of applications, newly engineered NPs are constantly being generated to meet the increased demand for smaller, lighter, faster, more efficient, and durable products [3]. The potential toxic and adverse health effects of these "new" NPs, however, are not fully understood [4]. Therefore, it is vital that efficient, model ways of analyzing the diverse physiochemical components of manufactured NPs are determined, in order to fully understand both the potential detrimental and advantageous effects of engineered NPs [5]. While many debates are ongoing during the constant attempt to gain such information, one specific discussion concerns how to adequately test all NPs efficiently and effectively in regard to their biological interaction.

2 Toxicology and Nanotoxicology

Toxicology is the study of the adverse effects of substances on living organisms [6]. One of the fundamental principles of toxicology is that the "risk" posed by a substance is a function of its potential to cause harm, or "hazard," and the amount of substance a biological system is "exposed" to. This is defined by the following principle:

$$\text{Risk} = \text{Hazard} \times \text{Exposure}$$

It is necessary to consider exposure levels of substances according to the principle that all materials are toxic if exposure occurs in sufficient quantities [6]. Most commonly this principle is referred to via the quote of *Paracelsus* (CH, 1493–1541). "*All things are poison and nothing is without poison, only the dose permits something not to be poisonous.*" These fundamental toxicology principles imply that substances with a low hazard generally pose a low risk. If, however, there is a high enough exposure to these low-hazard substances, even these can be harmful or even fatal.

Based on the principles of toxicology, nanotoxicology can be defined as "the study of the effects of NPs (either accidentally produced or engineered) on living organisms" and further described as a multidisciplinary science including material science, chemistry, physics, and medicine [7]. The field of nanotoxicology focuses upon gaining a thorough understanding of the relationship between properties such as particle size, surface area and reactivity, dose (concentration), material composition, and the potential toxicity of NPs. In relation to this, it is also proposed that nanotoxicology will develop and implement "nano-specific" protocols in order to investigate and gain the knowledge necessary to determine the suggested toxicity of the plethora of different NPs.

3 The Potential "Risk" Posed by NPs

There are numerous accounts of exposed individuals succumbing to disease following exposure to particles. Such points, for example, include welders (metal fume fever), coal miners (coal miner's pneumoconiosis), and silicosis amongst stone masons. If and how particulate or more aptly nanoparticulate can elicit such effects, or even the contribution that they have to the aforementioned human diseases, is a relatively unknown entity.

The first study to ever describe that a material when nanosized may elicit a significantly adverse biological response compared to its bulk counterpart was Ferin et al. [8]. In this landmark study, the authors reported that titanium dioxide (TiO_2) in a nanosized form could induce a significantly higher inflammatory reaction in vivo compared to micron-sized TiO_2. Briefly, rats were exposed, via inhalation, to both NP (21-nm diameter) and fine (250-nm diameter) TiO_2, as well as intratracheally instilled with TiO_2 particles of various sizes (12, 21, 230, and 250 nm in diameter) over a period of 12 weeks. The effects of treatment with each particle size were then observed over a 70-week post-exposure period. It was demonstrated that TiO_2 NPs promoted an acute inflammatory response following both intratracheal instillation and sub-chronic inhalation techniques compared with the larger particles (230 and 250 nm). The inflammation observed in exposed animals was subsequently found to reduce to control levels post-exposure (64 weeks), with a noted decrease (from peak levels) in the number of neutrophils present in the lung at that time. Ultrafine particles were also found to remain within the lung longer (501 days) than fine particles (174 days). The prolonged retention of TiO_2 NPs in the lung was suggested to be an effect of the finding that at equivalent masses, NPs were able to translocate to the pulmonary interstitium more efficiently than the larger TiO_2 particles. It was suggested that the translocation of NPs to the interstitium was due to the smaller particles (12 and 21 nm) not being taken up by alveolar macrophages and undergoing clearance from the alveoli via uptake by alveolar type-1 epithelial cells instead. In addition to this, it was found that an increased dose (increased number of particles and decreased particle size) promoted movement of particles within the pulmonary system. It was also observed that the number of particles present, particle size, delivered dose, and the delivered dose rate also had an effect on the translocation process. Ferin et al. [8] concluded that the observed inflammation was due to the exposure of the rat lung to NPs, impaired lung clearance, and NP redistribution.

In a subsequent publication, these findings were supported by Oberdorster et al. [9], who showed increased levels of inflammation present in the alveolar space of rats after instillation with 500 μg TiO_2 NPs (20 nm) over 24 h, compared with TiO_2 fine particles (250 nm) at the same mass dose. It was also found, in comparison to Ferin et al. [8], that NPs were able to enter the interstitium more readily than the larger TiO_2 particles. Oberdorster et al. [9] further suggested that the increased inflammatory response to acute NP exposure was not due to the movement of particles to the interstitium, but it was related to the larger surface area of the particles and their interaction with alveolar macrophages and interstitial cells. Focusing on alveolar macrophages, it was noted that they were effective inhibitors of the inflammatory response to NPs due to the active uptake of these particles. The findings of both Ferin et al. [8] and Oberdorster et al. [9] promoted interest in the effects of NPs in the lung, as well as the possible adverse effects on respiratory and cardiovascular function that exposure to NPs might pose.

In addition to laboratory investigations, epidemiological studies conducted over the past two decades have shown a positive correlation between the level of particulate air pollution and increased adverse health effects [10], including increased pulmonary diseases [11], as well as a rise in the number of deaths from cardiovascular disease [12]. It was specifically shown that these adverse health effects manifested predominantly in susceptible individuals who had pre-existing pulmonary or cardiovascular disease following increased exposure to areas of heightened particulate air pollution [13]. To determine the levels of particulate air pollution, or particulate matter (PM), these studies used the global sampling method of PM. This method collects all airborne PM that has an aerodynamic diameter of $\leq$10 μm, via a machine with a size-selective inlet of $\leq$10 μm and 50 % efficiency [14]. The composition of PM_{10} is complex and difficult to define exactly; thus, it is termed as a "particle cocktail," comprising a wide variation of particulate materials, including carbon, a complex mixture of metals, exhaust fumes, endotoxins, windblown dusts, and a mixture of chemicals ranging from >2.5 μm to <100 nm in size [15]. All of the components are thought to contribute to the observed toxicity and reducing effects of PM on human health [16]. Despite this, the NP component of PM_{10} has been held principally accountable for eliciting the adverse health effects that were highlighted in these epidemiological studies. This led to the generation of the "ultrafine hypothesis" by Seaton et al. [17], whereby NPs (<100 nm) were considered to be particularly hazardous and to drive the adverse effects to human health, specifically within the lung. It is also relevant to note that these findings and suggestions were in agreement with those of Ferin et al. [8] and Oberdorster et al. [9], whereby smaller particles were potentially of greater risk to health than larger particles of a similar material. This hypothesis was subsequently supported in an epidemiological study by Peters et al. [18], who reported that human respiratory ill health was associated with the number of ambient ultrafine (uf) particles inhaled. In the study by Peters et al. [18], 27 non-smoking asthmatics, who resided within a highly polluted city in Central Europe, were examined over a 6-month period. Analysis of the peak expiratory flow (PEF) of the lungs for each individual volunteer demonstrated the number of uf particles in the air to have a significant mean decrease on PEF, compared with the mass of larger particulates in the atmosphere. It was concluded, therefore, that the adverse health effects observed following exposure to PM_{10} could be related to the size distribution of particulates in the environment, thereby supporting the theory that NPs could potentially drive toxicity in the lung. Although there are a number of additional epidemiological studies that have studied particulate effects on human health [19, 20], further epidemiological studies are still required in order to support the findings of the plethora of laboratory-based investigations. Despite the shortage of epidemiological studies, the hypothesis of Seaton et al. [17] has been the basis of increased experimental, laboratory-based analysis over the past two decades. In a study by Li et al. [21], PM_{10} samples were examined in vivo and in vitro. Following intratracheal instillation of PM_{10} in rats for 6 h, an increase in inflammation (increased neutrophil count) and toxicity (increased levels of lactate dehydrogenase, LDH) was measured in the bronchoalveolar lavage (BAL) fluid, as well as an increase in BAL protein, indicative of increased epithelial permeability. Further analysis of the BAL fluid found depletion in glutathione, indicating an increase in oxidative stress. It was also observed that PM_{10}-treated BAL leucocytes in vitro produced a significant increase in the level of tumor necrosis factor (TNF)-α, suggestive of increased inflammation in PM_{10}-exposed lung cells. Subsequent analysis of the effects of uf carbon black (ufCB), due to the diverse range of materials in PM_{10}, is difficult to replicate in a laboratory environment for experimental purposes, and therefore, many studies have used surrogates of PM_{10}, such as fine (>100 nm) carbon black (CB) and its NP counterpart, ufCB; in vivo, following intratracheal instillation, demonstrated that these smaller particles were able to cause a greater level of inflammation compared with both fine CB and PM_{10}. Subsequent comparative analysis of CB and ufCB by Li et al. [22], in which a mass of 125 μg of each particle type was instilled into rats, also reported ufCB to induce greater proinflammatory effects than CB. A 50 % increase in neutrophils was observed in the alveolar region of the lung following BAL of rats exposed to ufCB, compared with CB particles after 6 h of exposure. Subsequent analysis of the BAL profile of the ufCB-instilled rats after 6 h, compared with rats exposed to CB, also showed an increase in LDH, indicative of cell damage (increased cell membrane permeability), as well as a rise in total BAL protein, suggestive of increased epithelial cell permeability. It was shown, however, that both particle sizes (ufCB and CB) caused a decrease in glutathione content in the lung, indicating that both ufCB and CB induce oxidative stress in the lung, although the highest levels of depletion were observed in rats treated with ufCB. In conclusion of these findings, Li et al. [22], in comparison to Oberdorster et al. [9], suggested that the biological reactivity of ufCB was related to the size and surface area of the particle type. The findings of both Li et al. [22] and Oberdorster et al. [9] were supported by Brown et al. [23] who examined the effects of the instillation of 64-, 202-, and 535-nm engineered polystyrene beads (PBs) in rats. It was discovered that treatment of 64-nm PBs induced an increased inflammatory response via an influx of neutrophils, LDH, and protein in BAL fluid from the rat lung compared with exposure to both the 202- and 535-nm-sized PBs. Further analysis also showed that the surface area of the particles instilled into the rats was relative to the level of inflammation in the rat lung, and therefore a proportional relationship between particle surface area and inflammation existed. Therefore, Brown et al. [23] concluded that as the 64-nm PBs had a greater surface area than both the 202- and 535-nm PBs, NPs have the ability to cause a greater level of inflammation than their larger particle counterparts at the same mass dose due to a greater surface area and reactivity. These findings were subsequently supported by Duffin et al. [24, 25], who demonstrated a number of low-toxicity particles (TiO_2, CB, and PBs), at different sizes in vivo, to cause a heightened inflammatory response, via an increased neutrophil count in BAL fluid from the rat lung between 18 and 24 h after exposure to particles <100 nm in size. Stoeger et al. [26] also showed the specific surface area of NPs to be related to their potential to induce proinflammatory cytokine production in mice. It was observed by Stoeger et al. [26] that mice instilled with 5, 20, and 50 μg of a series of different particles, including flame soot particles, spark-generated uf carbon particles (ufCPs), and diesel exhaust particles (DEP), ranging in their surface area from 30 to 800 m^2/g, caused increased levels of LDH, total protein, TNF-α, interleukin (IL)-1β, and macrophage inflammatory protein-2 (MIP-2) as well as polymorphonuclear leucocytes (neutrophils) (PMNs) in BAL fluid after 24-h exposure. Investigation demonstrated that the ufCPs were more potent than the other particles tested and that the inflammatory indicators were not detectable when the instilled dose was ≤ 20 cm^2. It was subsequently concluded by Stoeger et al. [26] that although the surface area of particles indicates their ability to cause inflammation, with uf particles causing an increased proinflammatory gene expression compared with larger particles, a threshold may exist in relation to the surface area of NPs and their relevant toxicity. The findings of these studies support the hypothesis that NPs have the potential to drive toxicity in the lung and that they are responsible for the subsequent adverse health effects observed [17]. In addition, although these studies are limited to only a few specific NP types (predominantly CB, PBs, and TiO_2), it is suggested that the specific characteristics of particle size, particle mass dose, surface area, and surface reactivity are associated with NP toxicity in general. Although it is well documented which characteristics influence NP toxicity, the precise mechanism by which this observed toxicity occurs is not fully understood. Despite this, as a result of increased laboratory-based investigations that have been highlighted in this section, a paradigm termed the "oxidative stress paradigm" was proposed, which considers how NPs may cause adverse effects to cells [16].

4 The Oxidative Stress Paradigm

Oxidative stress occurs when a greater number of oxidants than antioxidants are present within the cell, causing an oxidant/antioxidant imbalance [27]. Increased oxidation can occur within cells, such as macrophages following activation [16]. The activation of macrophage cells can cause the generation of the superoxide anion, which is readily converted into the hydroxyl radical ($^{\bullet}OH$) via the influence of superoxide dismutase [27]. The presence of the $^{\bullet}OH$, as well as the superoxide anion, which are examples of reactive oxygen species (ROS), can thus cause increased oxidation within the cell because these molecules possess unpaired electrons and are highly unstable [27]. Additionally, ROS can be produced via nicotinamide adenine dinucleotide phosphate (NADPH) oxidase, which is the most common form of ROS found in cells, and is usually produced when cells are performing the phagocytosis of xenobiotics [27]. Therefore, this suggests that although cells purposefully clear hazardous particles from the tissue, the phagocytosing cells can unintentionally or intentionally produce ROS. In addition, the potential production of ROS following encapsulation of particles via phagocytosis further emphasizes the necessity to understand the specific uptake mechanism of NPs, in order to determine their potential route within the cell, and how their uptake may relate to their toxicity. As it is not possible to explore in detail the entry mechanisms of NPs into cells, the reader is referred to the comprehensive review of Unfried et al. [28].

The production of ROS and subsequent oxidative stress in cells can be extremely deleterious, causing a reduction in cell metabolic competence via a reduction in mitochondrial respiration as well as an increase in the production of (pro)inflammatory mediators (i.e., cytokines and chemokines). The effects of ROS production

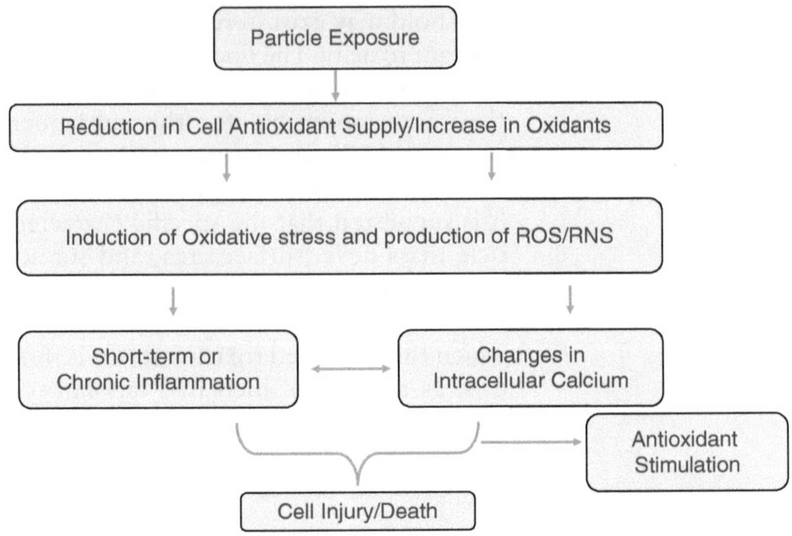

Fig. 1 Summary of the oxidative stress paradigm for nanotoxicological research testing strategies

and subsequent oxidative stress/inflammatory response can further be associated with diseases and illnesses such as pulmonary and cardiovascular diseases, including asthma, chronic obstructive pulmonary disease, as well as atherosclerosis and even tumor formation [16]. The manner in which oxidative stress production can elucidate deleterious effects is further discussed in this section and summarized within Fig. 1.

In a study by Stone et al. [29], the potential for NPs, specifically ufCB, to cause oxidative stress was assessed in A549 epithelial cells. Cells were measured for their glutathione content to determine the level of oxidative stress present. Glutathione is the most abundant nonprotein thiol present inside most cells, and it is essential for maintaining cell viability by detoxification of pathogens as well as by regulating cell cycle and gene expression (Droge et al.) [30]. A reduction in the glutathione content of cells, therefore, is known to indicate an oxidative stress environment [29]. It is also known that glutathione can have a protective role following the onset of oxidative stress, due to its antioxidant gene characteristics [27]. Antioxidant genes are common throughout cells, such as the c-*fos* gene. This gene is part of a series of growth- and differentiation-related genes that are expressed by cells in response to foreign materials [27]. It is thought that antioxidant genes are "switched on" following NP stimulus, which can thus overwhelm the NP effect [16]. Research into these effects, however, is limited and requires further investigation. In relation to the potential for NPs to induce oxidative stress, it was observed by Stone et al. [29] that following treatment of A549 epithelial cells with up to 0.78 μg mm^2 of ufCB and CB, the glutathione levels, as measured specifically in its reduced form (GSH), decreased after 2 h, with a significant decrease also found after 6-h exposure to ufCB, but not to CB. It was subsequently concluded by Stone et al. [29] that ufCB is more potent at inducing oxidative stress than its larger counterpart, CB. Li et al. [31] further studied the potential for NPs to induce oxidative stress in cells. In the study by Li et al. [31], the effects of ambient (25–10 μm), fine (<2.5 μm), and uf (<0.1 μm) particles on RAW 264.7 macrophage cells and BEAS-2B epithelial cells were assessed. It was observed that NPs, in comparison to both ambient and fine particles, induced an increased cellular expression of heme-oxygenase-1 (HO-1) expression in each cell line, indicative of oxidative stress, as well as a decrease in intracellular glutathione levels. High levels of ROS production were also demonstrated following dithiothreitol (DTT) analysis. Further examination of the different particle types by transmission electron microscopy (TEM) showed the NPs, as well as a small amount of larger particles, to predominantly locate within mitochondria, causing subsequent major structural damage thought to contribute to increased oxidative stress (decrease in GSH) and toxicity previously observed. It was therefore concluded that the increased biological potency of NPs can be associated with the redox cycling of organic chemicals due to their increased ability to damage the mitochondria, causing ROS and oxidative stress. Recently, Xia et al. [32] also showed ufCB, in comparison to TiO_2 and CB, to cause an increased production of ROS, as measured via the ROS quencher, furfuryl alcohol, as well as by assessment of the level of NADPH peroxidase with RAW 264.7 macrophage cells over a 4- and 16-h period. Subsequent analysis of the oxidative stress levels in these macrophage cells found depletion in GSH levels, as well as toxic oxidative stress after similar exposure periods. It was further illustrated that the toxic oxidative stress observed was specific to an injury to mitochondria due to increased cytosolic calcium (Ca^{2+}) production and uptake, causing structural damage to the organelle. Ca^{2+} is an essential and the most abundant mineral in the body [33]. Maintaining normal Ca^{2+} levels (~155 ± 9 nM) within the cell, also known as Ca^{2+} homeostasis, is essential for cell viability [34].The finding by Xia et al. [32] that an increased Ca^{2+} production (signaling) can occur in cells following exposure to NPs supports previous studies that have suggested the increase of cytosolic Ca^{2+} to be associated with the onset of increased ROS production and subsequent oxidative stress [16, 35–39]. Stone et al. [36] initially showed that ufCB exposure for 30 min to Mono Mac 6 (MM6) human macrophage cells induced a 1.6-fold increase in the resting cytosolic Ca^{2+} concentration, measured using the Ca^{2+} chelator Fura 2-AM [40] in MM6 cells at a concentration of 66 μg/m, while no changes were observed following treatment of MM6 cells with CB. Subsequent investigation by Stone et al. [35] examined the effects of CB, both fine and uf (33 μg/ml), and latex beads

(64, 202, and 535 nm in diameter) (1 mg/ml) on MM6 cells and primary rat macrophages. Analysis of the latex beads showed a 2.3-fold increase in cytosolic Ca^{2+} concentration (as determined by using the Fura 2-AM molecule) in response to thapsigargin stimulation following treatment with the 64-nm latex beads; however, no effects on MM6 cell Ca^{2+} concentration were observed following treatment with either the 202- or 535-nm latex beads. In this study, thapsigargin was used to assess the viability of the cells via Ca^{2+} signaling. This chemical stimulant causes release of Ca^{2+} from the endoplasmic reticulum (ER); a loss of cell viability, via apoptosis, can be related to a loss of Ca^{2+} in the ER store and therefore a reduced Ca^{2+} response to thapsigargin stimulation. Similarly, Stone et al. [35] showed a 2.6-fold increase in Ca^{2+} in BAL cells (>80 % macrophages) following stimulation with thapsigargin and after exposure to ufCB, but not to CB. These latter findings support those previously reported by Stone et al. [36] that ufCB can cause an increase in cytosolic Ca^{2+} concentration and further demonstrate that different macrophage types (MM6 and primary rat macrophages) can elicit similar responses following NP exposure. Further analysis by Stone et al. [35] demonstrated that these effects were attenuated when the MM6 cells were pretreated with either the antioxidant mannitol or nacystelyn, suggesting that the increased cytosolic Ca^{2+} concentrations observed following NP exposure could be mediated via ROS and oxidative stress. Stone et al. [35] performed further examination of the potential of all sizes of the latex beads to produce ROS, using the dye 2′,7′-dichlorofluorescin diacetate (DCFH-DA), which when oxidized converts into 2′,7′-dichlorofluorescin (DCFH) and shows an increasing fluorescence intensity relative to an increase in ROS production. It was found that the NP latex beads (64 nm) were more potent in causing ROS, with the 202- and 535-nm latex beads showing no oxidative effects after 10 min. These findings were subsequently supported by Brown et al. [37], who also measured cytosolic Ca^{2+} via the use of the fluorescent marker Fura 2-AM. It was reported that ufCB elicited a heightened cytosolic Ca^{2+} concentration in MM6 cells following treatment for 30 min. Subsequent analysis by Brown et al. [37], which examined the effects of transition metals, specifically Fe in the form of Fe chloride ($FeCl_2$), using the transition metal chelator, Desferal, found that these metals had no effect on cytosolic Ca^{2+} concentration after 30 min of exposure. In addition, Brown et al. [37] also investigated the inflammatory potential of the ufCB and CB particles as well as of the transition metals in vivo. It was observed that the ufCB particles, but not the CB particles, induced an increase in the number of PMNs present within the lungs of rats, with a significant increase in the number of neutrophils found within the BAL fluid after 24-h exposure. No inflammatory effects were found with $FeCl_2$. Following inductively coupled plasma–mass spectrometry, it was observed that the $FeCl_2$

particles were detectable in ng mg^{-1} concentrations within ufCB particles. It was subsequently concluded that ufCB does elicit an increase in cytosolic Ca^{2+} concentration, in support of Stone et al. [35, 36], and that ufCB particles induce an inflammatory response from cells, which was suggested to be independent of the transitional metal content of these NPs. The observation by Brown et al. [37] that ufCB can cause increased inflammation supports the suggestion that ROS production, oxidative stress, and altered Ca^{2+} signaling caused by these NPs can cause an activation of transcription pathways, via a cascade of events within the cell, including both the nuclear factor-kappa B (NF-κB) and activator protein-1 (AP-1) pathways [16]. Activation of these pathways was subsequently proposed to result in an escalated inflammatory response, with an increase in proinflammatory gene expression and proinflammatory mediator production, such as the cytokines IL-8 and TNF-α [16, 27]. Subsequent analysis by Brown et al. [38] further supported this hypothesis, demonstrating that ufCB particles, and not CB particles, cause an increase in resting cytosolic Ca^{2+} concentration (as assessed by using Fura 2-AM) in rat alveolar macrophages. Similar effects were also found on treatment of rat alveolar macrophages in a dose–response relationship (12.5–50 μg/ml) following thapsigargin stimulation. It is also worth noting that these effects are similar to the findings of Stone et al. [35, 36] and Brown et al. [37], further demonstrating the consistency between different forms of macrophage cells in assessing the toxicity associated with NP exposure. Additional study by Brown et al. [38] also showed that these effects were decreased following addition of antioxidants (Trolox and nacystelyn), further supporting the findings of Stone et al. [35], who concluded that the increased Ca^{2+} signaling observed in macrophage cells after NP exposure was mediated via ROS. As it was suggested, by Donaldson et al. [16], that subsequent inflammation could occur due to the activation of specific transcription pathways (such as NF-κB and AP-1) following the altered Ca^{2+} signaling caused by ROS and oxidative stress of NP-treated cells, Brown et al. [38] further studied the effects of ufCB on both NF-κB and AP-1 transcription pathways. Investigation of the NF-κB pathway showed ufCB (100 μg/ml) treated human monocyte cells to show increased fluorescence of the sub-units of NF-κB, p50, and p65 over a 4-h period compared with untreated monocyte cells. Additional analysis of the AP-1 pathway showed no significant increase in the intensity of the AP-1 protein following treatment with ufCB at 200 μg/ml, after 4 h in rat alveolar macrophages. The findings relative to both these pathways were found to be attenuated following the addition of antioxidants, further supporting the suggestion that these events are mediated via ROS production. Subsequent analysis of the inflammatory potential of ufCB in rat alveolar macrophages found the production of the proinflammatory cytokine TNF-α to be dose dependent (25–200 μg/ml) after 4-h exposure. It was subsequently concluded by Brown et al. [38] that uf particles can exert proinflammatory effects by altering Ca^{2+} signaling, activating transcription factors and causing the production of proinflammatory cytokines via ROS-mediated mechanism, thus supporting the proposed oxidative stress paradigm. Although the findings of these studies suggest that the toxicity observed following exposure to NPs is relative to increases in the production of ROS and subsequent oxidative stress, as well as an escalation in Ca^{2+} signaling and inflammation within cells, the oxidative paradigm is only a hypothesis, and further research is required to fully understand the mechanisms of NP toxicity and how they may relate to the many new and different types of NPs being manufactured. It is also prudent to highlight that the oxidative stress paradigm is flexible. In the previously discussed literature, it is evident that oxidative stress induces an inflammatory response which affects cell signaling. This sequence, while correct, is not always true for NPs. Any resultant effects can be initiated following the onset of oxidative stress following NP exposure occur either in the presence or absence of a state of oxidative stress.

5 Beyond the Oxidative Stress Paradigm

While the potential for NPs to cause oxidative stress has been the basis for increased research over the past two decades, an additional theory has been proposed which probes further the effects of oxidative stress on a biological system. The theory of genotoxicity [41] is predominantly based upon the oxidative stress paradigm; however, it moves on from determining an inflammatory response to assess what the stimulation of inflammatory mediators could induce to the biological system in regard to human health. The theory describes a two-tiered approach: (1) primary genotoxicity and (2) secondary toxicity. In regard to primary genotoxicity, it is suggested that NPs can cause genotoxicity following direct exposure to the biological system. Secondary genotoxicity however describes, initially, the oxidative stress paradigm (in theory: NP exposure = ROS/reactive nitrogen species (RNS) production (also oxidative stress (oxidant/antioxidant imbalance)) = chronic inflammatory response), which causes genotoxicity and (possibly) subsequent tumor formation. Secondary genotoxicity, however, may not be caused by the NPs alone; it may also be caused via interaction of the biological system and the chemicals contained within the NPs. It is suggested that the NPs might be completely inert but are able to penetrate the cellular membrane, possibly locating within the nucleus. At this stage, due to the highly acidic pH, the chemicals present within the NPs could be released (such as Fe within Fe platinum NPs), causing a toxic response. This form of secondary toxicity has also been referred to as the "Trojan horse" effect [42]. It is also possible, however, in relation to this theory that the cells might undergo cell death and thus not induce genotoxicity and tumor formation. Additionally, in response to the agonist of genotoxicity, the cells might proliferate or undergo repair of the deoxyribonucleic acid (DNA) chain and negate any NP effects [41].

Although both the oxidative stress paradigm and theory of genotoxicity can fit to any form or NP, they have predominantly been focused upon through the assessment of the biological response to spherical, crystalline, and non-fibrous NPs. However, since the portrayal that CNTs may cause "asbestos-like" effects in the lung [43], increased research has been performed using the theory of genotoxicity with fibrous NPs. While this is apt, specific, well-studied, and proposed paradigms already exist in regard to the biological effects following fibrous stimuli [44].

6 Investigating the Oxidative Stress Paradigm In Vitro

Although all the experimental analyses previously described in this chapter can be performed with cells obtained from in vivo experimentation (ex vivo) as well as in vitro, it is pertinent to highlight the constant debate on the advantages and disadvantages of both in vitro and in vivo testing. Although in vivo (specifically mice and rat models; however, also considering research on hamster, guinea pig, and monkey models), research is known to enable scientific research to observe almost "firsthand" the effects of a substance as they would occur in *Homo sapiens*, in vitro models provide the possibility to investigate toxic effects on human cells extensively, which cannot be conducted in vivo [45]. Cultured human and animal cells can be better controlled and therefore yield more reproducible data than in vivo systems; however, they require a high standardization to maximize reproducibility.

Due to the oxidative stress paradigm being the foundation for nanotoxicology research, assessment of the formation of ROS and oxidative stress can be achieved by an ever-increasing list of assays and diagnostic kits. Despite this, there are specific parameters that are constantly used in order to denote the production of ROS, the increase in oxidants, or the loss in antioxidants within a biological model. In regard to the production of ROS, the most common parameter used is the fluorescent dye 2′,7′-dichlorodihydrofluorescein diacetate (DCFH-DA), a reliable fluorogenic marker for ROS in living cells. This method assesses the production of ROS via the increase in fluorescence. Upon the production of ROS, the DCFH-DA molecule is cleaved, resulting in a DCFH molecule which is highly fluorescent. The fluorescence signals can be both imaged via CLSM and quantified using FACSTM analysis and fluorescent spectrophotometry. A similar form of testing procedure is also based upon the fluorescent dye dihydroethidium (DHE). Similar to the DCFH-DA reaction in which ROS production is determined by the level of fluorescence, DHE detects the extent of oxidative stress and ROS production in cells, by oxidizing to ethidium, which is highly fluorescent. These methods are extremely advantageous, as not only they can be performed via multiple approaches, but they can also be used in a cell-free environment. This is important because it provides important information pertaining to the NPs alone and thus their ability to cause ROS independent of a cellular environment. This, of course, is extremely advantageous in regard to assessing the ability of NPs to cause secondary genotoxicity. As previously mentioned, there are a number of kits available to measure ROS production and mostly these are based on the DCFH-DA reaction. An example of this is the Image-iT LIVE Green ROS Detection Kit (provided by Molecular Probes). The production of ROS is concomitant with a reduction in antioxidants, thus causing an oxidant imbalance within the cell and therefore oxidative stress. It is therefore pertinent that an assessment of the antioxidant capacity is performed in order to obtain a thorough and clear understanding of the oxidant-related effects of NPs. This form of analysis has commonly been performed in regard to an assessment of intracellular glutathione levels. Glutathione is an intracellular nonprotein thiol readily available in a wide range of living cells and is a key factor in a number of biological functions, including oxidative stress [46]. It is usually present in cells as GSH; however, when existing in cells as its oxidized glutathione (GSSG) form, glutathione is rapidly converted to GSH via an enzymatic reaction using glutathione reductase. In the method originally published by Hissin and Hilf [46], the levels of both GSH and GSSG were determined via their pH-sensitive reactions with the fluorescent reagent *o*-phthalaldehyde. There are of course many alternative ways to determine glutathione levels within cells, including diagnostic kits and microscopy-related assays. In addition, there are numerous ways to assess the antioxidant capacity of cells, such as the Trolox equivalent antioxidant capacity method (TEAC); however, this is extremely laborious and difficult to optimize. In addition to determining the antioxidant capacity, it is also possible to determine the level of oxidants present within the biological sample. This is notoriously difficult and time-consuming; however, it provides extremely valid and optimal results in regard to the oxidative potential of NPs. Such testing strategies include assessment of the $^{\bullet}OH$, as well as the superoxide anion via polymerase chain reaction (PCR). These can be determined via diagnostic kits; however, due to the intrinsic nature of these radicals/anions within the cellular system the most reliable method for assessing these aspects is via PCR.

In addition to the measurement of ROS and oxidative stress, it is also prudent to determine the ability for nitric oxide (NO) to be produced. Often, however, NO is not assessed in relation to the radical formation potential of NPs. Increased research should be directed towards this aspect, though, due to the fundamental role of NO in cellular function. A radical gas, NO, is derived from the guanidine nitrogen of the amino acid l-arginine and molecular oxygen in a reaction catalyzed by the enzyme nitric oxide synthase (NOS). There are at least two types of NOS. The calcium-dependent form is present constitutively in a variety of tissues and produces the physiological concentration of NO needed for maintenance of blood pressure. The other, calcium-independent form, is inducible (iNOS) in a number of cell types, including macrophages, hepatocytes, neutrophils, and endothelial cells, by a variety of immunological stimuli such as interferon-γ(IFN-γ) and bacterial lipopolysaccharide (LPS). iNOS is a high-output enzyme and nitrite concentrations are easily detected. It is possible to measure NO (or iNOS) via spectrophotometry using a colorimetric test known as the Griess reaction, for which numerous diagnostic kits are available. It is important to note, however, that problems have been reported when assessing NO in primary cells, as well as some cell lines (e.g., A549 epithelial cells). This test is, however, advantageous in assessing the ability of NPs only to produce NO (i.e., in a cell-free environment). In order to assess NO in cells therefore, it is also possible to use PCR analysis. This testing strategy, however, as with measurement of both the •OH and the superoxide anion, is extremely laborious and time-consuming.

Assessment of inflammation in vitro is usually achieved via the use of enzyme-linked immunosorbent assay (ELISA) diagnostic kits. Although expensive, these diagnostic kits provide valuable and reproducible testing strategies to measure the plethora of different proinflammatory cytokines and chemokines that have been associated with NP exposure to culture cells. In addition to ELISA kits, it is also possible to measure cytokine/chemokine production either via PCR analysis or flow cytometry; however, these methods are again laborious and, in comparison to the ELISA method, provide no further advantages. The influence of NPs on cell signaling, specifically upon Ca^{2+} signaling in cells, is a significant factor related to the oxidative stress paradigm. In relation to this, Ca^{2+} signaling is not as well researched as the ability of NPs to stimulate ROS and inflammatory cytokines/chemokines. Although it is possible to measure intracellular calcium levels via a number of means, again including diagnostic kits, the most commonly used indicator of changes in intracellular calcium is the fluorescent marker FURA 2-AM [40]. The function of this dye is similar to that of DCFH-DA, as during the incubation period with the cells (and xenobiotic), if there is a change in Ca^{2+}, intracellular esters are able to remove the acetoxymethyl (AM) group from the Fura 2-AM molecule, making the membrane impermeable to Ca^{2+}-sensitive fluorescent dye, Fura 2. It is then possible to determine the level of Ca^{2+} present in samples

Table 1
Example of how to combine the many different biological endpoint tests and microscopy techniques available in order to gain a clear and effective understanding of the NP-cell interaction

Cytotoxicity and cell viability	Oxidative stress and inflammation	ROS/RNS	Genotoxicity and DNA damage	Uptake and intracellular fate	Chemical analysis
MTT assay	Reduction in glutathione	DCFH-DA assay	Ames test	Confocal microscopy	EELS
LDH assay	Superoxide anion	Griess reaction test	Micronucleus assay	Electron microscopy	EFTEM
Apoptosis/necrosis	PCR analysis	PCR analysis	Comet assay	Fluorescent trackers for cell organelles	ICP-AES
BrdU assay	ELISA	Intracellular calcium	PCR analysis	Light microscopy	EDX analysis
TUNEL method	Western blots	Antioxidant stimulation	Various kits	FACS analysis	Mass spectroscopy

via spectrophotometry, as well as by CLSM. In addition to this, PCR is a very useful tool in assessing the effects of cell signaling. Although this method can identify specific proteins and genes (ribonucleic acid (RNA) transcription/isolation), it is laborious and requires increased optimization time in order to obtain valid and reliable results. It is obvious therefore that in order to fully understand how NPs interact with biological systems and how to measure how they may fit into the oxidative stress paradigm, there are numerous endpoint testing strategies. In a final note, it is important to point out to the reader that when strategically investigating how NPs interact with biological systems, a combination of the many different available tests is the most advantageous way forward. An example of this can be achieved not only for studying oxidative stress, but the complete NP-cell interaction (Table 1).

7 Conclusion

In summary, how NPs interact with biological systems is a complex field to study, and is based upon a plethora of factors that can all have a significant impact upon the final biological effect. While there is no specific or defined biological theory as to why NPs may elicit adverse effects to any biological system compared to their larger-sized counterparts, a number of paradigms/theories exist. The most widely accepted of these is the oxidative stress paradigm. This paradigm is complex and can fit a variety of different NPs, with only one prerequisite: they all cause an imbalance of oxidants to antioxidants within the biological system being investigated. In order to test this paradigm, a plethora of different biological endpoint tests are available. The most advantageous of which is debatable; however, it is accepted that through the combination of these many available tests, it promotes the best way forward to establishing effective and efficient testing strategies for determining the oxidative stress potential of all NPs being manufactured. It is clear though that with the hundreds to thousands of studies being performed and published within the field of nanotoxicology, while there is no definitive theory behind how NPs cause adverse biological effects, the oxidative stress paradigm is paramount in understanding how NPs may affect both human health and the environment.

References

1. Oberdorster G, Oberdorster E, Oberdorster J (2005) Nanotoxicology: an emerging discipline evolving from studies of ultrafine particles. Environ Health Perspect 113:823–839
2. Müller L, Gasser M, O'Raemy D, Herzog F, Brandenberger C, Schmid O, Gehr P, Rothen-Rutishauser B, Clift MJD (2011) Realistic exposure methods for investigating the interaction of nanoparticles with the lung at the air-liquid interface in vitro. InSci J (Nanotech) 1: 30–64
3. Maynard AD (2007) Nanotechnology: the next big thing, or much ado about nothing? Ann Occup Hyg 51:1–12
4. Clift MJ, Gehr P, Rothen-Rutishauser B (2011) Nanotoxicology: a perspective and discussion of whether or not in vitro testing is a valid alternative. Arch Toxicol 85:723–731
5. Maynard AD, Aitken RJ, Butz T, Colvin V, Donaldson K, Oberdorster G, Philbert MA, Ryan J, Seaton A, Stone V, Tinkle SS, Tran L, Walker NJ, Warheit DB (2006) Safe handling of nanotechnology. Nature 444:267–269
6. Timbrell J (1999) Principles of biochemical toxicology, 3rd edn. CRC, Philadelphia, PA, USA
7. Donaldson K, Stone V, Tran CL, Kreyling W, Borm PJ (2004) Nanotoxicology. Occup Environ Med 61:727–728
8. Ferin J, Oberdorster G, Penney DP (1992) Pulmonary retention of ultrafine and fine particles in rats. Am J Respir Cell Mol Biol 6: 535–542
9. Oberdorster G, Ferin J, Morrow PE (1992) Volumetric loading of alveolar macrophages (AM): a possible basis for diminished AM-mediated particle clearance. Exp Lung Res 18:87–104
10. Zanobetti A, Schwartz J, Samoli E, Gryparis A, Touloumi G, Peacock J, Anderson RH, Le Tertre A, Bobros J, Celko M, Goren A, Forsberg B, Michelozzi P, Rabczenko D, Hoyos SP, Wichmann HE, Katsouyanni K (2003) The temporal pattern of respiratory and heart disease mortality in response to air pollution. Environ Health Perspect 111: 1188–1193
11. Schwartz J (2004) The effects of particulate air pollution on daily deaths: a multi-city case crossover analysis. Occup Environ Med 61: 956–961
12. Medina S, Plasencia A, Ballester F, Mucke HG, Schwartz J (2004) Apheis: public health impact of PM10 in 19 European cities. J Epidemiol Community Health 58:831–836
13. Dockery DW, Pope CA 3rd, Xu X, Spengler JD, Ware JH, Fay ME, Ferris BG Jr, Speizer FE (1993) An association between air pollution and mortality in six U.S. cities. N Engl J Med 329:1753–1759
14. Donaldson K, Stone V, Clouter A, Renwick L, MacNee W (2001) Ultrafine particles. Occup Environ Med 58:211–216
15. Wilson MR, Lightbody JH, Donaldson K, Sales J, Stone V (2002) Interactions between ultrafine particles and transition metals in vivo and in vitro. Tox Appl Pharm 184:172–179
16. Donaldson K, Stone V, Borm PJ, Jimenez LA, Gilmour PS, Schins RP, Knaapen AM, Rahman I, Faux SP, Brown DM, MacNee W (2003) Oxidative stress and calcium signaling in the adverse effects of environmental particles (PM10). Free Radic Biol Med 34:1369–1382
17. Seaton A, MacNee W, Donaldson K, Godden D (1995) Particulate air pollution and acute health effects. Lancet 345:176–178

18. Peters A, Wichmann HE, Tuch T, Heinrich J, Heyder J (1997) Respiratory effects are associated with the number of ultrafine particles. Am J Respir Crit Care Med 155:1376–1383
19. Wichmann HE, Spix C, Tuch T, Wolke G, Peters A, Heinrich J, Kreyling WG, Heyder J (2000) Daily mortality and fine and ultrafine particles in Erfurt, Germany part I: role of particle number and particle mass. Res Rep Health Eff Inst 98:5–86, discussion 87–94
20. Schulz H, Harder V, Ibald-Mulli A, Khandoga A, Koenig W, Krombach F, Radykewicz R, Stampfl A, Thorand B, Peters A (2005) Cardiovascular effects of fine and ultrafine particles. J Aerosol Med 18:1–22
21. Li XY, Gilmour PS, Donaldson K, MacNee W (1997) In vivo and in vitro proinflammatory effects of particulate air pollution (PM10). Environ Health Perspect 105(Suppl 5): 1279–1283
22. Li XY, Brown D, Smith S, MacNee W, Donaldson K (1999) Short-term inflammatory responses following intratracheal instillation of fine and ultrafine carbon black in rats. Inhal Toxicol 11:709–731
23. Brown DM, Wilson MR, MacNee W, Stone V, Donaldson K (2001) Size-dependent proinflammatory effects of ultrafine polystyrene particles: a role for surface area and oxidative stress in the enhanced activity of ultrafines. Toxicol Appl Pharmacol 175:191–199
24. Duffin RTC, Clouter A, Brown DM, MacNee W, Stone V, Donaldson K (2001) The importance of surface area and specific reactivity in the acute pulmonary inflammatory response to particles. Ann Occup Hyg 4:242–245
25. Duffin R, Tran L, Brown D, Stone V, Donaldson K (2007) Proinflammogenic effects of low-toxicity and metal nanoparticles in vivo and in vitro: highlighting the role of particle surface area and surface reactivity. Inhal Toxicol 19:849–856
26. Stoeger T, Reinhard C, Takenaka S, Schroeppel A, Karg E, Ritter B, Heyder J, Schulz H (2006) Instillation of six different ultrafine carbon particles indicates a surface area threshold dose for acute lung inflammation in mice. Environ Health Perspect 114:328–333
27. MacNee W (2001) Oxidative stress and lung inflammation in airways disease. Eur J Pharmacol 429:195–207
28. Unfried K, Albrecht C, Klotz LO, von Mikecz A, Grether-Beck S, Schins RPF (2007) Cellular responses to nanoparticles: target structures and mechanisms. Nanotoxicology 1:52–71
29. Stone V, Shaw J, Brown DM, Macnee W, Faux SP, Donaldson K (1998) The role of oxidative stress in the prolonged inhibitory effect of ultrafine carbon black on epithelial cell function. Toxicol In Vitro 12:649–659
30. Droge W, Schulze-Osthoff K, Mihm S, Galter D, Schenk H, Eck HP, Roth S, Gmunder H (1994) Functions of glutathione and glutathione disulfide in immunology and immunopathology. FASEB J 8:1131–1138
31. Li N, Sioutas C, Cho A, Schmitz D, Misra C, Sempf J, Wang M, Oberley T, Froines J, Nel A (2003) Ultrafine particulate pollutants induce oxidative stress and mitochondrial damage. Environ Health Perspect 111:455–460
32. Xia T, Kovochich M, Brant J, Hotze M, Sempf J, Oberley T, Sioutas C, Yeh JI, Wiesner MR, Nel AE (2006) Comparison of the abilities of ambient and manufactured nanoparticles to induce cellular toxicity according to an oxidative stress paradigm. Nano Lett 6:1794–1807
33. Alberts B, Johnson J, Lewis J, Raff M, Roberts K, Walter P (2002) Molecular biology of the cell, 4th edn. Garland Science—Taylor and Francis Group, New York, USA
34. Abbas AK, Lichtman AH (2003) Cellular and molecular immunology, 5th edn. Elsevier Health Sciences, USA
35. Stone V, Brown DM, Watt N, Wilson M, Donaldson K, Ritchie H, MacNee W (2000) Ultrafine particle-mediated activation of macrophages: intracellular calcium signaling and oxidative stress. Inhal Toxicol 12:345–351
36. Stone V, Tuinman M, Vamvakopoulos JE, Shaw J, Brown D, Petterson S, Faux SP, Borm P, MacNee W, Michaelangeli F, Donaldson K (2000) Increased calcium influx in a monocytic cell line on exposure to ultrafine carbon black. Eur Respir J 15:297–303
37. Brown DM, Stone V, Findlay P, MacNee W, Donaldson K (2000) Increased inflammation and intracellular calcium caused by ultrafine carbon black is independent of transition metals or other soluble components. Occup Environ Med 57:685–691
38. Brown DM, Donaldson K, Borm PJ, Schins RP, Dehnhardt M, Gilmour P, Jimenez LA, Stone V (2004) Calcium and ROS-mediated activation of transcription factors and TNF-alpha cytokine gene expression in macrophages exposed to ultrafine particles. Am J Physiol Lung Cell Mol Physiol 286:L344–L353
39. Brown DM, Hutchison L, Donaldson K, MacKenzie SJ, Dick CA, Stone V (2007) The effect of oxidative stress on macrophages and lung epithelial cells: the role of phosphodiesterases 1 and 4. Toxicol Lett 168:1–6
40. Grynkiewicz G, Poenie M, Tsien RY (1985) A new generation of Ca2+ indicators with greatly improved fluorescence properties. J Biol Chem 260:3440–3450

41. Schins RP, Knaapen AM (2007) Genotoxicity of poorly soluble particles. Inhal Toxicol 19(Suppl 1):189–198
42. Limbach LK, Li Y, Grass RN, Brunner TJ, Hintermann MA, Muller M, Gunther D, Stark WJ (2005) Oxide nanoparticle uptake in human lung fibroblasts: effects of particle size, agglomeration, and diffusion at low concentrations. Environ Sci Technol 39: 9370–9376
43. Poland CA, Duffin R, Kinloch I, Maynard A, Wallace WA, Seaton A, Stone V, Brown S, Macnee W, Donaldson K (2008) Carbon nanotubes introduced into the abdominal cavity of mice show asbestos-like pathogenicity in a pilot study. Nat Nanotechnol 3:423–428
44. Donaldson K, Murphy FA, Duffin R, Poland CA (2010) Asbestos, carbon nanotubes and the pleural mesothelium: a review of the hypothesis regarding the role of long fibre retention in the parietal pleura, inflammation and mesothelioma. Part Fibre Toxicol 7:5
45. Rothen-Rutishauser B, Blank F, Muhlfeld C, Gehr P (2008) In vitro models of the human epithelial airway barrier to study the toxic potential of particulate matter. Expert Opin Drug Metab Toxicol 4:1075–1089
46. Hissin PJ, Hilf R (1976) A fluorometric method for determination of oxidized and reduced glutathione in tissues. Anal Biochem 74:214–226

Chapter 8

Evaluation of the Effect of Acute and Subacute Exposure to TiO_2 Nanoparticles on Oxidative Stress

Shichang Liu and Zhuo Yang

Abstract

The nanosized titanium dioxide (nano-TiO_2) is produced abundantly and used widely in the chemical, electrical/electronic, and energy industries because of its special photovoltaic and photocatalytic activities. Past reports have shown that the nano-TiO_2 can enter into the human body through different routes such as inhalation, ingestion, dermal penetration, and injection. The effects of nano-TiO_2 on different organs are currently being investigated. Oxidative stress is considered to play an important role in the oxidative potential of nanoparticles. Here we discuss the association between oxidative stress and the toxicity caused by exposure to nano-TiO_2 in different organs.

Key words Nanosized titanium dioxide (nano-TiO_2), Nanotoxicity, Reactive oxygen species (ROS), Oxidative stress

1 Introduction

Nanotechnology involves the creation and manipulation of materials at the nanoscale level to create unique products that exploit novel properties. Recently, nanomaterials such as nanotubes, nanowires, fullerene derivatives (buckyballs), and quantum dots have received enormous attention to create new types of analytical tools for biotechnology and life sciences [1, 2]. Nanomaterials range in size from 1 to 100 nm; the new physical and chemical properties of novel engineered nanoparticle make them extremely attractive for use in the applications such as medical science [3], drug applications, and agricultural and defense industries [4]. Despite the wide ranges of applications, there is a serious lack of information on the impact of the nanoparticles on human health and the environment.

Nanosized titanium dioxide (nano-TiO_2), one of the most widely used nanoparticles, has now been produced in a large industrial scale and used widely in paints, paper, and plastics, as well as

Donald Armstrong and Dhruba J. Bharali (eds.), *Oxidative Stress and Nanotechnology: Methods and Protocols*, Methods in Molecular Biology, vol. 1028, DOI 10.1007/978-1-62703-475-3_8,

in food additives and colorants [5, 6]. Nano-TiO_2 occurs in three different crystallographic structures: rutile, anatase, and brookite. Although variations in the photocatalytic properties and biological effects based on differences in the crystal structure have not been completely established, some studies reported that the anatase, the most commercially used form, showed higher activity than the rutile form, and anatase is usually used as an efficient photocatalyst and in printing ink. Brookite, which is difficult to synthesize, is reported to have higher photocatalytic activity [7]. The versatility of nano-TiO_2 is due in part to its useful properties, including being insoluble in aqueous solutions, chemically unreactive in biological systems, non-radioactive, not radio-opaque, not costly, easy to prepare in stable suspensions in fine particulate form, and easy to measure by simple chemical techniques [8]. With the large surface area and zeta potential, the nano-TiO_2 has a direct connection with each other and therefore tends to aggregate in solution. Figure 1 shows transmission electron microscopy (TEM) images (a) of primary particles and dynamic light scattering (DLS) images (b) of aggregates of nano-TiO_2.

Despite the wide ranges of applications, there is a serious lack of information on the impact of nano-TiO_2 on human health and the environment. The extensive uses in medical research and industrial applications highlight that there are many routes for the nano-TiO2 to potentially enter into human bodies, such as through inhalation (respiratory tract), ingestion (gastrointestinal tract), dermal penetration (skin), and injection (blood circulation) [9]. Up to now, nano-TiO_2 has been reported to cause many toxic effects, including inflammation, cytotoxicity, phytotoxicity, and genomic instability in mammals, plants, and microorganisms [7].

2 Oxidative Stress and Reactive Oxygen Species

Oxidative stress is caused by an imbalance between damaging oxidants and protective antioxidants. Oxidants include hydrogen peroxide and hydroxyl radicals (collectively known as reactive oxygen species or ROS); antioxidants are molecules such as vitamin C and glutathione that scavenge the unwanted oxidants [10]. ROS including hydroxyl radicals, superoxide anions, singlet oxygen, and hydrogen peroxide are generated as by-products of cellular metabolism [11, 12], and organisms have developed their own antioxidant mechanisms including low-molecular-weight antioxidant molecules, such as glutathiones (GSH), melatonin, and various antioxidant enzymes including superoxide dismutase (SOD), catalase (CAT), glutathione peroxidase (GSH-Px), and glutathione reductase [13]. SOD destroys the superoxide radical (O_2^-) and

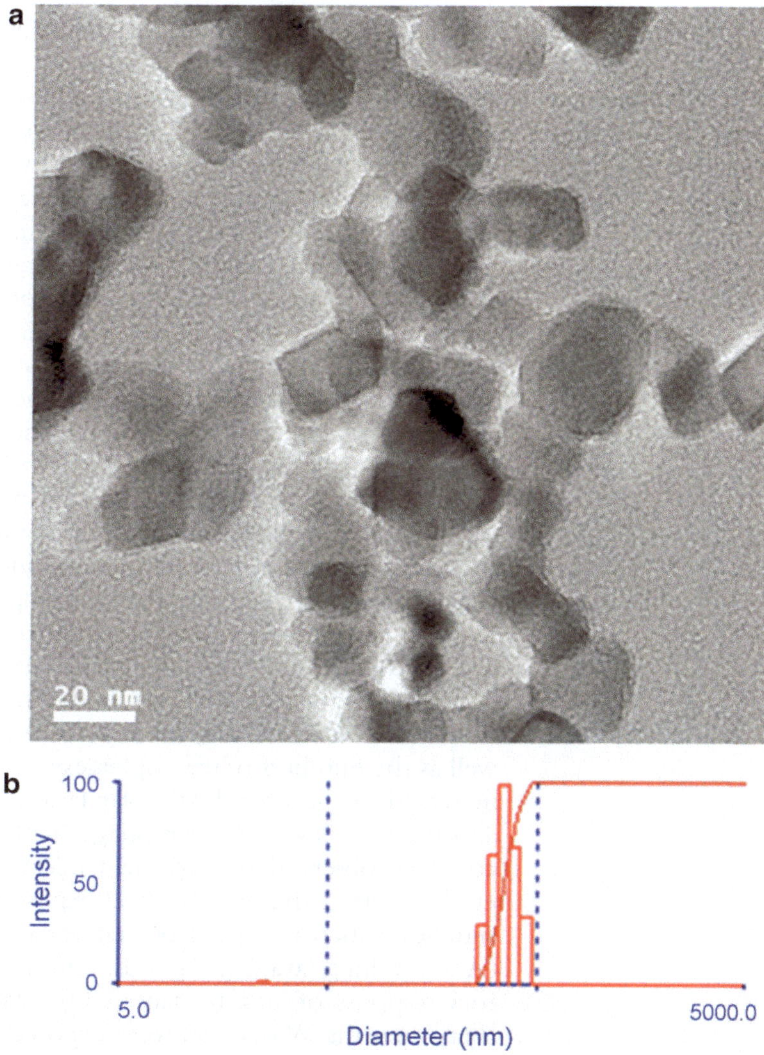

Fig. 1 Dispersion and characterization of the TiO_2 nanoparticles were characterized by TEM (**a**) and DLS (**b**). TEM images showed that the size of the nano-TiO_2 was distributed from 20 to 50 nm. And the DLS assay (**b**) stated that the particle size distribution had a wide range from 24 to 697 nm due to the aggregation, and the hydrodynamic diameter was 368.1 nm [48]

GSH-PX detoxifies H_2O_2 especially [6]. If oxidative stress persists, oxidative damage to critical biomolecules accumulates and eventually results in several biological effects such as alterations in signal transduction and gene expression for mitogenesis, mutagenesis, and cell death [14]. Because of the size of surface area, nano-TiO_2 has greater ability to produce reactive oxygen species (ROS) [15]. The various studies implicate that oxidative stress pathways play a special role in pathogenesis of nano-TiO_2-induced injuries [16].

3 The Effect of Exposure to Nano-TiO_2 Mediated by Oxidative Stress

3.1 The Influence of Acute and Subacute Exposure to Nano-TiO_2 on Lungs

During production, distribution, and use, nano-TiO_2 may become suspended in air. Therefore, manufactured nano-TiO_2 can become a component of indoor and outdoor environments and thus the air we breathe [17], and the respiratory tract becomes the primary target organ system for toxic effects of inhaled particles. The inhaled nano-TiO_2 particles in the respiratory tract are deposited in various ways, and the main mechanism for deposition is diffusion due to displacement when they collide with air molecules. Other deposition mechanisms such as inertial impaction, gravitational setting, and interception usually occur when the particles size is large [9]. The larger surface area enhanced the chemical reactivity and hence the pulmonary responses induced by inhaled nano-TiO_2 are considered to be greater than micron-sized particles. When mice were treated with a single intratracheal instillation of 0.1 mg nano-TiO_2, the lungs showed significant changes including disruption of the alveolar septa, alveolar enlargement (emphysematous change), type II pneumocyte proliferation, increased alveolar epithelial thickness, and the accumulation of particle-laden macrophages [18]. Moreover, when mice were exposed to nano-TiO_2 at the concentration of 0.77 or 7.22 mg/m^3 acutely, the numbers of total cells as well as the numbers of macrophages were found to be significantly increased in the bronchoalveolar lavage (BAL) fluid of the mice, and the large alveolar macrophages with phagocytized nano-TiO_2 could be observed by dark field microscopy (Fig. 2) [17]. Afaq et al. reported that intratracheal exposure of nano-TiO_2 caused a significant increase in the percentage of rat alveolar macrophages (AMs) in lung lavage [19], indicating that there was an inflammatory response of lung to nano-TiO_2. Ma-Hock et al. also showed that when the Wistar rats were exposed to aerosol of 50 mg/m^3 nano-TiO_2, there was a significant increase in lung weight, and the lung inflammation was associated with the increases of BAL fluid total cell and neutrophil counts, total protein content, enzyme activities, and levels of a number of cell mediators [20].

According to the previous studies, with increasing exposure time of nano-TiO_2, the pulmonary responses include bronchoalveolar hyperplasia and metaplasia, fibrosis, and pulmonary tumorigenesis. In animal study, long-term exposure of nano-TiO_2 to rats by inhalation revealed a dose-dependent dust cell accumulation, a foamy macrophage response, type II pneumocyte hyperplasia, alveolar proteinosis, alveolar bronchiolarization, cholesterol granulomas, focal pleurisy, bronchioloalveolar adenomas, and cystic keratinizing squamous cell carcinomas [21]. Heinrich et al. reported that long-term exposure to 10 mg/m^3 of nano-TiO_2 could induce a significantly increased lung tumor rate [22]. Moreover, exposure to nano-TiO_2 particles might stimulate an increase in the formation of reactive oxides. Schapira et al. had

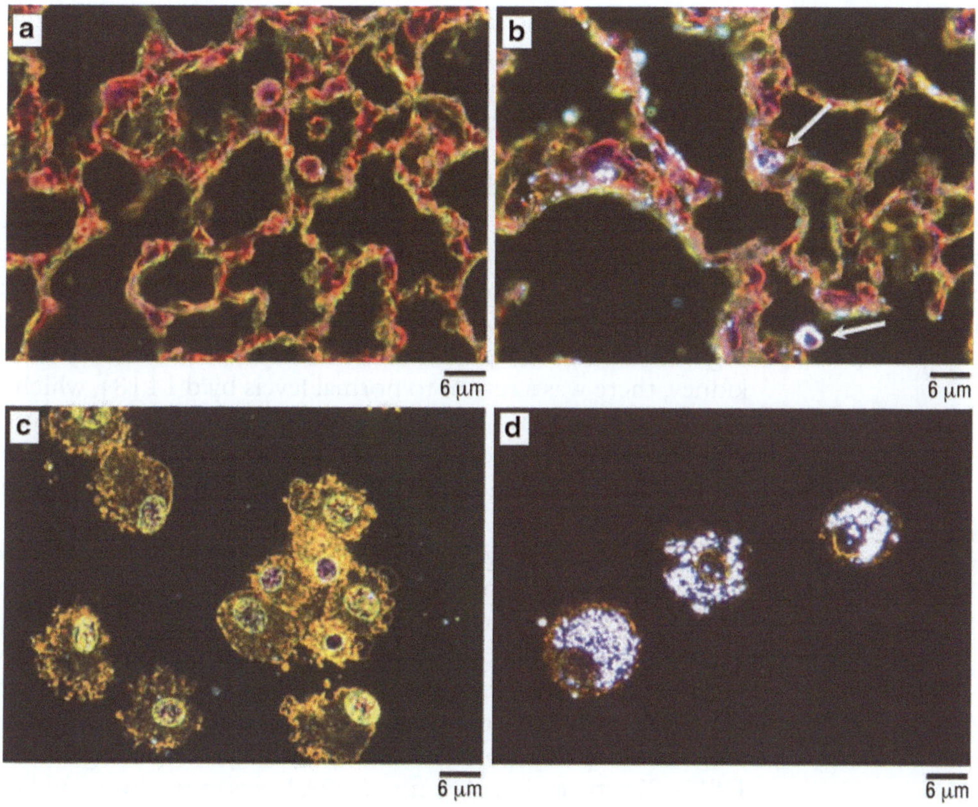

Fig. 2 Dark field micrographs of lung tissue with H&E staining (**a**, **b**) and alveolar macrophages prepared by cytospinning and H&E staining (**c**, **d**). *Arrows* point to nano-TiO_2 particle-laden macrophages [17]

shown the production of hydroxyl radicals from rat lung exposed to nano-TiO_2 particles [23]. The release of these ROS could oxidize polyunsaturated fatty acids located in the plasma membrane, resulting in the formation of the breakdown product malondialdehyde (MDA) of lipid peroxidation, which is in agreement with other mineral fibers [24]. Regarding the toxicity of nano-TiO_2, lungs showed significant changes in morphology and histology associated with pulmonary inflammation and emphysema, and long-term exposure can even lead to pulmonary tumors.

The potent biological activity of nano-TiO_2 was associated with high surface reactivity leading to ROS-mediated cytotoxicity and genotoxicity. By in vitro study, Bhattacharya et al. found that in human lung cells, nano-TiO_2 induced oxidative stress by generating elevated amounts of free radicals and therefore induced genotoxicity mainly by DNA adduct formation [25]. In human bronchial epithelial (BEAS-2B) cells, nano-TiO_2 induced the apoptotic process with the induction of ROS and the decreased level of intracellular GSH [26]. And the anatase-sized (20 and 20 nm) TiO_2 particles induced oxidative DNA damage, lipid peroxidation, and micronuclei formation and increased hydrogen peroxide and nitric oxide production [27]. Nano-TiO_2 particles also regulated key factors involved in the G2/M progression in the cell cycle by increasing the gene expression of cell division cycle 2 homologue A (cdc2a) and some cyclins [18].

3.2 The Influence of Acute and Subacute Exposure to Nano-TiO_2 Particles on Liver and Kidneys

Uptake of engineered nanoparticles into the human body has several different routes, and they can be distributed throughout the body. The liver is the major distribution site via uptake by Kupffer cells, followed by the spleen and kidneys [28, 29]. Moreover, after continuous exposure, nano-TiO_2 would be accumulating in these organs even at low exposure levels and was retained for 28 d. In the kidney, there was a return to normal levels by d 14 [8], which indicated that nano-TiO_2 might be slowly eliminated in the liver and kidneys. Wang et al. researched the impacts of nano-TiO_2 exposure by the route of oral administration in mice [30]. The results indicated that nano-TiO_2 particles mainly retained in the liver and kidneys, spleen, and lung. Also, the obvious hepatic injuries including hydropic degeneration around the central vein and the spotty necrosis of hepatocyte and renal lesion including proteinic liquids in the renal tubule and swelling in the renal glomerulus were observed in the histopathological examination. The exposure of nano-TiO_2 particles could also induce the presence of oxidative stress and lipid peroxidation response by decreasing SOD and GSH-PX activity and elevating MDA levels in the liver and kidneys [6]. Xiong et al. studied the acute exposure effect of nano-TiO_2 on zebrafish; the results showed that after the exposure for 96 h, lipid peroxidation was observed and the quantities of ·OH significantly increased, and the oxidative effects associated with SOD, CAT activities, and GSH contents were more severe in the livers [31], which indicated that nano-TiO_2 could induce the oxidative damage in the liver through superfluous ROS and the disruption of antioxidant systems. The in vitro studies confirmed the findings above. Nano-TiO_2 particles could lead to cellular morphological modifications and cause increased generation of ROS, depleted GSH, and reduced mitochondrial membrane potential in rat liver cells [32, 2]. In the kidneys, oxidative damage occurred likewise, with the exposure of 80 nm nano-TiO_2 particles on adult mice for 2 weeks; the renal tubule was filled with the proteinic liquids; and the serious swelling in the renal glomerulus can be observed [30]. Nano-TiO_2 could also induce nephric inflammation and impair nephric functions through ROS accumulation [33].

3.3 The Influence of Acute and Subacute Exposure to Nano-TiO_2 on Brain

Generally, the blood–brain barrier (BBB) separates blood from cerebrospinal fluid and is made of endothelial cells connected by tight junctions that limit the entry of many substances into the brain [34]. But with the ultrafine size and unusual properties, nanoparticles can enter the body and cross biological barriers such as BBB relatively unimpeded [4]; the nose-to-brain transport of exogenous materials is a potential route for bypassing the BBB [35]. Many studies have unequivocally showed that exposure to nano-TiO_2 could be translocated into the central nervous system (CNS) via the olfactory pathway and damage brain neurons and tissue in vitro and in vivo. After the exposure, the deposition of nano-TiO_2 could be observed in the sub-brain regions including olfactory bulb, cerebral cortex, and cerebellum, especially in the hippocampus region [36, 37]. When entered into the brain, nano-TiO_2 could cause obvious morphological changes of neurons in the cerebral cortex and significant disturbance of the monoamine neurotransmitter levels in the above sub-brain regions [38], but the damages in these regions did not occur simultaneously. Wang et al. reported that by intranasal instillation, the nano-TiO_2 particles first entered into the olfactory bulb and then deposited in the hippocampus and CNS as the time extended, and the pathological changes appeared after the exposure for 30 days [36]. Ma et al. also reported that after exposure for 14 days, a high dose of nano-TiO_2 accumulated in the ICR mouse brain and triggered a cascade of reactions such as lipid peroxidation, the decreases of the total antioxidation capacity and activities of antioxidative enzymes, the excessive release of nitric oxide, the reduction of glutamic acid, and the downregulated level of acetylcholinesterase activities, thereby inducing the oxidative stress and injury of the brain [39].

Wang et al. also found that the hippocampus had the highest Ti contents even after 30 days, indicating that the hippocampus would be greatly injured compared with that of the other parts of brain. As the hippocampus is one of the most important functional regions of the brain and hippocampal cells are greatly sensitive to ROS [40], the accumulation of nano-TiO_2 in the hippocampus could easily induce oxidative damage and affect synaptic plasticity subsequently, which was associated with certain high functions of the CNS such as learning and memory [41–43]. Hu et al. evaluated the ultrastructure of hippocampal neurons after intragastric administration with nano-TiO_2 for consecutive 60 days; the result showed a significant shrinkage of the nuclear membrane as well as chromatin marginalization with 50 mg/kg nano-TiO_2, which suggested that nano-TiO_2 could cause neuron damage in the hippocampus (Fig. 3) [41]. Furthermore, nano-TiO_2 had also been shown to induce apoptosis and oxidative stress in microglia, neurons [4, 44–46], Purkinje cells [47], and PC12 cells [48] by in vitro studies.

3.4 Oxidative Stress Damage of Skin by Exposure to Nano-TiO_2

With the unusual properties, nano-TiO_2 is included in some sunscreen formulations to physically block ultraviolet radiation and thereby has the possibility to get exposed to the skin directly. Most of the commercial micropigments based on TiO_2 have a mean particle size below 30 nm. Because of the small particle size, a penetration of ultrafine TiO_2 into human skin seems to be possible [49]. In order to evaluate whether nano-TiO_2 can penetrate intact skin, several

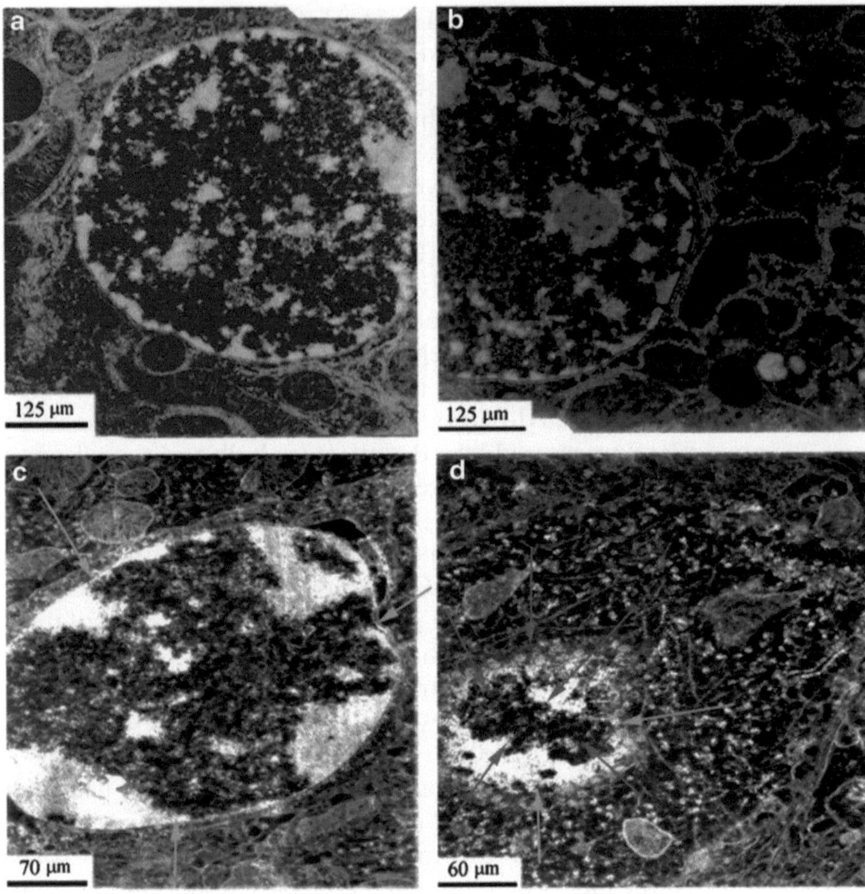

Fig. 3 Ultrastructure of hippocampal neuron in female mice caused by intragastric administration with nano-TiO_2 for consecutive 60 days. (**a**) Control, (**b**) 5 mg/kg BW nano-TiO_2, (**c**) 10 mg/kg BW nano-TiO_2, (**d**) 50 mg/kg BW nano-TiO_2. *Green arrows*, shrinkage of nucleus in the hippocampus CA neuron; *red arrows*, chromatin marginalization [41]

studies have been conducted and published. Tan et al. excised skin from human volunteers following 2- to 6-week applications of a sunscreen containing 8 % nano-TiO_2 for 2–6 weeks. The authors found elevated but not significant levels of TiO_2 in the skin of the test patients when compared to skin from cadavers; however, if one extreme value is removed from the cadaver group, a significant difference was detected with an ~50 % increase in skin TiO_2 levels [50]. Bennat et al. examined the penetration of microfine TiO_2 into human skin in vitro and in vivo [49]. The results indicated that TiO_2 penetrated the upper layers of the epidermis when applied as a 40 % suspension in oil, with less penetration from a water suspension. Lademann et al. reported that nano-TiO_2 particles could penetrate into the horny layer and the follicular orifice [51]. The above studies confirmed that nano-TiO_2 could penetrate the outermost layer of skin. Moreover, in vitro studies were also carried out to assess the uptake and cellular toxicity induced by exposure of nano-TiO_2. The study in human epidermal cells showed that nano-TiO_2 had a mild cytotoxic potential after 48 h of exposure, but nano-TiO_2 could induce ROS and oxidative stress leading to oxidative DNA damage and micronucleus formation [52]. In addition, when exposed to mouse fibroblast (L929) cells, nano-TiO_2 could induce significant cytotoxicity including cell adhesion, proliferation via lysosome enhancement, redox homeostasis damage, and DNA damage [53].

4 Conclusions

As one of the most common nanoparticles, nano-TiO_2 particle is widely used in many fields related to our daily life. Thus, more attention is needed on the impact of nano-TiO_2 on human health and the environment. Among various mechanisms in which the effect of nano-TiO_2 exposure is involved, oxidative stress seems to be an important mechanism which involves interaction between nano-TiO_2 particles and living organisms. All the presented results and theories demonstrated that nano-TiO_2 could induce oxidative stress damage by the generation of ROS, depletion of GSH, and reduction of mitochondrial membrane potential, thereby causing cell apoptosis; the possible schematic mechanism of nano-TiO_2-induced cellular toxicity in vitro condition was shown in Fig. 4. Otherwise, when the internal of exposure period changed, nano-TiO_2 could cause different effects on the organs. Generally, nano-TiO_2 induced more serious influence with the long-term exposure. In conclusion, the mechanism of oxidative stress occurring in different organs as well as the extent and type of subsequent injuries still need detailed and profound investigation. This would be helpful to realize the toxicity of nano-TiO_2 particles and make the application of nano-TiO_2 safely used.

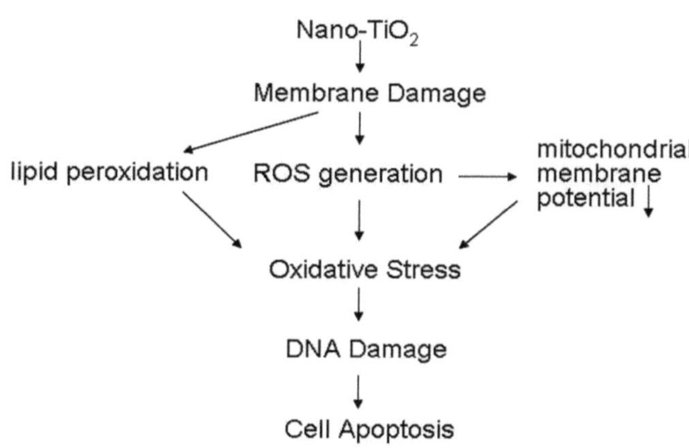

Fig. 4 The possible mechanism of cellular toxicity induced by nano-TiO_2

Acknowledgment

This work was partly supported by the National Natural Science Foundation of China (31070890).

References

1. Cui Y et al (2001) Nanowire nanosensors for highly sensitive and selective detection of biological and chemical species. Science 293:1289–1292
2. Hussain SM et al (2005) In vitro toxicity of nanoparticles in BRL 3A rat liver cells. Toxicol In Vitro 19:975–983
3. Emerich DF, Thanos CG (2003) Nanotechnology and medicine. Expert Opin Biol Ther 3:655–663
4. Long TC et al (2007) Nanosize titanium dioxide stimulates reactive oxygen species in brain microglia and damages neurons in vitro. Environ Health Perspect 115:1631–1637
5. Park JH et al (2006) Novel carbon-doped TiO_2 nanotube arrays with high aspect ratios for efficient solar water splitting. Nano Lett 6:24–28
6. Liang G et al (2009) Influence of different sizes of titanium dioxide nanoparticles on hepatic and renal functions in rats with correlation to oxidative stress. J Toxicol Environ Health A 72:740–745
7. Warheit DB et al (2007) Pulmonary toxicity study in rats with three forms of ultrafine-TiO_2 particles: differential responses related to surface properties. Toxicology 230:90–104
8. Fabian E et al (2008) Tissue distribution and toxicity of intravenously administered titanium dioxide nanoparticles in rats. Arch Toxicol 82:151–157
9. Oberdorster G et al (2005) Nanotoxicology: an emerging discipline evolving from studies of ultrafine particles. Environ Health Perspect 113:823–839
10. Stone V, Donaldson K (2006) Nanotoxicology: signs of stress. Nat Nanotechnol 1:23–24
11. Klamt F et al (2001) Imbalance of antioxidant defense in mice lacking cellular prion protein. Free Radic Biol Med 30:1137–1144
12. Crack PJ, Taylor JM (2005) Reactive oxygen species and the modulation of stroke. Free Radic Biol Med 38:1433–1444
13. Huang J et al (2008) Reactive oxygen species mediate oridonin-induced HepG2 apoptosis through p53, MAPK, and mitochondrial signaling pathways. J Pharmacol Sci 107:370–379
14. Ermak G, Davies KJ (2002) Calcium and oxidative stress: from cell signaling to cell death. Mol Immunol 38:713–721
15. Moller P et al (2010) Role of oxidative damage in toxicity of particulates. Free Radic Res 44: 1–46
16. Mocan T et al (2010) Implications of oxidative stress mechanisms in toxicity of nanoparticles (review). Acta Physiol Hung 97:247–255
17. Grassian VH et al (2007) Inhalation exposure study of titanium dioxide nanoparticles with a primary particle size of 2 to 5 nm. Environ Health Perspect 115:397–402
18. Chen HW et al (2006) Titanium dioxide nanoparticles induce emphysema-like lung injury in mice. FASEB J 20:2393–2395
19. Afaq F et al (1998) Cytotoxicity, pro-oxidant effects and antioxidant depletion in rat lung alveolar macrophages exposed to ultrafine titanium dioxide. J Appl Toxicol 18:307–312
20. Ma-Hock L et al (2009) Development of a short-term inhalation test in the rat using nano-titanium dioxide as a model substance. Inhal Toxicol 21:102–118
21. Lee KP et al (1985) Pulmonary response of rats exposed to titanium dioxide (TiO_2) by inhalation for two years. Toxicol Appl Pharmacol 79:179–192
22. Heinrich U, Fuhst R, Rittinghausen S (1995) Chronic inhalation exposure of Wistar rats and two different strains of mice to diesel engine exhaust, carbon black, and titanium dioxide. Inhal Toxicol 7:533–556
23. Schapira RM et al (1995) Hydroxyl radical production and lung injury in the rat following silica or titanium dioxide instillation in vivo. Am J Respir Cell Mol Biol 12:220–226
24. Arif JM et al (1993) Modulation of macrophage-mediated cytotoxicity by kerosene soot: possible role of reactive oxygen species. Environ Res 61:232–238
25. Bhattacharya K et al (2009) Titanium dioxide nanoparticles induce oxidative stress and DNA-adduct formation but not DNA-breakage in human lung cells. Part Fibre Toxicol 6:17
26. Park EJ et al (2008) Oxidative stress and apoptosis induced by titanium dioxide nanoparticles in cultured BEAS-2B cells. Toxicol Lett 180:222–229
27. Gurr JR et al (2005) Ultrafine titanium dioxide particles in the absence of photoactivation

can induce oxidative damage to human bronchial epithelial cells. Toxicology 213:66–73
28. Jani PU, McCarthy DE, Florence AT (1994) Titanium dioxide (rutile) particle uptake from the rat GI tract and translocation to systemic organs after oral administration. Int J Pharm 105:157–168
29. Sugibayashi K et al (2008) Safety evaluation of titanium dioxide nanoparticles by their absorption and elimination profiles. J Toxicol Sci 33:293–298
30. Wang J et al (2007) Acute toxicity and biodistribution of different sized titanium dioxide particles in mice after oral administration. Toxicol Lett 168:176–185
31. Xiong D et al (2011) Effects of nano-scale TiO_2, ZnO and their bulk counterparts on zebrafish: acute toxicity, oxidative stress and oxidative damage. Sci Total Environ 409:1444–1452
32. Du H et al (2012) Oxidative damage and OGG1 expression induced by a combined effect of titanium dioxide nanoparticles and lead acetate in human hepatocytes. Environ Toxicol 27:590–597
33. Zhao JF, Wang J, Wang SS, Zhao XY, Yan JY, Ruan J (2010) The mechanism of oxidative damage in the nephrotoxicity of mice caused by nano-anatase TiO_2. J Exp Nanosci 5:447–462
34. Begley DJ (1996) The blood–brain barrier: principles for targeting peptides and drugs to the central nervous system. J Pharm Pharmacol 48:136–146
35. Illum L (2000) Transport of drugs from the nasal cavity to the central nervous system. Eur J Pharm Sci 11:1–18
36. Wang J et al (2008) Time-dependent translocation and potential impairment on central nervous system by intranasally instilled TiO(2) nanoparticles. Toxicology 254:82–90
37. Gao X et al (2011) Effects of developmental exposure to TiO(2) nanoparticles on synaptic plasticity in hippocampal dentate gyrus area: an in vivo study in anesthetized rats. Biol Trace Elem Res 143(3):1616–1628
38. Zhang L et al (2011) Rutile TiO(2) particles exert size and surface coating dependent retention and lesions on the murine brain. Toxicol Lett 207:73–81
39. Ma L et al (2010) Oxidative stress in the brain of mice caused by translocated nanoparticulate TiO_2 delivered to the abdominal cavity. Biomaterials 31:99–105
40. Yang Z et al (2010) A review of nanoparticle functionality and toxicity on the central nervous system. J R Soc Interface 7(Suppl 4):S411–S422
41. Hu R et al (2011) Molecular mechanism of hippocampal apoptosis of mice following exposure to titanium dioxide nanoparticles. J Hazard Mater 191:32–40
42. Wang J et al (2008) Potential neurological lesion after nasal instillation of TiO(2) nanoparticles in the anatase and rutile crystal phases. Toxicol Lett 183:72–80
43. Hu R et al (2010) Neurotoxicological effects and the impairment of spatial recognition memory in mice caused by exposure to TiO_2 nanoparticles. Biomaterials 31:8043–8050
44. Long TC et al (2006) Titanium dioxide (P25) produces reactive oxygen species in immortalized brain microglia (BV2): implications for nanoparticle neurotoxicity. Environ Sci Technol 40:4346–4352
45. Yu Y et al (2008) Nanosize titanium dioxide cause neuronal apoptosis: a potential linkage between nanoparticle exposure and neural disorder. Neurol Res, http://www.ncbi.nlm.nih.gov/pubmed/18662496
46. Li XB, Xu SQ, Zhang ZR, Hermann JS (2009) Apoptosis induced by titanium dioxide nanoparticles in cultured murine microglia N9 cells. Chin Sci Bull 54:3830–3836
47. Takeda K, Suzuki KI, Ishihara A, Kubo-Irie M, Fujimoto R, Tabata M, Oshio S, Nihei Y, Ihara T, Sugamata M (2009) Nanoparticles transferred from pregnant mice to their offspring can damage the genital and cranial nerve systems. J Health Sci 55:95–102
48. Liu S et al (2010) Oxidative stress and apoptosis induced by nanosized titanium dioxide in PC12 cells. Toxicology 267:172–177
49. Bennat C, Muller-Goymann CC (2000) Skin penetration and stabilization of formulations containing microfine titanium dioxide as physical UV filter. Int J Cosmet Sci 22:271–283
50. Tan MH et al (1996) A pilot study on the percutaneous absorption of microfine titanium dioxide from sunscreens. Australas J Dermatol 37:185–187
51. Lademann J et al (1999) Penetration of titanium dioxide microparticles in a sunscreen formulation into the horny layer and the follicular orifice. Skin Pharmacol Appl Skin Physiol 12:247–256
52. Shukla RK et al (2011) ROS-mediated genotoxicity induced by titanium dioxide nanoparticles in human epidermal cells. Toxicol In Vitro 25:231–241
53. Jin CY et al (2008) Cytotoxicity of titanium dioxide nanoparticles in mouse fibroblast cells. Chem Res Toxicol 21:1871–1877

Chapter 9

Biomedical Applications of Nano-antioxidant

Geeta Watal, Aparna Watal, Prashant Kumar Rai, Devendra Kumar Rai, Gaurav Sharma, and Bechan Sharma

Abstract

For centuries now, antioxidants have been known to provide better health by neutralizing the free radicals which are continuously produced in the human body. In normal circumstances, self-antioxidant defense system of the human body is capable of quantitatively managing the free radicals. However, in certain cases, which are at the threshold of developing diseases like diabetes and Alzheimer's, the human body calls for an external source of antioxidants. Since orally delivered antioxidants are easily destroyed by acids and enzymes present in the human system, only a small portion of what is consumed actually gets absorbed. Hence, there is a recognized and urgent need to develop effective methods for efficiently delivering antioxidants to the required sites. This chapter provides an in-depth overview and analysis of two such methods and processes—nano-encapsulation and nano-dendrimers. Among the various nanoscale delivery mechanisms, nano-encapsulation has emerged as a key and efficient delivery process. Designed as a sponge-like polymer, nano-encapsulated antioxidants provide a protective vehicle which keeps antioxidants from being destroyed in the human gut and ensures their better absorption in the digestive tract. In fact, the nano-capsules bind themselves to the intestinal walls and pour antioxidants directly into the intestinal cells, which allow them to be absorbed directly into the blood stream. Another distinguished and popular mode for delivering antioxidants is that of nano-polymers known as dendrimers. Dendrimers involve multiple branches and sub-branches of atoms radiating out from a central core. Dendrimers afford a high level of control over their architectural design, including their size, shape, branching length or density, and surface functionality. Such flexibility makes these nanostructures ideal carriers in biomedical applications such as drug delivery, gene transfection, and imaging. Antioxidant dendrimers, made out of numerous units of antioxidants connected with each other in a branched fashion, provide numerous possible sites to couple with an active species and have enhanced free radicals scavenging potency. These dendrimer chains are biocompatible, biodegradable with nontoxic degradation products, and well suited for targeted drug delivery and other biomedical applications. Recent successes in simplifying and optimizing the synthesis of dendrimers, such as the "lego" and "click" approaches, provide a large variety of structures while at the same time reducing the cost of their production. The use of these highly branched, nanometer-sized, polymeric materials as nano-antioxidants for prevention and treatment of human diseases, associated with oxidative stress, is of immense public health relevance globally.

Key words Dendrimer, Antioxidant, Drug delivery, Polymeric micelles, Liposome, Nanomedicine

Donald Armstrong and Dhruba J. Bharali (eds.), *Oxidative Stress and Nanotechnology: Methods and Protocols*, Methods in Molecular Biology, vol. 1028, DOI 10.1007/978-1-62703-475-3_9, © Springer Science+Business Media New York 2013

1 Introduction

Nanotechnology is an emerging science that demands an understanding of the physicochemical properties of nanomaterials that may lead to adverse biological outcomes. Although impressive from the perspective of material science, the novel properties of nanoparticles could lead to adverse biological effects. The unique physicochemical properties of engineered nanoparticles are attributable to their small size, large surface area, chemical composition, surface reactivity, charge, shape, and media interactions [1–4]. The application of nanotechnology in the health-care sector, in imaging, diagnostics, drug delivery, and therapeutics, also referred to as nanomedicine, has gained ground over the past 5 years. This can be observed from the increase in the USA budget for nanomedicine research as well as an increase in the number of nanopharmaceutical patents [5].

2 Significance of Physicochemical Properties of Nanoparticles on the Efficiency of Drug Delivery

A drug delivery system is defined as a formulation or a device that enables the introduction of a therapeutic substance in the body and improves its efficacy and safety by controlling the rate, time, and location of release of drugs in the body. When material is at a nanometer size range, it acquires unique physical and chemical properties. More specifically, the parameters that play a major role in drug delivery and have significant pharmacological effects are particle size, surface area, hydrophobicity, surface charge, and crystallinity.

3 Particle Size

The submicron size of nanoparticles offers a number of distinct advantages over microparticles particularly in drug delivery due to the fact that these particles are in the size region of macromolecules. This physical characteristic, particularly for particles less than 100 nm in size, allows these particles to reach virtually all tissues in the body [5].

4 Surface Charge

The surface charge in nanoparticles reflects the electrical potential of particles and is influenced by the chemical composition of the particle and the medium in which it is dispersed. Nanoparticles with a high positive or negative zeta potential have been shown to be stable in suspension, as the surface charge prevents aggregation of the particles [6, 7].

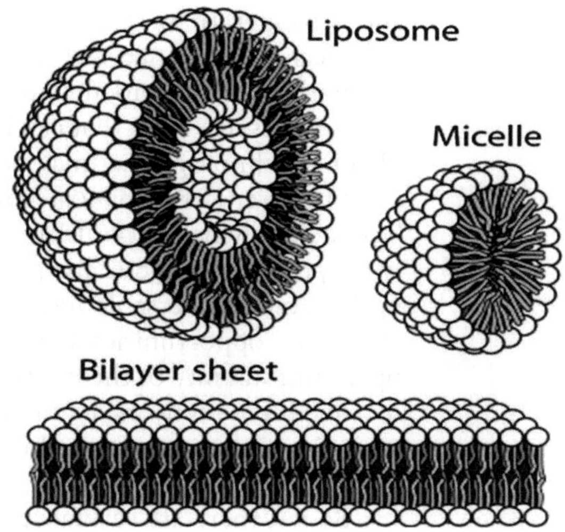

Fig. 1 Schematic illustration of liposomes and polymeric micelles

4.1 Polymeric Micelles

Polymeric micelles are a supramolecular assembly of amphiphilic block copolymers or polymer-lipid conjugates or other surface-active molecules that self-assemble in aqueous media to form structures with a hydrophobic core as illustrated in Fig. 1. Their advantages include high drug-loading capacity of the inner core and the possibility of their modulation to respond to various stimuli such as pH and temperature [8]. The ability to functionalize the micelles as well as tailor the disintegration behavior by varying the copolymer composition is a beneficial parameter in making them drug carriers of choice. Their small size (1–50 nm) makes them ideal for intravenous delivery. In addition they are also more stable, when compared to liposomes due to the ability to design them to be chemically stable and biocompatible [9].

5 Why Is the Interest in Nanomedicine Research?

The potential advantages of the clinical trials of the nanoscale research are well known and are fairly obvious because the small size allows extraordinary access to target areas in the body. For instance, the assay of very small amounts of bioanalyte, treatment or repair inside diseased cells and tissues, nanostructures, and devices for imaging are more sensitive diagnosis for the nanomedicine. The rapidly growing interest in nanotechnology is for the reason that the nature of some materials is altered in unexpected ways as size is reduced, called "quantum effects," thereby producing changes in properties such as color, reactivity, electrical conductivity, and strength. It can also be understood with a good example of carbon and steel. Carbon in the form of graphite being soft and malleable in nature when changed into carbon nanotubes (~1.5 nm in diameter) becomes more flexible, resilient, and stronger than steel along with increased fluorescence and electrical conductance with virtually no resistance [10, 11].

Further advancement must be made in encapsulation of glucose sensors in a form that can be embedded in the body. An example of a technology that may be appropriate for this is electrostatic layer-by-layer (LBL) nanoassembly of capsules composed of multilayers of polymer films. LBL encapsulation has been achieved by two approaches; these are providing tunable permeability to be both stable and versatile modes. LBL deposition of nanofilms may also offer opportunities for improving the biocompatibility and operating stability of the existing needle-type implantable amperometric glucose sensors, by incorporation of anti-inflammatory and antioxidant agents, extracellular matrix mimics, and other factors in the layers.

So far, there are several applications involved in nanomedicine but only few researches have been carried out in the field of oxidative stress. Antioxidants are compounds that protect cells against reactive oxygen cells or free radicals in the body. Although they are created as part of the body's normal metabolic functions, free radicals react with other cells and may interfere with their ability to function. Free radicals are unstable atoms or molecules with free outer electrons. This makes them highly reactive because free electrons always strive to form a stable bond. This stabilization involves gaining an electron from another molecule, triggering a chain reaction. Such reactions are omnipresent in the human body, but under certain circumstances can damage biomolecules. Whether nanoparticles are intracellular taken up leading to the activation of free radical production is currently being discussed. Various in vitro and in vivo studies show that free radical formation can be triggered by nanoparticles, viz., fullerenes, carbon nanotubes, quantum dots, and emission particles. Nanoparticles can be taken up actively (phagocytosis) by certain cells (macrophages) and initiate reactive oxygen species (ROS) formation. Ongoing studies are investigating whether the amount of free radicals formed on the surface of nanoparticles is sufficient to induce cellular effects. This dossier provides an overview about what free radicals are, how they originate, why organisms need them, how they are neutralized, and what we know about the connection between nanoparticles and free radical production.

References

1. Finkel T (2003) Oxidant signals and oxidative stress. Curr Opin Cell Biol 15:247–254
2. Liu H, Colavitti R, Rovira II, Finkel T (2005) Redox-dependent transcriptional regulation. Circ Res 97:967–974
3. McCord JM, Edeas MA (2005) SOD, oxidative stress and human pathologies: a brief history and a future vision. Biomed Pharmacother 59:139–142
4. Maren S, Baudry M (1995) Properties and mechanisms of long-term synaptic plasticity in the mammalian brain: relationships to learning and memory. Neurobiol Learn Mem 63: 1–18

5. Mishra B, Patel BB, Tiwari S (2010) Colloidal nanocarriers: a review on formulation technology, types and applications toward targeted drug delivery nanomedicine: nanotechnology. Biol Med 6:9–24
6. McNeil SE (2005) Nanotechnology for the biologist. J Leukoc Biol 78:585–594
7. Reynolds IJ, Hastings TG (1995) Glutamate induces the production of reactive oxygen species in cultured forebrain neurons following NMDA receptor activation. J Neurosci 15: 3318–3327
8. Narayan RJ, Kumta PN, Sfeir C, Lee DH, Olton D, Choi D (2004) Nanostructured ceramics in medical devices: applications and prospects. J Min Met Mat Soc 56:38–43
9. Curtis ASG, Dalby M, Gadegaard N (2006) Cell signaling arising from nanotopography:implications for nanomedical devices. Nanomedicine 1:67–72
10. O'Connell MJ, Bachilo SH, Huffman CB (2002) Band gap fluorescence from individual single walled carbon nanotubes. Science 297: 593–596
11. Dresselhaus MS, Dresselhaus G, Jorio A (2004) Unusual properties and structure of carbon nanotubes. Ann Rev Mater Res 34: 247–278

Part II

Protocols

Chapter 10

Lipid Peroxidation Due to In Vitro and In Vivo Exposure of Biological Samples to Nanoparticles

Anca Dinischiotu, Loredana Stanca, Daniela Gradinaru, Sorina Nicoleta Petrache, Mihaela Radu, and Andreea Iren Serban

Abstract

The increasing use of nanomaterials in biological applications raises numerous concerns about the dangers they might pose to living organisms. The rise in oxidative stress is usually the most readily observed effect induced by nanoparticles, with the measurement of lipid peroxidation levels being one of the most frequently used biological markers for its evaluation. Here, we describe the spectrophotometric and fluorimetric methods for determining the modifications of the malondialdehyde (MDA) level induced by many types of nanoparticles in in vitro and in vivo biological systems.

Key words Malondialdehyde, Nanoparticles, Spectrophotometric method, Fluorimetric method

1 Introduction

The rapid development of nanotechnologies with medical, agricultural, industrial, environmental, domestic, and cosmetic use has raised considerable and increasing health concerns as far as the public and governments worldwide are concerned [1]. Such widespread use, alongside the potential entry of nanoparticles (NP) through dermal, ingestion, and inhalation routes, suggests that nanosized particles could pose an exposure risk to humans and animals [2]. Once NP have entered the animal body and translocated to organs and tissues, they pass the cell membranes. Depending on their chemical composition and properties, the pathways involved in the internalization of NP into cells can be clathrin-mediated transport [3], caveolae-dependent endocytosis [4], macropinocytotic uptake [5], as well as other mechanisms in which adhesive interactions due to electrostatic forces, van der Waals, or steric interactions [6, 7] are involved. While recent studies highlight simple diffusion as a pathway for NP uptake into cells [8], a potential route for inhaled NP uptake could be the paracellular pathway down the gradient for physiological water absorption [9].

Donald Armstrong and Dhruba J. Bharali (eds.), *Oxidative Stress and Nanotechnology: Methods and Protocols*, Methods in Molecular Biology, vol. 1028, DOI 10.1007/978-1-62703-475-3_10,

Free radicals are continuously produced by enzymatic and nonenzymatic mechanisms in cells and tissues. The enzymatic reactions involving free radicals are those of the respiratory chain, phagocytosis, prostaglandin synthesis, and some xenobiotic compounds biotransformation in the cytochrome P_{450} system [10].

Nowadays, it is widely accepted that NP are generators of reactive oxygen species (ROS) by chemical and/or enzymatically catalyzed reactions. ROS are chemical species possessing one or more unpaired electrons oxygen species capable of independent existence.

For example, some types of NP can act as photosensitizers, causing the production of both singlet oxygen and superoxide from ground-state molecular oxygen under the influence of light [11]. In the case of transitional metals containing NP, their ions may be released in the cells and catalyze Fenton-type reactions [12]. Alternatively, these ions could also replace Fenton ions such as Fe^{2+}, Cu^{2+}, Mn^{2+} from cellular storage protein, generating hydroxyl radicals [13]. Additionally, infinitesimal quantities of transitional metal ions released from these NP in the cytosol could also affect the electron transport chain of the mitochondrial inner membrane, thus generating superoxide leakage [14].

For combustion-born NP, the released organic compounds (e.g., guinones or polyaromatic hydrocarbons) could undergo redox cycling as part of phase I metabolism, thus yielding redox active compounds [15].

NP uptaken by cells could interact with NADPH oxidase in the plasma membrane, thus generating superoxide in the extracellular space [16], which would enter the cell through an anion channel [17]. Furthermore, these nanoparticles could interact with a recently identified NADPH oxidase isoform located in the endoplasmic reticulum [18], consequently forming the oxygen radical endogenously.

It is also possible that the nanoparticle-induced damage of the mitochondria and endoplasmic reticulum could determine Ca^{2+} release into the cytosol, where several Ca^{2+}/calmodulin-dependent enzymes, such as nitrogen monoxide synthase isoforms, become activated, resulting in production of nitric oxide and peroxynitrite. The latter can yield hydroxyl radicals due to a homolysis reaction [19].

Cellular redox homeostasis is carefully maintained by an elaborate antioxidant defense system which includes antioxidant enzymes, proteins, and low-molecular-weight scavengers. Excessive ROS production or weakening of the antioxidant defense could lead to oxidative stress, which can be related to the etiopathogenesis of several chronic diseases and aging [20, 21]. Free radicals formation and lipid peroxidation are physiologically used by the organism to change organic or morphologic structures via oxidative pathways [22]. Lipid peroxidation results subsequently to a free radical attack on a double covalent bond within a polyunsaturated fatty acid or fatty acid chains encompassed by several biological structures [23]. In the presence of molecular oxygen, chain reactions lead to numerous compounds with different structures such as conjugated dienes, lipid hydroperoxides, aldehydes, and isoprostanes [24]. Due to the fact that lipid peroxides are unstable compounds, they tend to degrade rapidly in a variety of subproducts, mainly aldehydes which increase oxidative insults.

Since the discovery in 1948 of the reaction between thiobarbituric acid (TBA) and organic peroxides by F. Bernheim, numerous reliable methods have been developed for the study of lipid peroxidation processes in the biological systems [25]. Aldehydes, which include malonic dialdehyde (MDA), result from decomposing hydroperoxides and can react with TBA with subsequent formation of colored or fluorescent compounds [26]. The assay known as TBARS (*thiobarbituric acid-reactive substances*) is based on the aforementioned reaction, which is easy to carry out but lacks specificity [27].

MDA can potentially be derived from fatty acids with less than three double bonds and from primary carbonyl compounds which undergo a secondary oxidation [28]. Moreover, MDA can also be generated in vivo by enzymatic reactions involved in thromboxane A_2 biosynthesis [29]. This compound has a high stability and reactivity and is largely responsible for many types of cellular damage associated with oxidative stress.

4-Hydroxy-2-nonenal (HNE) is another aldehyde, a toxic product of lipid peroxidation and at the same time a toxic mediator of free radicals [30]. Isoprostanes, which are particularly stable and specific lipid peroxidation products, are newer markers for lipid peroxidation assessment [31].

Long regarded as a key mechanism in the damaging alterations of biological structures under reactive species action, lipid peroxidation has attracted a lot of interest. Additionally, this process has been widely used as an oxidative stress biomarker, in both in vitro and in vivo experimental models [32].

The present work aims at describing simple methods that could be used for assessing lipid peroxidation in noncellular and cellular in vitro systems as well as in in vivo ones. This type of experiments could indicate the overall way of action of nanoparticles, namely, prooxidant or antioxidant, in biological systems.

The most rapid and simple methods for MDA determination are the colorimetric [25] and fluorimetric [33] assays.

2 Materials

Prepare all solutions using ultrapure water (prepared by purifying deionized water to attain a resistivity of 18 M Ω cm at 25 °C) and analytical-grade reagents. Prepare in advance and store all reagents at 4–8 °C (unless indicated otherwise).

2.1 Materials and Reagents for the Colorimetric Method

Biological samples: human erythrocyte suspension; potassium phosphate buffer: 0.1 N K_2HPO_4, 0.1 N KH_2PO_4, pH 7.4; 0.15 M NaCl; Drabkin reagent: 0.05 % KCN, 0.14 % KH_2PO_4 0.2 % $K_3[Fe(CN)_6]$ (dilute 1:10 before use); phosphate-buffered saline (PBS) 10 mM, pH 7.4; 10 mM H_2O_2 in PBS; PBS–azide solution (PBSA) 100:1; 1 % TBA solution prepared in water (TBA) (*see* **Note 1**); 28 % trichloroacetic acid (TCA) with 0.1 N sodium arsenite; MDA standard stock solution: 10 μM 1,1,3,3-tetramethoxy propane in 0.1 N HCl.

2.2 Materials and Reagents for the Fluorimetric Method

Biological samples: whole tissue extract or cell lysate; 0.1 N HCl; 0.025 M TBA solution prepared in 0.1 N HCl (*see* **Note 1**); tissue lysis buffer: 0.1 M Tris–HCl, 5 mM EDTA, pH 7.4; MDA standard stock solution: 1 μM 1,1,3,3-tetramethoxy propane in 0.1 N HCl; bovine serum albumin (BSA) solution prepared in lysis buffer having a concentration to match the protein concentration of the biological samples (*see* **Note 2**).

3 Methods

3.1 Colorimetric Method

The method of evaluating the human erythrocyte susceptibility to lipid peroxidation is based on the generation, under oxidative stress (in the presence of H_2O_2) of malonyldialdehyde (MDA) evaluated by the use of the reaction with TBA [34, 35]:

1. Standardize the erythrocyte suspension to a final hemoglobin concentration of 1 g/100 mL. You may determine hemoglobin content by the Drabkin and Austin method [36].
2. Use tubes with screw caps to prepare the reaction mixture consisting in 2 mL of erythrocytes in PBSA +2 mL 10 mM of H_2O_2 solution. In our experiment, 25 μL suspensions of either Fe_2O_3, Fe_3O_4, or SiO_2 nanoparticles were added to this reaction mixture, resulting in a final concentration of 25 μg/mL.
3. Incubate 2 h at 37 °C using a water bath.
4. Precipitate the proteins with 3 mL of 28 % (w/v) TCA solution (*see* **Note 3**).
5. Centrifuge the mixture at 3,000 ×*g* for 10 min at room temperature and collect the supernatant.
6. To achieve a calibration curve, make dilutions of 1,1,3, 3-tetramethoxypropane (10 μM stock solution) to a final volume of 2 mL with 0.1 N HCl in the range of 0–10 μM (*see* Fig. 1).
7. Add 0.5 mL of 1 % TBA solution to 2 mL of supernatant or standard dilutions and incubate at 95 °C for 15 min.
8. Read the absorbance at 532 nm.
9. The results are expressed in nmoles MDA/g Hb using the calibration curve (*see* Fig. 2).

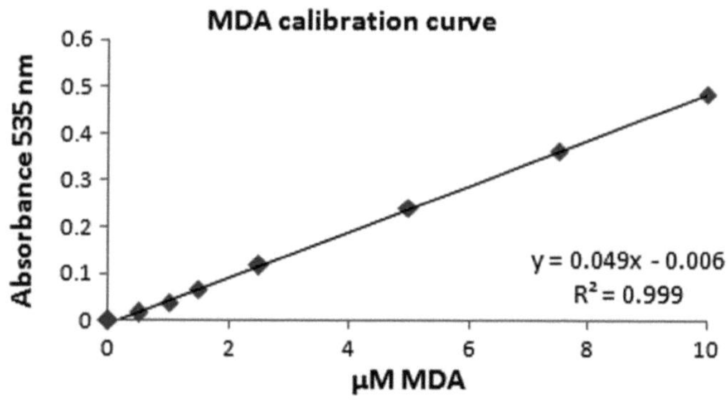

Fig. 1 Typical data obtained when reading the MDA calibration curve as described in the colorimetric method. One should use this data for reference only. This data should not be used to interpret actual assay results

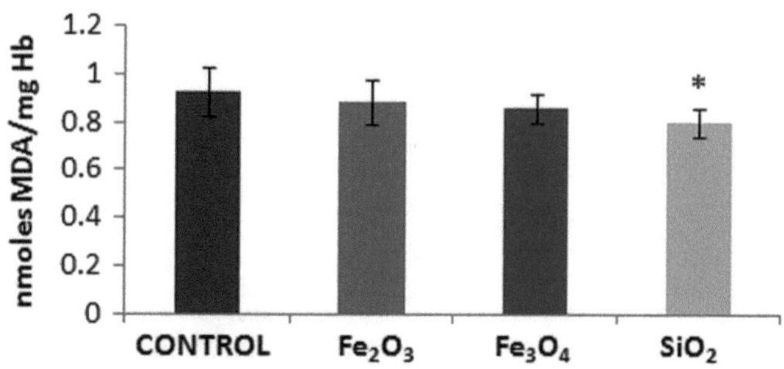

Fig. 2 Lipid peroxide levels expressed as nmoles MDA/mg Hb in isolated human erythrocytes after 2 h of incubation at 37 °C in the presence of 25 μg/mL of Fe_2O_3, Fe_3O_4, SiO_2 nanoparticles

3.2 Fluorimetric Method

Lipid peroxidation induced in hepatic and red muscular tissue of crucian crap (*see* **Note 4**) or in MRC-5 cultured cells (*see* **Note 5**) by the exposure to nanoparticles was expressed as MDA concentration, measured by the fluorimetric method:

1. Measure the protein concentration of the samples by Bradford or Lowry methods [37, 38]. Dilute them in order to equalize the protein concentration (*see* **Note 6**).
2. To achieve a calibration curve, make dilutions of 1,1,3,3-tetramethoxypropane (1 μM stock solution) to a final volume of 200 μL with 0.1 N HCl in the range of 0–1 μM (*see* Fig. 3).
3. Mix 200 μL of each diluted sample (*see* **Note 6**) or standards dilutions with 700 μL 0.1 N HCl.
4. Homogenize and incubate the samples or standards for 20 min at room temperature.
5. Mix each sample or standards with 900 μL of 0.025 M TBA.
6. Homogenize and incubate for 65 min at 37 °C.

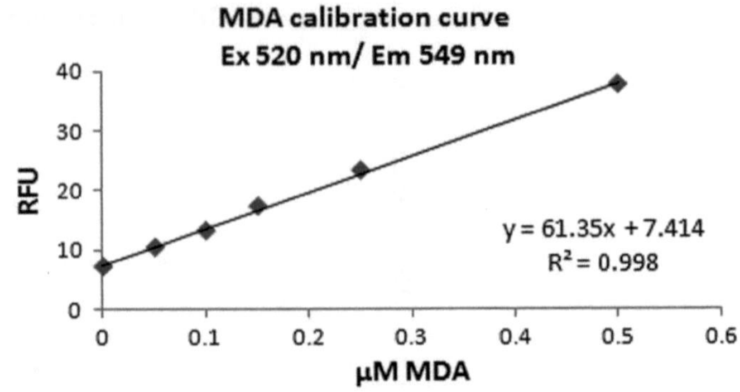

Fig. 3 Typical data obtained when reading the MDA calibration curve as described in the fluorimetric method. One should use this data for reference only. This data should not be used to interpret actual assay results

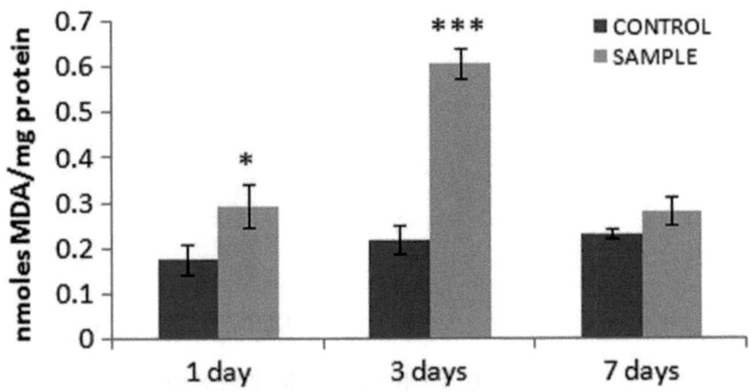

Fig. 4 MDA levels in the liver of individuals of freshwater crucian carp intraperitoneally injected with 2 mg silicon-based quantum dots/kg body weight, compared to controls intraperitoneally injected with the vehicle. Data are calculated as means ± SD ($n = 5$ in each group at each time point); *$P < 0.05$, ***$P < 0.001$

7. Mix each sample with 400 μL of lysis buffer and the standards with 400 μL of BSA solution (*see* **Note 2**).
8. Read the fluorescence at Ex 520 nm/Em 549 nm.
9. Extrapolate the MDA concentration of the samples on the calibration curve and express it as nmoles/mg protein (*see* Figs. 4, 5, and 6).

4 Notes

1. For best results, use only freshly prepared thiobarbituric reagent. To improve the solubility of TBA in water or 0.1 N HCl, it is necessary to heat the solution using a water bath.
2. We observed a direct correlation between protein concentration and the increase of the fluorescence level $Ex_{520\ nm}/Em_{549\ nm}$.

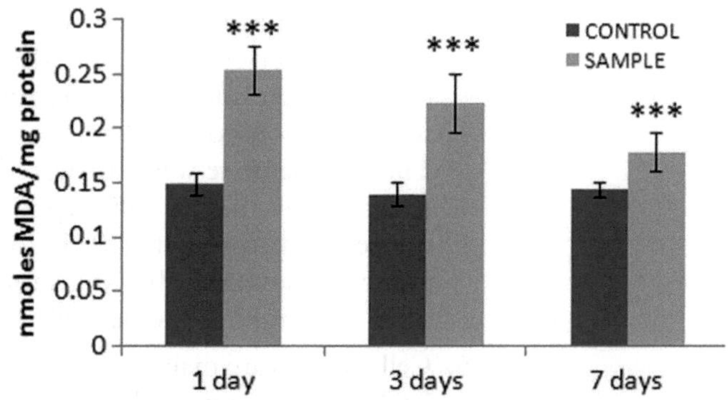

Fig. 5 MDA levels in the red muscle of individuals of freshwater crucian carp intraperitoneally injected with 2 mg silicon-based quantum dots/kg body weight, compared to controls intraperitoneally injected with the vehicle. Data are calculated as means ± SD ($n = 5$ in each group at each time point); $^{*}P < 0.05$, $^{***}P < 0.001$

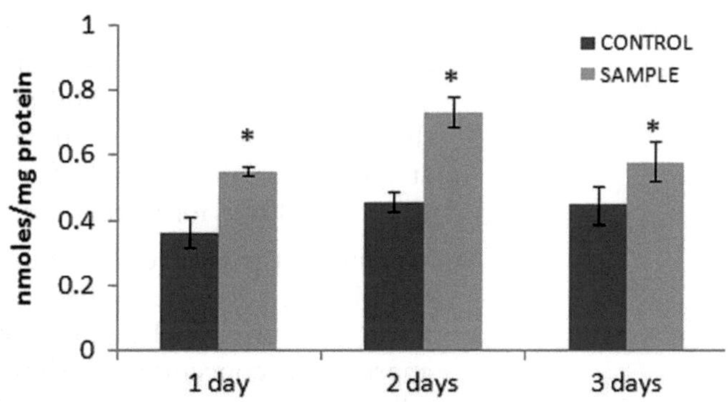

Fig. 6 The variation of MDA levels in MRC-5 cells exposed to (6.3×10^5 nanoparticles per individual cell) SiO_2 nanoparticles for 1, 2, 3 days compared to controls. Data are calculated as means ± SD ($n = 5$ in each group at each time point); $^{*}P < 0.05$

Sample fluorescence is cumulative and is due to MDA–TBA adducts and also due to the proteins presence. To ensure that the fluorescence of the calibration curve is not reduced by the absence of proteins (comparative to the samples), we add a freshly prepared BSA solution in the same volume and with the same concentration as that of the samples.

3. The presence of NaN_3 in PBSA is required for its inhibitory effect on erythrocyte catalase to prevent the decomposition of the H_2O_2.

4. Specimens of *Carasius auratus gibelio* intraperitoneally injected with 2 mg Si/SiO_2 nanoparticles/kg body weight (nanoparticles suspension stock of 20 mg/mL in 7‰ NaCl) were sacrificed by sectioning of the vertebral spine in the cervical region, and tissues were prelevated and stored at −80 °C. Fragments of red muscle and hepatic tissue were weighed, placed in a 1.5 mL tube that contained 0.1 M Tris–HCl, 5 mM EDTA buffer,

pH 7.4 in a ratio of 1:10 (w:v). The tissues are homogenized using the Retsch Mixer Mill MM301 (2 grinding balls Ø 4 mm per tube were added and a frequency of 16 radial vibrations per second was used for 2 min) and immediately placed on ice. The homogenate was transferred into another tube and centrifuged for 30 min at 8,000 rpm at 4 °C. Supernatants were aliquoted and stored at −80 °C for 2 weeks maximum. To avoid tissue degradation, thawing should be reduced as much as possible by weighing the partially frozen samples instead.

5. Cell cultures are maintained in an appropriate medium with 10 % fetal bovine serum. Cells are grown as a monolayer in a humidified 5 % CO_2 air atmosphere at 37 °C in culture flasks. They are seeded at a density of 2.5×10^5 cells/mL. The stock suspensions of SiO_2 nanoparticles (25 mg/mL) are sterilized before use. In each experiment, the stock suspension is sonicated and freshly diluted to appropriate concentrations in the cell medium. Controls without treatment are performed for each experiment and time point. After trypsinization, the cells are washed in PBS and resuspended in 200–500 μl PBS. To obtain the cell lysate, the cells are sonicated three times for 20 s on ice and centrifuged at $5{,}000 \times g$, 10 min at 4 °C. Supernatants are collected and stored at −80 °C for further use.
6. We diluted the samples to about 0.5 mg protein/mL, which we found to be adequate for hepatic and muscular tissue, as well as for the cultured cell lysate. For other types of samples, the appropriate concentration should be determined experimentally. If your samples appear pink or purple at the end of this assay, you should make further dilutions. You should not be able to see any visible coloration after the 65 min incubation.

Acknowledgement

This study was financially supported by the National Research Council of Higher Education, Romania, grant number 127TE/2010.

References

1. Yang Z, Liu ZW, Allaker P, Reip P, Oxford J, Ahmad Z, Ren G (2010) A review of nanoparticle functionality and toxicity on the central nervous system. J R Soc Interface 7:S411–S422
2. Long TC, Tajuba J, Sama P, Saleh N, Swartz C, Parker J, Hester S, Lowry GV, Veronesi B (2007) Nanosize titanium dioxide stimulates reactive oxygen species in brain microglia and damage neurons in vitro. Environ Health Perspect 115:1631–1637
3. Stearns R, Paulauskis JD, Godleski JJ (2001) Endocytosis of ultrafine particles by A549 cells. Am J Respir Cell Mol Biol 24:108–115
4. Műhlfeld C, Rothen-Rutishauser B, Blank F, Vanhecke D, Ochs M, Gehr P (2008) Interactions of nanoparticles with pulmonary structures and cellular responses. AJP Lung Physiol 294:L817–L829
5. Shukla R, Bansal V, Chaudhary M, Basu A, Bhonde RR, Sastry M (2005) Biocompatibility

of gold nanoparticles and their endocytotic fate inside the cellular compartment: a microscopic overview. Langmuir 21:10644–106654

6. Green FHY, Gehr P, Lee MM, Schürch S (2000) The role of surfactant in disease associated with particle exposure. In: Gehr P (ed) Particle-lung interaction. Marcel Dekker, New York, pp 533–576
7. Rothen-Rutishauser B, Schurch S, Gehr P (2007) Interaction of particles with membranes. In: Donaldson K, Borm P (eds) Toxicology of particles. CRC, Boca Raton, FL, pp 139–160
8. Taylor U, Klein S, Petersen S, Kues W, Barcikowski S, Rath D (2010) Nonendosomal cellular uptake of ligand-free, positively charged gold nanoparticles. Cytometry Part A 77A:439–446
9. Miserocchi G, Sancini G, Mantegazza F, Chiappino G (2008) Translocation pathways for inhaled asbestos fibers. Environ Health 7:8 p.
10. Droge W (2002) Free radicals in the physiological control of cell function. Physiol Rev 82:47–95
11. Qian HS, Guo HC, Ho PC, Mahendran R, Zhang Y (2009) Mesoporous-silica-coated up-conversion fluorescent nanoparticles for photodynamic therapy. Small 5:2285–2290
12. Niu H, Zhang D, Zhang S, Zhang X, Meng Z, Cai Y (2011) Humic acid coated Fe3O4 magnetic nanoparticles as highly efficient Fenton-like catalyst for complete mineralization of sulfathiazole. J Hazardous Mat 190:559–565
13. Jiin S-J, Chern C-L, Shen J-Y, Lin T-H (2000) Cadmium-induced liver, heart and spleen lipid peroxidation in rats and protection by selenium. Biol Trace Elem Res 78:219–230
14. Radu M, Munteanu MC, Petrache S, Serban AI, Dinu D, Hermenean A, Sima C, Dinischiotu A (2010) The depletion of intracellular glutathione and increase of lipid peroxidation mediate the hematite nanoparticles-induced cytotoxicity in MRC-5 cells. Acta Biochem Pol 57:355–360
15. Klotz LO, Sies H (2009) Cellular generation of oxidants: relation to oxidative stress. In: Jacob C, Winyard PG (eds) Redox signaling and regulation in biology and medicine. Wiley-VCH, Weinheim, pp 45–61
16. Dhaunsi G, Paintlia MK, Kauer J, Turner R (2004) NADPH oxidase in human lung fibroblasts. J Biomed Sci 11:617–622
17. Bedard K, Krause KH (2007) The NOX family of ROS-generating NADPH oxidases: physiology and pathophysiology. Physiol Rev 87:245–313
18. Chen K, Kirber MT, Xiao H, Yang Y, Keaney JF (2008) Regulation of ROS signal transduction by NADPH oxidase 4 localization. J Cell Biol 181:1129–1139
19. Josephy PD, Mannervik B (2006) Toxicology of oxygen in molecular toxicology. Oxford University Press, London
20. Ceconi C, Boraso A, Cargnoni A, Ferrari R (2003) Oxidative stress in cardiovascular disease: myth or fact? Arch Biochem Biophys 420:217–221
21. Sohal RS, Mockett RJ, Orr WC (2002) Mechanism of aging: an appraisal of the oxidative stress hypothesis. Free Radic Biol Med 33:575–586
22. Gutteridge JMC (1989) Free radicals in disease processes: a compilation of cause and consequence. Free Radic Res Commun 19:141–158
23. Guéraud F, Atalay M, Bresgen N, Cipak A, Eckl PM, Huc L (2010) Chemistry and biochemistry of lipid peroxidation products. Free Radic Res 44:1098–1124
24. Catala A (2010) A synopsis of the process of lipid peroxidation since the discovery of the essential fatty acids. Biochem Biophys Res Commun 399:318–323
25. Niki E (2009) Lipid peroxidation: physiological levels and dual biological effects. Free Radic Biol Med 47:469–484
26. Spiteller G (2010) Is lipid peroxidation of polyunsaturated acids the only source of free radicals that induce aging and age-related diseases? Rejuvenation Res 13:91–103
27. Dmitriev LF, Titov VN (2010) Lipid peroxidation in relation to ageing and the role of endogenous aldehydes in diabetes and other age-related diseases. Ageing Res Rev 9: 200–210
28. Grotto D, Santa ML, Valentini J, Paniz C, Schmitt G, Garcia C, Pombum VJ, Rocha JBT, Farina M (2009) Importance of the lipid peroxidation biomarkers and methodological aspects for malondialdehyde quantification. Quim Nova 32:169–174
29. Hecker M, Ullrich V (1989) On the mechanism of prostacyclin and thromboxane A2biosynthesis. J Biol Chem 264:141–150
30. Esterbauer H, Schaur RJ, Zollner H (1991) Chemistry and biochemistry of 4-hydroxynonenal, malonaldehyde and related aldehydes. Free Rad Biol Med 11:81–128
31. Halliwell B, Lee CY (2010) Using isoprostanes as biomarkers of oxidative stress: some rarely considered issues. Antioxid Redox Signal 13:145–156
32. Møller P, Jacobsen NR, Folkmann JK, Danielsen PH, Mikkelsen L, Hemmingsen JG (2010) Role of oxidative damage in toxicity of particulates. Free Radic Res 44:1–46

33. Del Rio D, Pellegrini N, Colombi B, Bianchi M, Serafini M, Torta F, Tegoni M, Musci M, Brighenti F (2003) Rapid fluorimetric method to detect total plasma malondialdehyde with mild derivatization conditions. Clin Chem 49:690–692
34. Stocks J, Gutteridge JMC, Sharp RJ, Dormandy TL (1974) Assay using brain homogenate for measuring the antioxidant activity of biological fluids. Clin Sci Mol Med 47:215–222
35. Rusu C, Borsa C, Gradinaru D, Ionescu C (1990) The assessment of rat liver, brain and kidney homogenates peroxidation potentials, superoxide dismutase activity and oxidoreduction capacity in reference to age. Rom J Gerontol Geriat 11:125–136
36. Drabkin DL, Austin JH (1935) Spectrophotometric constants for common hemoglobin derivatives in human, dog and rabbit blood. J Biol Chem 112:51–65
37. Bradford MM (1976) A rapid and sensitive method for the quantitation of microgram quantities of protein utilizing the principle of protein-dye binding. Anal Biochem 72: 248–254
38. Lowry OH, Rosenbrough NJ, Farr AL, Randall BJ (1951) Protein measurement with Folin phenol reagent. J Biol Chem 193: 265–275

Chapter 11

Testing Nanomaterial Toxicity in Unicellular Eukaryotic Algae and Fish Cell Lines

Alexandra Kroll, Dana Kühnel, and Kristin Schirmer

Abstract

Nanoecotoxicology as a sub-discipline of ecotoxicology aims to identify and predict effects elicited on ecosystems by nano-sized materials (NM). Two key groups of model organisms in this context are algae and fish. In this chapter, we present considerations for testing NM with respect to their impact on unicellular algae and cell lines derived from various organs of fish.

Based on currently available literature on NM effects in unicellular algae and fish cell lines, and our own experience, we provide guidance on test design, including principle test considerations, materials, NM presentation to cells, exposure, bioavailability, and effect assessment. Assessment needs to be based on a meaningful choice of exposure scenario(s) related to the research question. As a first step, one needs to address whether effects of NMs are to be investigated under environmentally relevant or probable conditions, which may include processes such as agglomeration, or whether NM effects from mono-dispersed particles are of interest, which may require special steps to ensure stable NM suspension. Moreover, whether effects on cells are to be studied in the short- or long-term is important with regard to experimental design. Preparation of NM suspensions, which can be done in aqueous media different from the exposure medium, is addressed with regard to energy input, sterility (as required for algae and fish cell exposure) and particle purity.

Specified for the two model systems, algae and fish cell lines, availability and choice of culture media are presented and discussed with regard to impact on NM behavior. Light, temperature, and agitation, which are variables during exposure, are discussed. We further provide guidance on the characterization of the NM in the chosen aqueous exposure media regarding size, zeta potential and electrophoretic mobility. The state of NM in exposure media is decisive for their bioavailability and therefore for potential particle effects. Therefore, we present ways of deriving a mass balance and quantitative/qualitative information on the uptake and distribution of NM in cells.

As NM have a high surface-to-volume ratio and possess specific physical-chemical properties, which make them prone to interfere with various compounds and certain types of toxicity tests, potential interferences and appropriate controls are introduced. Furthermore, different types of dose metrics, which is still a strongly debated issue in nanotoxicology, are highlighted. We also consider laboratory safety regarding NM handling and disposal.

Key words Ecotoxicology, Algae, Fish cells, Nanomaterial properties, Nanoparticle characterization, Test design, Exposure scenario, Toxicity, Uptake, Accumulation

Donald Armstrong and Dhruba J. Bharali (eds.), *Oxidative Stress and Nanotechnology: Methods and Protocols*, Methods in Molecular Biology, vol. 1028, DOI 10.1007/978-1-62703-475-3_11, © Springer Science+Business Media New York 2013

1 Introduction

Nanoecotoxicology has recently emerged as a subdiscipline of ecotoxicology and aims to identify and predict effects elicited on ecosystems by nano-sized materials. According to a recent definition by the EU, "nanomaterial" (NM) means a natural, incidental, or manufactured material containing particles in an unbound state or as an aggregate or as an agglomerate and where, for 50 % or more of the particles in the number size distribution, one or more external dimensions is in the size range 1–100 nm [1]. In the following, we refer to spherical NM as nanoparticles (NP). Agglomerates of NM are based on weak interactions, whereas chemical bonds between NM are formed in aggregates [2]. To achieve its aims, nanoecotoxicology needs to take into consideration the entry routes and fate of NM in the abiotic and biotic environment to define exposure. It moreover needs to identify those interactions of NM with biota that alter the proper function of cells comprising an organism, thus impacting populations, which in turn can lead to changes in community structure and function. For nanoecotoxicology concerned with the aquatic environment, key species of consideration are algae and fish. Algae are important primary producers in aquatic ecosystems and represent the group of photoautotrophic organisms. In contrast, fish are heterotrophs further up the food chain; with many conserved functions in vertebrates, they are also considered early indicators of impacts on water quality on higher vertebrates, including humans.

In this chapter, we present key considerations for testing NM with respect to their impact on algae and fish using cellular models, specifically, unicellular algae and cell lines derived from various organs of fish. Currently, algae are a paraphyletic group of eukaryotes that conduct photosynthesis in chloroplasts and that differ from plants in, a.o., reproduction and anatomy (e.g., ref. 3). They fall into different groups according to the origin and architecture of their chloroplasts including pigmentation. Cyanobacteria are not included in this definition. Here we focus on unicellular algae as opposed to colonial or filamentous algae. A selection of unicellular algae has become widely used model organisms. Algae form the base of the food web in aquatic ecosystems. In toxicity tests, algal species often prove to be more sensitive than other model organisms [4]. Algae can be isolated from their natural habitat and cultured at standardized conditions in the laboratory [5]. The challenge is to generate cultures of a single species that is not contaminated by bacteria or fungi (axenic). Some algae species cannot be kept axenically as they depend on symbiotic bacteria (e.g., in *Rhopalodia gibba*, a diatom [6]). If the culture conditions are appropriate, algae can be cultured indefinitely. Fish cell lines can be derived from primary cultures of cells, tissues, or organs taken directly from organisms. If a primary culture can be divided into new culture vessels and successfully propagated, it becomes a cell line. A cell line may be propagated a limited number of times, in which case it is finite or, indefinitely, in which case it becomes an immortal or continuous (or permanent) fish cell line [7, 8]. Cell lines from a variety of different species of fish and from different tissues and organs have been established in this way. In contrast to mammalian cells, many fish cell lines arise spontaneously without any specific immortalization treatment, possibly owing to a high telomerase activity [7]. A wide range of both unicellular algae and fish cell lines are commercially available through dedicated repositories, such as the Culture Collection of Algae and Protozoa (CCAP [9, 10]) and the American Type Culture Collection for fish cell lines (ATCC, e.g., refs. [7, 11]), or they can be obtained from other research laboratories.

Owing to their small size and ease to produce them in rather large numbers, unicellular algae and fish cells are very useful to study a variety of NM in small-scale assays on integrative parameters, such as cell survival and reproduction, but as well on mechanisms of NM uptake and toxicity. Small scale refers to small flasks, holding a few milliliters of medium, down to multi-well plates with micro- to nanoliter volumes of exposure medium, making these tests amenable not only to high throughput screening but also to reduced material resources and, consequently, reduced waste. However, the small scale also bears problems, such as the high surface-to-volume ratio of test containers or little material for characterizing NM as they are present in the exposure chambers.

A summary of currently available publications on NM effect studies using unicellular algae and fish cell lines is presented in Tables 2 and 3. Based on these studies, and taking our own experience into consideration, we would like to provide guidance on test design in terms of principle test considerations, NM presentation to cells, exposure, bioavailability, and effect assessment.

2 Materials

Materials specifically recommended for working with NM in algae or fish cell culture environments are as follows:

2.1 Equipment

1. High-precision, antistatic scale to weigh NM provided as powder.
2. Ultrasonic bath or sonication cup to fit on a sonication tip for indirect sonication of NM dispersions (direct sonication may add impurities to your sample!).

Table 1
Possible interferences of NM with cell-based toxicity tests and adequate controls

Interference	Control
Quenching of fluorescence emitted from test reaction product	Measure known concentrations of NM and reaction product in combination
Quenching of fluorescence emitted from organism (e.g., algal autofluorescence)	Difficult to control as change in fluorescence is the end point; wash organisms to ± remove NM
Light absorption at the wavelength used in a fluorescence or light absorbance measurement	Measure light absorption spectrum of NM at known concentrations
Autofluorescence (in the respective wavelength required for detection in the assay)	Measure fluorescence of NM suspension
Conversion of assay substrates, e.g., by catalysis or ROS formation	Incubate NM with assays substrates and measure end point taking into account possible optical interference and sorption (see above)
Binding of assay compounds	Difficult to control (separate them from cells and prevent necessary interactions)
Interference with an enzyme (inhibition, enhancement) and/or adsorption of assay substrates or products	Incubate known concentrations of enzyme, substrate, and NM and measure end point, taking into account possible optical interference and sorption (see above)
Interaction with RNA/DNA, PCR mixes	Treat known concentrations of RNA/DNA with NM dispersions and perform PCRs in the presence of NM to check for interference/destruction
Shading	Use external shading
pH is influenced by NM–e.g., change in fluorescence of fluorophores	Check pH in the presence of NM
Nutrient or growth factor depuration (NM adsorb compounds from the media)	Analyze compounds associated with NM

3. Stir plates with exact (digital) displays for reproducible stirring of NM dispersions.
4. Room or incubator with controlled light and temperature conditions for the generation of dispersions and exposure of test organisms.

2.2 Reagents and Supplies

Please refer to specific literature on NM characterization (e.g., ref. 12) and on algae (e.g., ref. 5) and fish cell culture (e.g., ref. 13, 14), as well as the small-scale assays cited in Tables 1 and 2, for details on specific infrastructure, instruments, consumables, and chemicals. Be sure to protect yourself and your colleagues from exposure to NM (*see* **Note 1**).

Table 2
Reports on toxicity tests of NM in algae

Type of NM	Test organism	Division	NM dispersions	NM characterization	NM concentration (mg/L)	Temperature (°C)	Light ($\mu Em^2/s$)	Light–dark cycle	Agitation	Exposure time (h)	End points 1 OECD 201 2 ISO 8692 3 DIN38412-33 4 US EPA	Effects expressed per…	Reference
Ag NP	*C. reinhardtii*	Chl		1, 2, 3, 4	10–10^5 nM	25	267	x	x	1–5	A	Total mass, [Ag^+]	[37]
Ag NP, 0.3 % PVP	*T. weissflogii*	Het	Fulvic acid, 1 week in Teflon, filtration	1, 5, 6	2.12×10^{-10} to 1.03×10^{-7}	20	170	x		48	A, B, C, E, EPS	[Ag^+]	[54]
Al_2O_3 NP	*P. subcapitata*	Chl	10 % (w/w) in H_2O, dilution in medium	7, 5	100	20	100		x	4.5	A	Total mass	[20]
Au NP	*S. subspicatus*	Chl			NPs/cell			x	x	24	B, D, F	Total mass	[55]
C60	*P. subcapitata*	Chl	Stirring (stock, 60 d; dilutions, 5 d)	1, 8	6–10	20	80–90				B, ISO 8692	Total mass	[56]
C60	*P. subcapitata*	Chl	Stirring in H_2O, dilution in medium	7, 5, 8	100	20	100		x	4.5	A	Total mass	[20]
CdTe QD	*P. tricornutum*	Het	Sonication, resuspension	2, 4, 8	0.1–0.3	20	3,000 Lux		x	24–144	B, E	Total mass	[48]
CeO_2 NP	*Anabaena* CPB4337	Cy	Sonication, dilution, stirring	1, 2, 4, 6, 9	0–100	28	65			72+96	D, F, G, ATP	Total mass	[57]
CeO_2 NP	*P. subcapitata*	Chl	Sonication, dilution, stirring	1, 2, 4, 6, 9	0–100	22	100			72+96	B	Total mass	[57]

(continued)

Table 2
(continued)

Type of NM	Test organism	Division	NM dispersions	NM characteri-zation	NM concentration (mg/L)	Temperature (°C)	Light (μEm²/s)	Light–dark cycle	Agitation	Exposure time (h)	End points 1 OECD 201 2 ISO 8692 3 DIN38412-33 4 US EPA	Effects expressed per…	Reference
CeO_2 NP	*P. subcapitata*	Chl	Sonication, shaking, filtration	1, 3, 4, 5, 10[a]	0–100	24	70			72	B, membrane	Total mass	[40]
CeO_2 NP	*P. subcapitata*	Chl		1, 2, 11	0–32	25	70			72	B, 1	Total mass, BET	[58]
CeO_2 NP	*P. subcapitata*	Chl	10 % (w/w) in H_2O, dilution in medium	7, 5	100	20	100		x	4.5	A	Total mass	[20]
$CuZnFe_4O_4$ NP	*P. subcapitata*	Chl	Rotation, filtration		0.1–100						B	Total mass	[59]
CuO NP	*P. subcapitata*	Chl	Sonication		0 to ~6.5	24					B, F, 1	Total mass	[60]
CuO-PS NP	*C. reinhardtii*	Chl		1, 4	0–20	23	200			6	A, F, G, ROS	Total mass	[61]
Er_2O_3 NP	*P. subcapitata*	Chl	Rotation, filtration		0.1–100						B	Total mass	[59]
g-Al_2O_3 NP	*P. subcapitata*	Chl	Sonication	2, 3, 6	10–200					24	B	Total mass	[62]
Ho_2O_3 NP	*P. subcapitata*	Chl	Rotation, filtration		0.1–100						B	Total mass	[59]

In_2O_3 NP	*P. subcapitata*	Chl	Rotation, filtration		0.1–100						B	Total mass	[59]
MWCNT	*D. tertiolecta*	Chl	Stock: filtered, washed, autoclaved		0–10	19	120	x		24	A, B, D, GSH	Total mass	[63]
MWCNT	*P. subcapitata*	Chl	Sonication in presence of NOM	2	0.1–100	24	80 ± 5	x		96	A, B, 1	Total mass	[43]
MWCNT	*C. vulgaris*	Chl	Sonication in presence of NOM	2	0.1–100	24	80 ± 5	x		96	A, B, 1	Total mass	[43]
$Ni_{0.5}Zn_{0.5}Fe_2O_4$ NP	*P. subcapitata*	Chl	Rotation, filtration		0.1–100						B	Total mass	[59]
NIPAM/BAM NP	*P. subcapitata*	Chl	Dispersed on ice after freeze-drying	1, 2, 3, 6	25–1,000	20	10^4 lx	x		72	B, 1	Total mass	[64]
PMMA NP	*P. subcapitata*	Chl	10 % (w/w) in H_2O, dilution in medium	7, 5	100	20	100		x	4.5	A	Total mass	[20]
PNIPAM NP	*P. subcapitata*	Chl	Dispersed on ice after freeze-drying	1, 2, 3, 6	25–1,000	20	10^4 lx	x		72	B, 1	Total mass	[64]
QD	*C. reinhardtii*	Chl			0–100	25	55		x	24 + 120	B, C, ox. stress	Total mass	[65]
QD	*P. subcapitata*	Chl			0–50 ppbm					96	B, F, G, 4	Total mass	[21]
QD	*Chlamydomonas sp.*	Chl		1, 2	0.05–5 ppm	RT				2	D, G, [CO_2]	Total mass	[22]

(continued)

Table 2 (continued)

Type of NM	Test organism	Division	NM dispersions	NM characterization	NM concentration (mg/L)	Temperature (°C)	Light (μEm²/s)	Light–dark cycle	Agitation	Exposure time (h)	End points 1 OECD 201 2 ISO 8692 3 DIN38412-33 4 US EPA	Effects expressed per…	Reference
Sm_2O_3 NP	*P. subcapitata*	Chl	Rotation, filtration		0.1–100						B	Total mass	[59]
SiO_2 NP	*C. kessleri*	Chl			0–5 % w/v	25	3.4 W/m²	x	x	120	B, G, F	Total mass	[66]
SiO_2 NP	*P. subcapitata*	Chl	Stirring	1, 2, 4	0–460	25	70			72	B, D, 1	Total mass	[67]
$SrFe_{12}O_{19}$ NP	*P. subcapitata*	Chl	Rotation, filtration		0.1–100						B	Total mass	[59]
SWCNT	*P. subcapitata*	Chl	Stirring in H_2O, dilution in medium	7, 5	100	20	100		x	4.5	A	[C]	[20]
TiO_2 NP	*A. variabilis*	Cy	Sonication	2, 6	0–150	20				24	D, F, ROS	Total mass	[68]
TiO_2 NP	*D. subspicatus*	Chl	Sonication, ± light		12.5–50	20				72	G, 1, 2, 3	Total mass	[42]
TiO_2 NP	*D. tertiolecta*	Chl	Sonication, vortexing, dilution	2, 8	0–1	15	100–200	x		96	B	Total mass	[69]
TiO_2 NP	*I. galbana*	Chl	Sonication, vortexing, dilution	2, 8	0–1	15	100–200	x		96	B	Total mass	[69]
TiO_2 NP	*P. subcapitata*	Chl	Sonication	2, 3, 6	10–200					24	B	Total mass	[62]
TiO_2 NP	*P. subcapitata*	Chl	Sonication		0–120	24					B, F, 1	Total mass	[60]

TiO_2 NP	*P. subcapitata*	Chl	Stirring stock for at least 30 min	12		25	50–100	x		96	B, 4	Total mass	[70]
TiO_2 NP	*P. subcapitata*	Chl	Stirring, sonication, varying conditions	2, 3, 14	3.7–100	22	95		x		B, 1	Total mass	[27]
TiO_2 NP	*P. subcapitata*	Chl	10 % (w/w) in H_2O, dilution in medium	7, 5, 14	100	20	100		x	4.5	A	Total mass	[20]
TiO_2 NP	*P. subcapitata*	Chl	Rotation, filtration		0.1–100						B	Total mass	[59]
TiO_2 NP	*S. marinoi*	Het	Sonication, vortexing, dilution	2, 8	0–1	15	100–200	x		96	B	Total mass	[69]
TiO_2 NP	*T. pseudonana*	Het	Sonication, vortexing, dilution	2, 8	0–1	15	100–200	x		96	B	Total mass	[69]
TiO_2 (P25) NP	*A. carterae*	Dino	Directly mixed with algal culture	12, 13, 6, 8, 1	500	25	UV–Vis			72	B	Total mass	[71]
TiO_2 (P25) NP	*C. reinhardtii*	Chl		2, 14, 15	0–100	25	55		x	24 + 120	B, C, ox. stress	Total mass	[65]
TiO_2 (P25) NP	*P. subcapitata*	Chl		2, 6, 12, 15	0–100	23.8	9,500 lx		x	72	B, 1	Total mass	[72]
TiO_2 (P25) NP	*T. suecica*	Chl	Directly mixed with algal culture	1, 6, 8, 12, 13	500	25	UV–Vis			72	B	Total mass	[71]

(continued)

Table 2
(continued)

Type of NM	Test organism	Division	NM dispersions	NM characteri-zation	NM concentration (mg/L)	Temperature (°C)	Light (μEm²/s)	Light–dark cycle	Agitation	Exposure time (h)	End points 1 OECD 201 2 ISO 8692 3 DIN38412-33 4 US EPA	Effects expressed per…	Reference
TiO_2-Ag NP	*A. carterae*	Dino	Directly mixed with algal culture	1, 6, 8, 12, 13	500	25	UV–Vis			72	B	Total mass	[71]
TiO_2-Ag NP	*T. suecica*	Chl	Directly mixed with algal culture	1, 6, 8, 12, 13	500	25	UV–Vis			72	B	Total mass	[71]
$Y_3Fe_5O_{12}$ NP	*P. subcapitata*	Chl	Rotation, filtration		0.1–100						B	Total mass	[59]
ZnO NP	*C. gracilis*	Het	Sonication	1, 2, 4, 12	0–80	17	120	x			B	Total mass	[47]
ZnO NP	*D. tertiolecta*	Chl	Sonication, vortexing, dilution	1, 4, 8	0–1	15	100–200	x		96	B	Total mass	[69]
ZnO NP	*I. galbana*	Chl	Sonication, vortexing, dilution	1, 4, 8	0–1	15	100–200	x		96	B	Total mass	[69]
ZnO NP	*P. tricornutum*	Het	Sonication	1, 2, 4, 12	0–80	17	120	x			B	Total mass	[47]
ZnO NP	*P. subcapitata*	Chl	Sonication	–	0–0.5	24					B, F, 1	Total mass	[60]
ZnO NP	*P. subcapitata*	Chl	± Sonication, ± filtration	1, 2, 4	0.025–0.6					72	B	Total mass $[Zn^{2+}]$	[73]
ZnO NP	*S. marinoi*	Het	Sonication, vortexing, dilution	1, 4, 8	0–1	15	100–200	x		96	B	Total mass	[69]

ZnO NP	*T. pseudonana*	Het	Filtration	1, 2, 4	9.50×10^{-9} to 1.52×10^{-7} M Zn^{2+}	20	50	x			B, A, C	[Zn^{2+}]	[74]
ZnO NP	*T. pseudonana*	Het	Sonication, vortexing, dilution	1, 4, 8	0–1	15	100–200	x		96	B	Total mass	[69]
ZnO NP	*T. pseudonana*	Het	Sonication	1, 2, 4, 12	0–80	17	120	x			B, D, E	Total mass	[47]
ZrO_2 NP	*P. subcapitata*	Chl	10 % (w/w) in H_2O, dilution in medium	7, 5	100	20	100		x	4.5	A	Total mass	[20]

Chl (*Chlorophyceae*), Cy (*Cyanobacteria*), Dino (*Dinoflagellatae*), Het (*Heterokontophytae*)
A Photosystem II quantum yield
B Growth rate/inhibition/cell number
C Chlorophyll a content
D NM localization
E Bioaccumulation
F Morphology
G Fluorescence/luminescence
1 TEM
2 DLS
3 Electrophoretic mobility
4 Solubility/ICP-MS
5 EDS
6 BET
7 AFM
8 Spectrophotometry
9 TXRF
10 PEELS
11 XANES
12 XRD
13 XPS
14 Concentration
15 Visual observation
Empty cells not specified
[a]ROS formation, peroxidation of linoleic acid

3 Methods

This chapter guides through important elements in designing experiments to assess the toxicity of NMs in unicellular algae and fish cell lines. Inasmuch as the dominant route of exposure to NMs toward these organisms in the aquatic environment is via NM dispersions (aside from uptake in fish through food), we will focus this chapter on exposure via liquid media and NMs dispersed into it. A general design scheme is presented in Fig. 1. This scheme assumes a prior selection of cells. This selection needs to be decided upon according to the question at hand, availability of desired cell types, and other potential considerations as briefly summarized below.

3.1 Considerations for Selection of Cells

3.1.1 Unicellular Eukaryotic Algae

The three NM-specific considerations are the algal cell wall, algal class, and habitat. Algal cell walls are composed of silica in diatoms and of polysaccharides and/or glycoproteins in other groups. Consequently algal cell walls have different chemical and physical properties that will possibly influence the availability of NM at the cell membrane. The silica-based cell wall or frustule of diatoms has pores and slits of defined sizes which filter material from the surrounding medium. A study on two diatom species revealed a lower size limit of 40 nm in frustule pores, possibly limiting the diffusion of larger particles to the cell membrane [15]. Other classes may have additional routes of uptake for NM, such as dinoflagellates

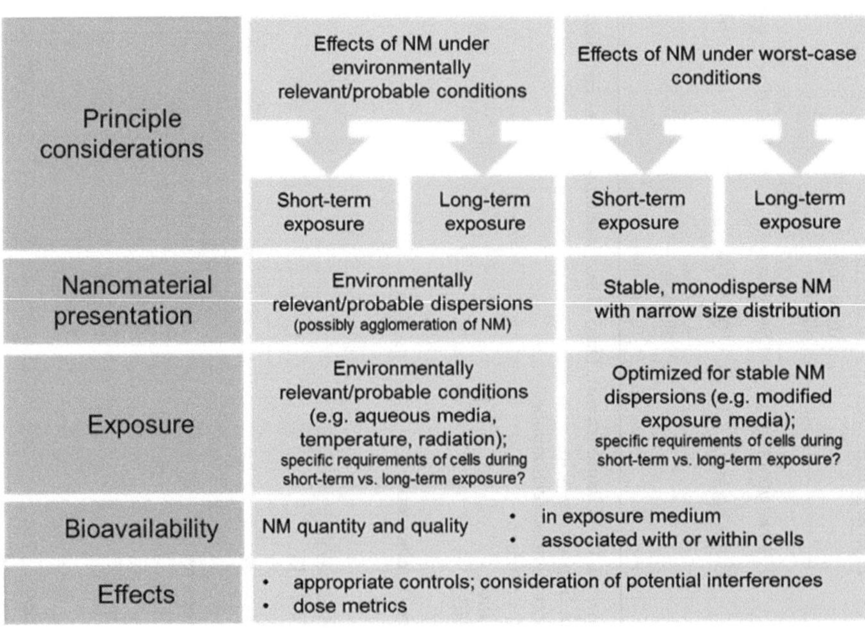

Fig. 1 Scheme for the design of toxicity tests of nanomaterials

that can ingest food particles (e.g., ref. 16). The original habitat and thus the required medium will influence the fate of NM during exposure. Marine algae, for example, require high salt medium as opposed to freshwater algae. The ionic strength strongly influences the stability of NM dispersions.

3.1.2 Fish Cell Lines

Fish cell lines are not as diverse in structure as algae cells, but the species and its associated habitat and the organ or tissue of origin are likewise important considerations. Inasmuch as fish cell lines generally thrive in temperatures closely mimicking the environment of their donor, cells from warm and cold water fish have very different temperature requirements. For example, cell lines from rainbow trout (*Oncorhynchus mykiss*) can be cultured between 4 and 25 °C, whereas cell lines from warm water fish, such as from zebra fish (*Danio rerio*), grow at temperatures also above 25 °C (23–34 °C). The organ or tissue of origin is important for consideration of specific target sites. For example, if one wants to mimic the exposure scenario at the gill, exposure media closely mimicking the fresh- or saltwater environment may be applied [11]. Moreover, in analogy to dinoflagellates, macrophage cells have additional routes of NM uptake, i.e., phagocytic routes, not of such relevance to other cell types.

3.2 Principle Considerations for Test Design

The assessment of NM toxicity needs to be based on a meaningful choice of exposure scenario(s) related to the research questions. The two possible "extreme" cases are sketched in Fig. 1. If the research focus is on the effects of NM under environmentally relevant or probable conditions, the use of natural water chemistry and low NM concentrations is implied. Due to the lack of measured data, different authors have estimated or modeled NM concentrations (such as TiO_2-, CeO_2-, Ag-based nanoparticles (NP)) which average in the ng/L range and below in surface waters, meaning several orders of magnitude below most reported test concentrations (Tables 2 and 3) [17–19]. To investigate NM effects in a "worst-case" exposure scenario, establishing a dose–response curve is necessary. However, the use of high NM concentrations may be limited by interference with toxicity assays (*see* Subheading 3.6.1), whereas low concentrations of NM are difficult to quantify and characterize (*see* Subheading 3.5).

Furthermore, the exposure duration needs to be considered (Fig. 1). In the literature, exposure is typically grouped in short-term (hours to days) and long-term (days to weeks) approaches. As the life cycle of NM in the environment has not been studied yet, we do not have any data to base our selection on. From other anthropogenic inputs into surface waters, we know that there may be pulsed and long-term/chronic exposure. Both may thus apply to NM, too. Reported exposure times in nanotoxicity studies range from 4.5 h [20] to 5 d [21] in algae and from 3 to 72 h [22]

Table 3
Reports on toxicity tests of NM in fish cell lines

Type of NM	Species	Cell type/cell line	Exposure condition/ scenario	NM characterization	NM concentration (mg/l)	Exposure time (h)	Temperature (°C)	Assay	End point	Effects expressed per…	Reference
Ag NP	Medaka (*Oryzias latipes*)	OLHNI2, adult fin tissue	DMEM w 20 % FBS	1,2,3,7,9,11	0.05–5 μg/ cm^2			A3.3, C2	A3, C	Total mass	[75]
Ag NP	Rainbow trout (*Oncorhynchus mykiss*)	Gill cells, RTgill-W1	L-15 with 10 % FBS	1,6		40, 48	37 (?)	A3.2, A3.4.	A, A3	Total mass	[76]
Ag nano-composite	Rainbow trout (*Oncorhynchus mykiss*)	Gill epithelial cells, RT-W1	Dispersion in distilled water, L-15 + 10 % FBS, alginic acid	1,2,4,7,10,11	0–25	6, 12, 24	24	A2.4, A3.1, B1.2	A3, B	Total mass	[77]
Au NP	Rainbow trout (*Oncorhynchus mykiss*)	Gill cells, RTgill-W1	L-15 with 10 % FBS	1,6		40, 48	37 (?)	A3.2, A3.4	A, A3	Total mass	[76]
Au NP polymer coated	Rainbow trout (*Oncorhynchus mykiss*)	Gill cells, RTgill-W1	L-15 with 10 % FBS	1,2,8,11	1–100	24	20	A1.3	A1, D	Total mass	[78]
CdO NP	Rainbow trout (*Oncorhynchus mykiss*)	Gill cells, RTgill-W1	L-15 with 10 % FBS	–		40, 48	37 (?)	A3.2, A3.4	A, A3	Total mass	[76]
CNT	Rainbow trout (*Oncorhynchus mykiss*)	Gill cells, RTgill-W1	L-15 with 10 % FBS	1,6		40, 48	37 (?)	A3.2, A3.4	A, A3	Total mass	[76]

NIPAM-NIPAM/BAM copolymers	Rainbow trout (*Oncorhynchus mykiss*)	Gonad cells (RTG-2)	Dispersed in Milli-Q, DMEM with 5 % FBS	2,3,5	25–1,000			A1.2, A2.2	A1, A2	Total mass [79]
TiO_2 NP, anatase	Goldfish (*Carassius auratus*)	Skin cells (GFSk-S1)	Suspended in sterilized PBS, final dilution in growth media, w/o UVA	–	0.1–1,000	24		A2.2, B2.1, C1	A2, B, C	Total mass [80]
TiO_2 (P25) NP	Rainbow trout (*Oncorhynchus mykiss*)	Gonad cells (RTG-2), passage 13–25	Suspension in dist. H_2O, sonication 12 h, exposure in PBS vs. MEM w/o UVA	1	0–50			A2.2, C1, C3	A2, C	Total mass [81]
WC NP	Rainbow trout (*Oncorhynchus mykiss*)	Gill cells, RTgill-W1	Dispersion in distilled water, L-15 + 10 % FBS, L-15, L-15ex	1,2,3,4,5	0–30	3, 72	19	A1.2, A2.2, A2.3, D1	A1, A2, D	Total mass [22]
WC-Co NP	Rainbow trout (*Oncorhynchus mykiss*)	Gill cells, RTgill-W1	Dispersion in distilled water, L-15 + 10 % FBS, L-15, L-15ex	1,2,3,4,5	0–33	3, 72	19	A1.2, A2.2, A2.3, D1	A1, A2, D	Total mass [22]

(continued)

Table 3
(continued)

Type of NM	Species	Cell type/cell line	Exposure condition/ scenario	NM characterization	NM concentration (mg/l)	Exposure time (h)	Temperature (°C)	Assay	End point	Effects expressed per...	Reference
ZnO NP	Topminnow fish (*Poeciliopsis lucida*)	Hepatocellular carcinoma cell line, PLHC-1	Serum-free medium, 1 min vortexing, a-MEM with 5 % FBS	1,2,7	0.00078–0.1	24	30	A1.1, A2.1, A2.2, B1.1	A1, A2, B1	Total mass	[82]

End points
A Cytotoxicity/viability
A1 Metabolic activity
A2 Membrane integrity
A3 Cell number/protein content
B ROS
C Genotoxicity
D NM uptake and localization
Assays/tests
A1.1 MTT
A1.2 alamarBlue
A1.3 Resazurin
A2.1 LDH
A2.2 Neutral red
A2.3 CFDA, AM, fluorescein retention
A2.4 Propidium iodide
A3.1 Hoechst 33324
A3.2 Protein content (sulforhodamine B, Bradford)
A3.3 Colony-forming assay

A3.4 Impedance and resistance measurements
B1.1 DCF
B1.2 MitoSOX red
B2.1 Electron spin resonance
C1 COMET
C2 Chromosome aberration
C3 Micronucleus assay
D1 Microscopy
Characterization
1 TEM
2 DLS
3 Electrophoretic mobility/zeta potential
4 EDS
5 BET
6 AFM
7 ICP-MS
8 Spectrophotometry
9 UV–Vis spectroscopy
10 XRD
11 Concentration
Empty cells not specified

in fish cell lines. Recovery from NM-induced adverse effects was shown in cell culture and animal tests [23], but has not yet been investigated in ecotoxicological tests. Strikingly, many studies in algae are based on 72 h exposures, often conducted according to the standard procedures OECD 201 [24], ISO 8692 [25], or DIN 38412-33 [26] (Table 2). Using these standard protocols may imply an easy comparison to data on the toxicity of chemicals. However, these protocols may not be suitable to answer individual research questions on NM toxicity. We further discuss this point in Subheadings 3.4.1 and 3.6.1.

3.3 Nanomaterial Presentation

3.3.1 Generation of NM Dispersions

NMs are available as powders or dispersions depending on the production process and posttreatment. Both types need to be dispersed in liquid media for exposure. As described in Fig. 1, dispersions should be generated according to the research approach. The dispersion method can greatly influence the outcome of NM toxicity tests and should thus be carefully chosen [27]. Relevant parameters are dispersion medium chemistry, pH, energy input, time, temperature, and light (*see* **Notes 4–7, 9–12**).

Stable, monodisperse NM suspensions with a narrow size distribution are needed to understand the specific effects of a defined size range of the NM under controlled conditions. Dispersion parameters need to be adapted to the respective NM. Depending on the NM surface properties (charged/uncharged, hydrophilic/hydrophobic), which are influenced by surface modifications and the exposure medium (*see* Subheading 3.4.1), more or less energy input during the dispersion process is required. Dispersion methods ranging from low energy input, like slow stirring or shaking, to high energy input, like vigorous stirring, shaking, or sonication, have been used (*see* Tables 2 and 3, *see* **Note 6**). The use of surfactants and temperature and/or pH different from exposure conditions may be required to achieve monodispersed NMs. Filtration or centrifugation may help to narrow the size distribution. Stock dispersions may be generated in a medium different from the exposure medium to ensure optimal dispersion (e.g., ref. 27).

Environmentally relevant or probable conditions may be contrary to the conditions needed to generate stable dispersions. Agglomeration (or possibly aggregation) of the NM may be the consequence (see, e.g., dispersion of Ag NPs in river water [28]). To design environmentally probable conditions, one should determine a relevant type of site in the environment and adapt the dispersion parameters mentioned above accordingly.

3.3.2 Sterility

Contamination of the NMs by microorganisms or their toxins may affect the test organisms and thus confound toxicity test results [15]. Verifying the sterility of NM for a nonspecialized laboratory is challenging. A simple but superficial approach is to test NM for growth of nonspecialized bacteria in standard media (e.g., lysogeny broth (LB)). If available, NM can be tested aerobically and anaerobically for sterility, applying fluid and solid media under long-term cultivation [15, 23]. The presence of common bacterial toxins (e.g., endotoxins) can be ruled out using available standard kits. Be sure to exclude interferences of the NM with the assay (*see* Subheading 3.6.1). Sterility of NM powders or suspensions can be achieved by autoclaving [22, 29], γ-irradiation [30], or possibly filtration. However, one has to verify that these processes do not change NM properties or concentration. We have previously used X-ray photoelectron spectroscopy (XPS) to verify that γ-irradiation did not change the surface properties of diverse NM [30].

3.3.3 Purity

Impurities originating from the production process (or side products formed during preparation of suspensions, e.g., by solvents) may result in measurable toxicity that is independent of actual NM effects and may confound the results [31]. The chemical purity of the NM should thus be determined. If impurities exist, washing steps (if possible) or appropriate controls in the toxicity tests (e.g., using the detected contaminants alone) are recommended.

3.4 Exposure

Depending on the research approach (Fig. 1), exposure conditions need to be adapted either to correspond as closely as possible to environmental conditions or to guarantee maintenance of stable, monodisperse NM suspensions with a narrow size distribution. The following sections describe the influence of different parameters on NM dispersions. They serve as basis to design the exposure conditions according to the desired scenario.

3.4.1 Exposure Media

Both algae and fish cells need growth media containing essential nutrients in order to ensure optimal growth and fitness. The constituents of these media will influence NM properties and thus their effects on the test organisms. If dispersion properties are to be controlled, e.g., to generate a stable monodisperse suspension or to minimize dissolution of metallic NM or loss of surface modifiers, the medium may have to be adapted. We recommend to model the speciation of media components including NM (e.g., with Visual MINTEQ, available from KTH, Department of Land and Water Resources Engineering). This helps to understand the fate of NM in the medium and allows optimizing the medium composition if necessary.

Typical liquid minimal algae media contain essential metals (sodium, potassium, calcium, magnesium), chloride, a source of nitrogen (nitrate or ammonium) and inorganic carbon (carbonate), trace metals (copper, zinc, cobalt, manganese, molybdenum, iron, possibly boron), and phosphate. Some algae require an organic carbon source (e.g., citrate), vitamins, or silicon (diatoms). A pH buffer may be necessary and a chelator (e.g., EDTA) to make the metals more bioavailable. Generally, bivalent cations such as calcium and magnesium tend to increase the agglomeration of NM; phosphate and carbonate may stabilize NM depending on their surface charge and modification. Ionic strength, pH, and organic compounds will also influence the stability of NM dispersions (TiO_2 [30, 32], CeO_2 NPs [30, 33], polyethylene glycol (PEG)-coated quantum dots (QDs) (CdSe/ZnS) [34], Ag [28]). Media properties may also change the solubility of NM and possible contaminants which in turn may influence the biological effects. The coating of, e.g., Ag NPs influences their dissolution in different media [35]. Metal-binding compounds such as EDTA or citrate may change the bioavailability of metal ions dissolved from NM. NaCl has been shown to enhance dissolution of Ag NPs [36]. Thus, a medium without chloride has been used to expose *C. reinhardtii* to Ag NPs [28, 37].

As mentioned above, many studies on NM effects on algae have been conducted according to standard test guidelines. This implies the use of standardized culture media. As described in this section, certain components found in standard media influence the dispersibility of NMs or may affect dissolution of NMs. The application of standard protocols should thus be carefully evaluated with respect to the research questions.

For the cultivation of fish cells, different commercially available standard media are in use, e.g., Leibovitz-15 (L-15), minimal essential medium (MEM), or medium 199 (M199). These media contain amino acids, vitamins, inorganic salts, and sugars essential for the growth of heterotrophic cells and differ only slightly in their formulations. Other than mammalian cells, fish cells are usually maintained in CO_2-free systems, hence the media does not contain sodium bicarbonate as a buffer, but is buffered by the use of Hank's salts, which buffer at atmospheric CO_2 concentrations. Additionally, a higher amino acid content, and complement of galactose is used (e.g., L-15 medium). To allow cell growth and division, the addition of serum is generally needed and the most frequently used is fetal bovine serum (FBS). Serum contains essential growth factors and proteins (e.g., albumin) but may influence NM behavior and hence the toxicity assessment. For example, we and others have found the addition of serum to stabilize NP in the media, hence preventing NP agglomeration [22, 38]. For short-term assays not requiring cell division, serum is not essential and even very simple buffers may be applicable, such as L-15ex [22, 39]. The absence or presence of serum leads to very different particle states and may modulate the particle toxicity [22]. In media without serum, NP may agglomerate due to the high salt content. Depending on tissue of origin, fish cells are capable to grow under hypo- and hyperosmotic conditions (e.g., ref. 11); hence, media with different salt contents can be designed in order to control NM states. The addition of antibiotics to prevent growth of bacteria in the media is common. To date, there are no reports on interference of antibiotics with NM; however, during exposures, the use of antibiotics is generally not needed and thus should be avoided.

3.4.2 Choice of Light, Temperature, and Agitation

Photoactive NMs such as TiO_2- or CeO_2-based NP generate reactive oxygen species (ROS) in the presence of UV and even visible light [40]. They may thus change exposure conditions by catalyzing the oxidation of media components (indirect damage due to change of nutrients). The light-induced generation of ROS may also result into phototoxicity of NM [41]. Photoreduction/oxidation of NM or ions in the medium may change NM dissolution, surface properties, and stability. The choice of light/dark conditions during exposure is therefore an important factor. Light conditions may be influenced locally due to NM shading the cells. This is especially relevant in studies based on photosynthetic organism [42, 43] for carbon nanotube-exposed algae, although only extremely high exposure concentrations resulted in this effect.

In previous studies, algae have been exposed under continuous light and light–dark cycles (Table 2). Continuous light produces a more or less homogenous culture and constant conditions, whereas light–dark cycles generate synchronized cultures similar to synchronization by the natural day–night cycle. Fish cells are usually cultivated in the dark and are only exposed to light during work steps performed under the sterile bench (splitting, exposure procedure, toxicity assays). If working with photoactive NM and if desired by the test design, light exposure during these steps can be minimized. On the other hand, it is reasonable to consider exposure under controlled light conditions for cell cultures that represent fish organs naturally exposed to light, such as gills, skin, or eyes [39, 44, 45].

In a non-agitated exposure system, stable-dispersed NMs move by Brownian motion which is dependent on, i.e., the temperature of the system. Consequently, the temperature influences NM behavior (e.g., speed, agglomeration, and contact with surfaces) and possibly interaction with organisms. Fish cells are typically cultured at 15–34 °C, depending on the species from which fish cells are derived (*see* Subheading 3.1.2), while temperatures reported for algae cultures range from 15 to 26 °C.

3.4.3 Agitation

Agitation (e.g., stirring, shaking, flowing) influences the stability of the NM dispersion and the way and frequency of organism–NM encounter. Fish cells grown in monolayers are cultivated in non-agitated systems; however, suspension cultures may need agitation. Algae are typically agitated by CO_2 flow-through, stirring, or shaking.

3.5 Bioavailability

To understand toxic effects of NM, it is important to characterize the NM in the exposure system with regard to size and charge as detailed as possible (Fig. 1). Furthermore, the distribution of the NM in the exposure system in the different compartments, i.e., the mass balance including possible dissolution, is an essential information for the interpretation of results. The NM characterization techniques need to be applicable to relevant NM concentrations. An overview of characterization methods is provided below. For a more detailed review, please refer to the literature [12, 46].

3.5.1 Characterization of NM in Exposure Media

Size

The size, among other properties, will determine the behavior of NM in suspension and thus its interaction with organisms. The most commonly used techniques for size characterization in NM toxicology are scanning electron microscopy (SEM) or transmission electron microscopy (TEM) or dynamic light scattering (DLS). The use of nanoparticle tracking analysis (NTA) is increasing. While EM provides images of typically dry material in high vacuum, DLS and NTA measure the hydrodynamic diameter in liquids, and calculations are based on spherical representations of the NM. Details and limitations of the methods can be found in specific literature (*see* **Notes 2**, **3**, and **13**).

In general, designing environmentally relevant or probable exposure scenarios may be contrary to the optimal sample properties for these techniques. For example, natural additives, like humic or fulvic acid or dissolved organic matter (DOC), can influence the measurements. All three methods have their limitations regarding the detectable NM concentration range. As environmentally relevant concentrations may be very low (e.g., up to μg/L instead of mg to g/L), characterizing NM size at these concentrations may be impossible or unfeasible with these methods to date. Field-flow fractionation (FFF) coupled to elemental analysis is applicable to very low NM concentrations but entails the removal of the NM from the exposure medium. The same holds true for the detection of single particles by inductively coupled plasma mass spectrometry (ICP-MS). The surface plasmon resonance (SPR) of metallic NP, especially silver and gold, can be measured by UV–Vis spectrophotometry and provide size information as the SPR peak wavelength depends on the NP size. DLS is not applicable to very polydisperse or unstable suspensions which might, however, represent the NP state in the environment. Large and/or sticky NM/agglomerates that adhere to the surfaces of the NTA sample compartment may also make reliable measurements challenging. Agglomerated NM are a common issue depending on the culture media and dispersion method (*see* Subheadings 3.6.1 and 3.6.2).

In summary, the more monodisperse the sample is and within the optimal concentration range of the methods, the more reproducible and meaningful are the results. We usually employ DLS, NTA, and EM in combination to profit from the complementary information gathered.

Zeta Potential and Electrophoretic Mobility

The zeta potential of a particle system is a measure for its surface charge and is indicative of particle stability and interactions with components of the suspension (molecules, other particles, surfaces) and test organisms. It can be derived mathematically from the electrophoretic mobility of particles within an electric field. Most publications list the derived zeta potential, but providing electrophoretic mobility data helps to compare results independent of the assumptions made when calculating zeta potential. Like size measurements in liquids, the optimal conditions for electrophoretic mobility analysis may not be fulfilled by environmentally relevant NM dispersions. The electrophoretic mobility/zeta potential strongly depends on the pH of a solution; hence, information on the pH value of a solution/medium is required. Shifts in pH (e.g., by oxidation processes or due to cell metabolism) during an exposure may lead to the presence of different particle states in an exposure period; usually this is prevented by using a buffer system with suitable capacity.

3.5.2 Mass Balance and Bioaccumulation of NM

Dispersed NM will partition to different parts of the exposure system depending on its material properties, exposure conditions (solution chemistry, agitation, temperature, illumination), material properties and surface area of the containers used, and the properties and number of the test organisms. Possible redistribution processes from the dispersion of both NM and ions or surface modifiers dissolved from the NM are:

1. Sorption to abiotic and biotic surfaces
2. Uptake by test organisms
3. Sedimentation.

The amount of NM in these different compartments should be quantified to understand the distribution of NM during the exposure. Quantifying NM adsorbed to abiotic surfaces may be difficult to determine but can be deduced from the initial total concentration of NM and the amount present on/in the organisms, in suspension, and possibly in the sediment.

The possible dissolution, coating or surface modification (engineered or by media components), and speciation of a NM change the bioavailability of its components and may influence its effects on test organisms. Thus, these properties should be analyzed under exposure conditions to understand which factor actually elicited measured effects.

The chemical characterization of nonorganic carbon-based NM is typically done via ICP-MS, ICP-AES (atomic emission spectroscopy), or ICP-OES (optical emission spectrometry) following adapted acidic digestion of the NM. When working in culture media, be sure to use matrix-adapted calibration curves. Organic carbon-based NM can be assessed via, e.g., OCEC (organic carbon–elemental carbon) analyzers based on infrared spectrometry.

Dissolved ions may be separated by, e.g., ultrafiltration, diffusive gradients in thin films (DGT) devices, or dialysis prior to chemical analysis or detected by ion selective electrodes. Surface composition may be determined by XPS.

Metal analysis (e.g., by ICP-MS) of entire cells (or possibly subcellular fractions) following acidic digestion informs about bioaccumulation of the NM within cells but does not differentiate between dissolved and particulate material. At the moment, determining intracellular mass balance and speciation is difficult or impossible, depending on the NM and test organism. One reported approach was based on radioactive NPs to facilitate detection and assess bioavailability of ZnO NP in algae [47]. To differentiate between Ag NP and Ag^+ adsorbed to the surface of cells and intracellular silver, one study used several wash steps including cystein to bind Ag^+ [37].

3.5.3 Intracellular Localization of NM

For the interpretation of toxicity test results, information on the localization of NM in or on cells is helpful. The method for detection of NM associated with and in cells depends on the NM and the organism. A combination of different methods should be used to account for the limitations of individual techniques. Optically dense NMs can be visualized by light microscopy (LM); however, it does not provide proof of the chemical identity of the detected particles. As fish cells are transparent, this type of visualization is feasible. LM can be unusable for algae with a cell wall and/or an optically dense cytosol. Electron microscopy of fixed and dried samples provides high-resolution images and can be coupled to, e.g., energy-dispersive X-ray spectroscopy (EDX) to give evidence of the NM chemistry. Preparation artifacts may distort the localization of NMs in the sample. Alternatives such as environmental SEM/TEM which do not require vacuum in the specimen chamber and thus do not require a dry sample are currently being refined. Fluorescence or preferably confocal laser scanning microscopy (CLSM) are valuable to visualize fluorescent NMs [48]. Depending on the fluorescence wavelength emitted by the NMs, algal autofluorescence may interfere with NM detection. Metal-containing NMs exhibit size-dependent surface plasmon resonance when interacting with light and can thus be visualized by CLSM in reflection mode [49]. For ions released from metallic NM, metal specific fluorescent probes may be used to detect dissolved fractions of NM within cells. This approach is limited by the detection limit of the probes and cross-reactivity with other metals. One possibly feasible approach is coupling CLSM with Raman spectroscopy to resolve subcellular chemistry in the nanometer range.

3.6 Effects

Algae and fish cells can be explored for NM effects in many different ways. One common approach is indicator dyes that can be assessed either by fluorescence, luminescence, or by absorbance;

a large variety of such indicator dyes is commercially available. Among them are indicator dyes to assess viability, ATP content, or formation of ROS (Fig. 1). Taking some limitations due to autofluorescent biomolecules, such as chlorophylls, into account, these indicator dyes can generally be applied to both algal and fish cells (*see* Tables 1 and 2 but as well as [13, 14] as examples for effect assessment in fish cells and [50, 51] for examples of effect assessment in unicellular algae). Also fluorescence-based are methods to quantify photosynthetic activity of NM-exposed algae, such as via pulse-amplitude modulated (PAM) fluorometry. PAM provides a measure of the relative activity of photosystem II reaction centers at a given time and information on which part of the electron transport chain is affected in response to a stressor. In principle, one can make use of the variety of effect measurements established also for exposure of cells to chemicals, including genotoxicity, gene and protein expression, and impact on enzyme activity (*see* **Notes 14–18** for critical points in choosing end points).

The key to successfully applying effect-focused assays is to include appropriate controls to take potential interferences by NMs into account. For example, carbon-based NMs have a high sorption capacity also for hydrophobic fluorescence indicator dyes, which may render some of these dyes impossible to use to measure NM cellular effects [52]. Another important aspect is how to express the effect assessment results in terms of exposure or dose. These two aspects, interferences/controls and dose metrics, are briefly discussed below.

3.6.1 Potential Interferences and Appropriate Controls

NMs have a high surface-to-volume ratio and possess specific physicochemical properties which make them prone to interfere with certain types of toxicity tests [30]. Possible interferences that need to be ruled out by appropriate cell-free controls are listed in Table 1 (also *see* **Note 8**). In certain cases it may be difficult to distinguish between the different types of interferences. Different surface modifications of NM may induce different interferences. Media components may also interact with the reaction substrate, product, or enzyme. The medium should thus be tested in the same way as NM dispersions.

With regard to the NMs, two types of controls should be considered: reference NMs to rank the test results and make them better comparable to other studies and positive controls to verify the functionality of the toxicity test. A repository of representative NMs has recently been established by the Institute for Health and Consumer Protection (IHCP) of the European Commission Joint Research Centre (EC JRC) [53]. However, round robin tests based on these materials have not yet been published. To distinguish between the effects of ions and particulate material of soluble NMs, appropriate characterization methods help to know the extent of ion dissolution or impurities from the production process. Additionally, controls with ion chelators and metal salts are recommended. Surface modifications may also induce effects independent of the NMs. It may thus be reasonable to test surface modifiers independent of the NM.

3.6.2 Dose Metrics

Dose metrics is a widely debated issue in nanotoxicology. Possible bases to express measured effects are the total mass, the surface area of the dry material, or the number concentration. Furthermore, these can be expressed per cell surface, cell volume, cell number, or total exposure volume. The most common choice is total mass per total exposure volume, mostly given in mass NM/volume medium (*see* Tables 2 and 3). The total mass is the easiest to measure but does not account for size distribution. NM surface area and number have to be derived from other information (most often size distribution by intensity) which may be biased toward smaller or larger particles and also do not account for size distribution. Additionally, in case of NM agglomeration (or possibly aggregation) in the test media, both size and free surface area may change. Certain assumptions such as uniform shape have to be made. Cell number and exposure volume are easy to determine, while cell surface and volume have to be derived also in making certain assumptions. In summary, none of the combinations is a perfect solution. Using total mass per exposure volume and providing a thorough NM dispersion characterization and detailed description of exposure conditions is one approach to increase comparability of reported results because it also allows for recalculating different dose metrics. For soluble NMs it is recommended to use the dissolved fraction as well as total concentration as basis to compare toxicity of different treatments to see whether free ions can explain the measured toxicity and whether it is identical or different for different forms of the same material.

4 Notes

1. Laboratory safety: Be sure to protect yourself and your colleagues from exposure to NM. If possible, dedicate a specific area in the laboratory to handling NM. In particular, a dedicated hood should be used for powdered material. We additionally keep track of the NM handled in the nano-laboratory and the involved persons. Handling NM waste is an important issue and should be discussed with the personnel responsible for chemical waste.
2. Containers: Consider choosing your dispersion and exposure containers depending on the tested NM, e.g., using plastics to minimize metal adsorption during exposure.

3. Glass ware: Soak in diluted HNO_3 to minimize adsorption of metals to the surface during experiments; muffle in case you need to determine organic carbon.
4. Sonication: Be sure to use indirect sonication to avoid metal contamination by direct sonication.
5. NM may have to be cleaned every time before use (e.g., ref. 42).
6. NPs may change agglomeration state in the presence of cells [68].
7. Properties of purchased NM may vary between different lots. This may be a cause of variable responses in the toxicity tests.
8. Test results may prove to be irreproducible in independent experiments or show high standard deviations (e.g., ref. 56), which requires troubleshooting. One cause of this variability may be differences in NM dispersions or nonintuitive factors such as frequency of opening and closing of incubators during growth and exposure.
9. Due to the variation in NM preparation and of exposure conditions in available reports, results (*see* Table 1) may differ from reported values even if the same NM and same organism was used.
10. DLS: Discard all DLS data not meeting software quality criteria.
11. NTA: Carefully adjust the data processing settings to the sample analyzed.
12. The methods for NM characterization that are currently available are limited regarding very low (but likely environmentally relevant) concentrations of NM.
13. Most of the toxicity end points reported in the literature to date are descriptors of the general state of cell cultures (such as growth rate, membrane integrity, photosynthetic activity).
14. Most methods were adopted from the testing of chemicals.
15. Direct interaction with intra- and extracellular molecules (e.g., DNA, specific enzymes, and extracellular polymeric substances) has not been studied.
16. The influence of NMs on the cell cycle and biochemical pathways is still unexplored.
17. We are far from understanding the mechanisms of interaction between the considered biological models and NM.
18. To minimize interference with subsequent measurements, cells should be washed, taking into account the chemistry and charge of extracellular substances and of the NM (e.g., slightly acidic buffer to neutralize extracellular negative charges and thus weaken NM-cell interaction).

References

1. European Commission (2012-01-22) http://europa.eu/rapid/pressReleasesAction.do?reference=IP/11/1202&format=HTML&aged=0&language=EN&guiLanguage=en
2. Jiang J, Oberdörster G, Biswas P (2009) Characterization of size, surface charge, and agglomeration state of nanoparticle dispersions for toxicological studies. J Nanoparticle Res 11:77–89
3. Lewis LA, McCourt RM (2004) Green algae and the origin of land plants. Am J Bot 91: 1535–1556
4. Weyers A, Sokull-Klüttgen B, Baraibar-Fentanes J, Vollmer G (2000) Acute toxicity data: a comprehensive comparison of results of fish, daphnia, and algae tests with new substances notified in the European Union. Environ Toxicol Chem 19:1931–1933
5. Lee RE (2008) Phycology (4 ed.) Cambridge University Press, Cambridge, UK.
6. Kneip C, Voss C, Lockhart PJ, Maier UG (2008) The cyanobacterial endosymbiont of the unicellular algae Rhopalodia gibba shows reductive genome evolution. BMC Evol Biol 8:30
7. Bols NC, Dayeh VR, Lee LEJ, Schirmer K (2005) Use of fish cell lines in the toxicology and ecotoxicology of fish. In: Mommsen TP, Moon TW (eds) Biochemistry and molecular biology of fishes. Elsevier Science, Amsterdam
8. Schirmer K (2006) Proposal to improve vertebrate cell cultures to establish them as substitutes for the regulatory testing of chemicals and effluents using fish. Toxicology 224: 163–183
9. Surek B (2008) Meeting report: algal culture collections 2008. An international meeting at the culture collection of algae and protozoa (CCAP), Dunstaffnage Marine Laboratory, Dunbeg, Oban, UK, June 8–11, 2008, Protist 159:509–517.
10. Gachon CM, Day JG, Campbell CN, Proschold T, Saxon RJ, Kupper FC (2007) The culture collection of algae and protozoa (CCAP): a biological resource for protistan genomics. Gene 406:51–57
11. Lee LE, Dayeh VR, Schirmer K, Bols NC (2009) Applications and potential uses of fish gill cell lines: examples with RTgill-W1. In Vitro Cell Dev Biol Anim 45:127–134
12. Meissner T, Kuhnel D, Busch W, Oswald S, Richter V, Michaelis A, Schirmer K, Potthoff A (2010) Physical-chemical characterization of tungsten carbide nanoparticles as a basis for toxicological investigations. Nanotoxicology 4:196–206
13. Dayeh VR, Schirmer K, Lee LEJ, Bols NC (2005) Rainbow trout gill cell line microplate cytotoxicity test. In: Blaise C, Férard JF (eds) Small-scale freshwater environment toxicity test methods. Kluwer, Norwell, MA, pp 473–503
14. Dayeh VR, Bols NC, Schirmer K, Lee LE (2003) The use of fish-derived cell lines for investigation of environmental contaminants. Curr Protoc Toxicol Chapter 1, Unitl 5.
15. Losic D, Rosengarten G, Mitchell JG, Voelcker NH (2006) Pore architecture of diatom frustules: potential nanostructured membranes for molecular and particle separations. J Nanosci Nanotechnol 6:982–989
16. Naustvoll L-J (1998) Growth and grazing by the thecate heterotrophic dinoflagellate Diplopsalis lenticula (Diplopsalidaceae, Dinophyceae). Phycologia 37:1–9
17. Gottschalk F, Sonderer T, Scholz RW, Nowack B (2009) Modeled environmental concentrations of engineered nanomaterials (TiO_2, ZnO, Ag, CNT, fullerenes) for different regions. Environ Sci Technol 43:9216–9222
18. Luoma SN (2008) Silver nanotechnologies and the environment: old problems or new challenges? In: Project on emerging nanotechnologies report, PEN 15. Woodrow Wilson Centre, Washington, DC, pp 1–67
19. Tiede K, Hassellöv M, Breitbarth E, Chaudhry Q, Boxall ABA (2009) Considerations for environmental fate and ecotoxicity testing to support environmental risk assessments for engineered nanoparticles. J Chromatogr A 1216:503–509
20. Velzeboer I, Hendriks AJ, Ragas AMJ, van de Meent D (2008) Nanomaterials in the environment aquatic ecotoxicity tests of some nanomaterials. Environ Toxicol Chem 27: 1942–1947
21. Bouldin JL, Ingle TM, Sengupta A, Alexander R, Hannigan RE, Buchanan RA (2008) Aqueous toxicity and food chain transfer of quantum dots™ in freshwater algae and Ceriodaphnia dubia. Environ Toxicol Chem 27:1958–1963
22. Kühnel D, Busch W, Meißner T, Springer A, Potthoff A, Richter V, Gelinsky M, Schirmer K (2009) Agglomeration of tungsten carbide nanoparticles in exposure medium does not prevent uptake and toxicity toward a rainbow trout gill cell line. Aquat Toxicol 93:91–99
23. Landsiedel R, Ma-Hock L, Kroll A, Hahn D, Schnekenburger J, Wiench K, Wohlleben W (2010) Testing metal-oxide nanomaterials for human safety. Adv Mater 22:2601–2627
24. OECD (2011),Test No. 201: Freshwater Alga and Cyanobacteria, Growth Inhibition Test, OECD Guidelines for the Testing of Chemicals, Section 2, OECD Publishing. doi: 10.1787/9789264069923-en

25. ISO (2012) Water quality—Fresh water algal growth inhibition test with unicellular green algae (ISO 8692:2012). International Organization for Standardization.
26. German standard methods for the examination of water, waste water and sludge; bio-assays (group L); determining the tolerance of green algae to the toxicity of waste water (Scenedesmus chlorophyll fluorescence test) by way of dilution series (L 33), (DIN 38412-33:1991-03)
27. Hund-Rinke K, Schlich K, Wenzel A (2010) TiO2 nanoparticles—relationship between dispersion preparation method and ecotoxicity in the algal growth test. Umweltwissenschaften und Schadstoff-Forschung 22:517–528
28. Piccapietra F, Sigg L, Behra R (2011) Colloidal stability of carbonate-coated silver nanoparticles in synthetic and natural freshwater. Environ Sci Technol 46:818–825
29. Hildebrand H, Kühnel D, Potthoff A, Mackenzie K, Springer A, Schirmer K (2009) Evaluating the cytotoxicity of palladium/magnetite nano-catalysts intended for wastewater treatment. Environ Pollut 158:65–73
30. Schulze C, Kroll A, Lehr CM, Schäfer UF, Becker K, Schnekenburger J, Schulze Isfort C, Landsiedel R, Wohleben W (2008) Not ready to use—overcoming pitfalls when dispersing nanoparticles in physiological media. Nanotoxicology 2:51–61
31. Hull MS, Kennedy AJ, Steevens JA, Bednar AJ, Weiss JCA, Vikesland PJ (2009) Release of metal impurities from carbon nanomaterials influences aquatic toxicity. Environ Sci Technol 43:4169–4174
32. Sharma V, Shukla RK, Saxena N, Parmar D, Das M, Dhawan A (2009) DNA damaging potential of zinc oxide nanoparticles in human epidermal cells. Toxicol Lett 185:211–218
33. Van Hoecke K, De Schamphelaere KAC, Van der Meeren P, Smagghe G, Janssen CR (2011) Aggregation and ecotoxicity of CeO_2 nanoparticles in synthetic and natural waters with variable pH, organic matter concentration and ionic strength. Environ Pollut 159: 970–976
34. Slaveykova VI, Startchev K (2009) Effect of natural organic matter and green microalga on carboxyl-polyethylene glycol coated CdSe/ZnS quantum dots stability and transformations under freshwater conditions. Environ Pollut 157:3445–3450
35. Odzak N, Behra R, Sigg L (2013) Dissolution of metal and metal oxide nanoparticles in aqueous media. Manuscript in preparation.
36. Kent RD, Vikesland PJ (2012) Controlled evaluation of silver nanoparticle dissolution using atomic force microscopy. Environ Sci Technol 46(13):6977–6984
37. Navarro E, Piccapietra F, Wagner B, Marconi F, Kaegi R, Odzak N, Sigg L, Behra R (2008) Toxicity of silver nanoparticles to Chlamydomonas reinhardtii. Environ Sci Technol 42:8959–8964
38. Deguchi S, Yamazaki T, Mukai SA, Usami R, Horikoshi K (2007) Stabilization of C(60) nanoparticles by protein adsorption and its implications for toxicity studies. Chem Res Toxicol 20:854–858
39. Schirmer K, Chan AG, Greenberg BM, Dixon DG, Bols NC (1997) Methodology for demonstrating and measuring the photocytotoxicity of fluoranthene to fish cells in culture. Toxicol In Vitro 11:107–119
40. Rogers NJ, Franklin NM, Apte SC, Batley GE, Angel BM, Lead JR, Baalousha M (2010) Physico-chemical behaviour and algal toxicity of nanoparticulate CeO_2 in freshwater. Environ Chem 7:50–60
41. Ma H, Kabengi NJ, Bertsch PM, Unrine JM, Glenn TC, Williams PL (2011) Comparative phototoxicity of nanoparticulate and bulk ZnO to a free-living nematode Caenorhabditis elegans: the importance of illumination mode and primary particle size. Environ Pollut 159:1473–1480
42. Hund-Rinke K, Simon M (2006) Ecotoxic Effect of Photocatalytic Active Nanoparticles (TiO2) on Algae and Daphnids. Environ Sci Pollut Res 13:225–232
43. Schwab F, Bucheli TD, Lukhele LP, Magrez A, Nowack B, Sigg L, Knauer K (2011) Are Carbon Nanotube Effects on Green Algae Caused by Shading and Agglomeration? Environ Sci Technol 45:6136–6144
44. Laycock NL, Schirmer K, Bols NC, Sivak JG (2000) Optical properties of rainbow trout lenses after in vitro exposure to polycyclic aromatic hydrocarbons in the presence or absence of ultraviolet radiation. Exp Eye Res 70: 205–214
45. Schirmer K, Chan AG, Greenberg BM, Dixon DG, Bols NC (1998) Ability of 16 priority PAHs to be photocytotoxic to a cell line from the rainbow trout gill. Toxicology 127:143–155
46. Hassellöv M, Readman J, Ranville J, Tiede K (2008) Nanoparticle analysis and characterization methodologies in environmental risk assessment of engineered nanoparticles. Ecotoxicology 17:344–361
47. Peng X, Palma S, Fisher NS, Wong SS (2011) Effect of morphology of ZnO nanostructures on their toxicity to marine algae. Aquatic Toxicology 102:186–196
48. Xu M, Deng G, Liu S, Chen S, Cui D, Yang L, Wang Q (2010) Free cadmium ions released from CdTe-based nanoparticles and their cytotoxicity on Phaeodactylum tricornutum. Metallomics 2:469–473

49. Tsai S-W, Chen Y-Y, Liaw J-W (2008) Compound Cellular Imaging of Laser Scanning Confocal Microscopy by Using Gold Nanoparticles and Dyes. Sensors 8:2306–2316
50. Nestler H, Groh KJ, Schonenberger R, Behra R, Schirmer K, Eggen RI, Suter MJ (2012) Multiple-endpoint assay provides a detailed mechanistic view of responses to herbicide exposure in Chlamydomonas reinhardtii. Aquat Toxicol 110–111:214–224
51. Szivák I, Behra R, Sigg L (2009) Metal-induced reactive oxygen species production in chlamydomonas reinhardtii (Chlorophyceae)1. J Phycol 45:427–435
52. Kroll A, Pillukat MH, Hahn D, Schnekenburger J (2012) Interference of engineered nanoparticles with in vitro toxicity assays, *Arch Tox* 86(7):1123–1136
53. European Commission Joint Research Centre. (2012-01-22). http://ihcp.jrc.ec.europa.eu/our_activities/nanotechnology/nanomaterials-repository
54. Miao A-J, Schwehr KA, Xu C, Zhang S-J, Luo Z, Quigg A, Santschi PH (2009) The algal toxicity of silver engineered nanoparticles and detoxification by exopolymeric substances. Environ Pollut 157:3034–3041
55. Renault S, Baudrimont M, Mesmer-Dudons N, Gonzalez P, Mornet S, Brisson A (2008) Impacts of gold nanoparticle exposure on two freshwater species: a phytoplanktonic alga (Scenedesmus subspicatus) and a benthic bivalve (Corbicula fluminea). Gold Bull 41:116–126
56. Baun A, Sørensen SN, Rasmussen RF, Hartmann NB, Koch CB (2008) Toxicity and bioaccumulation of xenobiotic organic compounds in the presence of aqueous suspensions of aggregates of nano-C60. Aquat Toxicol 86: 379–387
57. Rodea-Palomares I, Boltes K, Fernández-Piñas F, Leganés F, García-Calvo E, Santiago J, Rosal R (2011) Physicochemical characterization and ecotoxicological assessment of CeO_2 nanoparticles using two aquatic microorganisms. Toxicol Sci 119:135–145
58. Van Hoecke K, Quik JTK, Mankiewicz-Boczek J, Schamphelaere KACD, Elsaesser A, Meeren PVD, Barnes C, McKerr G, Howard CV, Meent DVD, Rydzyński K, Dawson KA, Salvati A, Lesniak A, Lynch I, Silversmit G, Samber BRD, Vincze L, Janssen CR (2009) Fate and effects of CeO_2 nanoparticles in aquatic ecotoxicity tests. Environ Sci Technol 43:4537–4546
59. Blaise C, Gagné F, Férard JF, Eullaffroy P (2008) Ecotoxicity of selected nano-materials to aquatic organisms. Environ Toxicol 23:591–598
60. Aruoja V, Dubourguier H-C, Kasemets K, Kahru A (2009) Toxicity of nanoparticles of CuO, ZnO and TiO_2 to microalgae Pseudokirchneriella subcapitata. Sci Total Environ 407:1461–1468
61. Saison C, Perreault F, Daigle J-C, Fortin C, Claverie J, Morin M, Popovic R (2010) Effect of core–shell copper oxide nanoparticles on cell culture morphology and photosynthesis (photosystem II energy distribution) in the green alga, Chlamydomonas reinhardtii. Aquat Toxicol 96:109–114
62. Li M, Czymmek KJ, Huang CP (2011) Responses of Ceriodaphnia dubia to TiO_2 and Al_2O_3 nanoparticles: a dynamic nano-toxicity assessment of energy budget distribution. J Hazardous Mater 187:502–508
63. Wei L, Thakkar M, Chen Y, Ntim SA, Mitra S, Zhang X (2010) Cytotoxicity effects of water dispersible oxidized multiwalled carbon nanotubes on marine alga, Dunaliella tertiolecta. Aquat Toxicol 100:194–201
64. Naha PC, Bhattacharya K, Tenuta T, Dawson KA, Lynch I, Gracia A, Lyng FM, Byrne HJ (2010) Intracellular localisation, geno- and cytotoxic response of polyN-isopropylacrylamide (PNIPAM) nanoparticles to human keratinocyte (HaCaT) and colon cells (SW 480). Toxicol Lett 198:134–143
65. Wang J, Zhang X, Chen Y, Sommerfeld M, Hu Q (2008) Toxicity assessment of manufactured nanomaterials using the unicellular green alga Chlamydomonas reinhardtii. Chemosphere 73:1121–1128
66. Fujiwara K, Suematsu H, Kiyomiya E, Aoki M, Sato M, Moritoki N (2008) Size-dependent toxicity of silica nano-particles to Chlorella kessleri. J Environ Sci Health Part A 43: 1167–1173
67. Van Hoecke K, De Schamphelaere KAC, Van der Meeren P, Lcucas S, Janssen CR (2008) Ecotoxicity of silica nanoparticles to the green alga pseudokirchneriella subcapitata: importance of surface area. Environ Toxicol Chem 27:1948–1957
68. Cherchi C, Chernenko T, Diem M, Gu AZ (2011) Impact of nano titanium dioxide exposure on cellular structure of Anabaena variabilis and evidence of internalization. Environ Toxicol Chem 30:861–869
69. Miller RJ, Lenihan HS, Muller EB, Tseng N, Hanna SK, Keller AA (2010) Impacts of metal oxide nanoparticles on marine phytoplankton. Environ Sci Technol 44:7329–7334
70. Hall S, Bradley T, Moore JT, Kuykindall T, Minella L (2009) Acute and chronic toxicity of nano-scale TiO_2 particles to freshwater fish, cladocerans, and green algae, and effects of organic and inorganic substrate on TiO_2 toxicity. Nanotoxicology 3:91–97
71. Rodríguez-González V, Alfaro SO, Torres-Martínez LM, Cho S-H, Lee S-W (2010) Silver–TiO_2 nanocomposites: synthesis and

harmful algae bloom UV-photoelimination. Appl Catal B: Environ 98:229–234

72. Warheit DB, Hoke RA, Finlay C, Donner EM, Reed KL, Sayes CM (2007) Development of a base set of toxicity tests using ultrafine TiO_2 particles as a component of nanoparticle risk management. Toxicol Lett 171:99–110
73. Franklin NM, Rogers NJ, Apte SC, Batley GE, Gadd GE, Casey PS (2007) Comparative toxicity of nanoparticulate ZnO, Bulk ZnO, and $ZnCl_2$ to a freshwater microalga (Pseudokirchneriella subcapitata): the importance of particle solubility. Environ Sci Technol 41:8484–8490
74. Miao A-J, Zhang X-Y, Luo Z, Chen C-S, Chin W-C, Santschi PH, Quigg A (2010) Zinc oxide-engineered nanoparticles: dissolution and toxicity to marine phytoplankton. Environ Toxicol Chem 29:2814–2822
75. Wise JP Sr, Goodale BC, Wise SS, Craig GA, Pongan AF, Walter RB, Thompson WD, Ng AK, Aboueissa AM, Mitani H, Spalding MJ, Mason MD (2010) Silver nanospheres are cytotoxic and genotoxic to fish cells. Aquat Toxicol 97:34–41
76. Hondroulis E, Liu C, Li CZ (2010) Whole cell based electrical impedance sensing approach for a rapid nanotoxicity assay. Nanotechnology 21:315103
77. George S, Lin S, Ji Z, Thomas CR, Li L, Mecklenburg M, Meng H, Wang X, Zhang H, Xia T, Hohman JN, Lin S, Zink JI, Weiss PS, Nel AE (2012) Surface defects on plate-shaped silver nanoparticles contribute to its hazard potential in a fish gill cell line and Zebrafish embryos. ACS Nano 6(5):3745–3759
78. Van Hoecke K, De Schamphelaere KA, Ali Z, Zhang F, Elsaesser A, Rivera-Gil P, Parak WJ, Smagghe G, Howard CV, Janssen CR (2013) Ecotoxicity and uptake of polymer coated gold nanoparticles. Nanotoxicology 7(1): 37–47
79. Naha PC, Casey A, Tenuta T, Lynch I, Dawson KA, Byrne HJ, Davoren M (2009) Preparation, characterization of NIPAM and NIPAM/BAM copolymer nanoparticles and their acute toxicity testing using an aquatic test battery. Aquat Toxicol 92:146–154
80. Reeves JF, Davies SJ, Dodd NJ, Jha AN (2008) Hydroxyl radicals (OH) are associated with titanium dioxide (TiO(2)) nanoparticle-induced cytotoxicity and oxidative DNA damage in fish cells. Mutat Res 640:113–122
81. Vevers WF, Jha AN (2008) Genotoxic and cytotoxic potential of titanium dioxide (TiO_2) nanoparticles on fish cells in vitro. Ecotoxicology 17:410–420
82. Fernández-Cruz ML, Lammel L, Connolly M, Conde E, Barrado AI, Derick S, Perez Y, Fernandez M, Furger C, Navas JM (2012) Comparative cytotoxicity induced by bulk and nanoparticulated ZnO in the fish and human hepatoma cell lines PLHC-1 and Hep G2 Nanotoxicology early online 1–18.

Chapter 12

Gold Nanoparticles and Oxidative Stress in the Blue Mussel, *Mytilus edulis*

Sara Tedesco, Hugh Doyle, Daniela Iacopino, Irene O'Donovan, Sarah Keane, and David Sheehan

Abstract

Relatively little is known about how gold nanoparticles (GNP) might interact *in vivo* with aquatic organisms. Gold nanoparticles (GNP) of defined average diameter may be synthesized and used to challenge test organisms held in aquaria. The blue mussel, *Mytilus edulis*, is a popular sentinel species in environmental toxicology. This chapter describes how mussels exposed to GNP (average diameter ~5 nm) may be dissected and extracts of digestive gland can be analyzed for oxidative stress. Protein thiols are labeled with 5′-iodoacetamide-fluorescein (IAF), and proteins are separated in one-dimensional electrophoresis. After scanning for IAF-associated fluorescence, gels are stained with colloidal coomassie. A ratio of fluorescence to protein stain is calculated revealing that thiol oxidation increases with GNP treatment.

Key words Gold nanoparticle, Digestive gland, Oxidative stress, Carbonylation, Thiol oxidation, Mussel, Mytilus

Abbreviations

1-DE	One-dimensional electrophoresis
DMSO	Dimethyl sulfoxide
EDTA	Ethylenediaminetetraacetic acid
GSH	Reduced glutathione
GNP	Gold nanoparticle
IAF	5′-Iodoacetamide-fluorescein
TEMED	*N,N,N,N′*-tetramethyl-ethylenediamine
PMSF	Phenylmethylsulfonyl fluoride
ROS	Reactive oxygen species
SDS PAGE	Sodium dodecyl sulfate polyacrylamide gel electrophoresis

Donald Armstrong and Dhruba J. Bharali (eds.), *Oxidative Stress and Nanotechnology: Methods and Protocols*, Methods in Molecular Biology, vol. 1028, DOI 10.1007/978-1-62703-475-3_12,

1 Introduction

Nanoparticles have one dimension <100 nm [1] and have a very large surface to volume ratio which may be responsible for toxicological risk [2]. In particular, recent reviews have focused strongly on the potential of nanomaterials to cause oxidative stress [2–6]. This occurs when cellular defenses against reactive oxygen species (ROS) are exceeded and redox homeostasis is lost [7]. Antioxidant defenses include enzymes (e.g., catalase and superoxide dismutases) and low M_r thiols (e.g., glutathione; GSH). GSH, the principal cytosolic antioxidant, is maintained in a reduced state by the GSH reductase/thioredoxin system [8]. Ultimately, antioxidant defense is supported by NADPH reducing equivalents supplied mainly from the pentose phosphate pathway [9]. Nanoparticles can contribute to oxidative stress at a number of biologically relevant levels [5]. ROS can be produced directly due to nanoparticles' large proportional surface area [2], or nanoparticles can mechanically disrupt cell structures such as mitochondria [10]. Alternatively, nanoparticles can become coated with a "corona" of proteins [11, 12] perhaps providing routes of passage through biomembranes (e.g., via receptors or endocytosis) with potential for effects on cell signaling pathways [13, 14].

Relatively little is known about the fate of nanoparticles in the environment or within complex biological systems, but, undoubtedly, some nanoparticles cross important biobarriers (e.g., skin, gut wall, blood–brain barrier) [4]. Nanomaterials could be released in large quantities into the environment [15], and they can be toxic to a wide range of biota including bacteria, algae, invertebrates, fish, and mammals. However, much of the currently available ecotoxicological data are limited to species used in regulatory testing or freshwater organisms [16]. Gold nanoparticles (GNP) have widespread commercial uses [17, 18]. Although supposedly an inert material, nanoscale gold exhibits properties that change according to size, surface coating, shape, oxidation state, and synthesis method. Studies have appeared on the biodistribution and toxicity of Au (I) and Au (III) compounds [19, 20], but few have evaluated in vivo toxicology of colloidal nanoscale Au (0). Furthermore, only a few studies have been carried out on filter-feeding aquatic organisms such as the blue mussel, *Mytilus edulis*, exposed to GNP [21–25].

Bivalve molluscs are popular sentinel organisms in environmental monitoring programs because they are widely distributed, sedentary, relatively tolerant to a wide range of environmental conditions, and are filter feeders that pump large volumes of water and bioconcentrate many chemicals in their tissues [26]. Mussels can absorb xenobiotics present in water directly through the gills and indirectly through the digestive system when xenobiotics are adsorbed on the small grain-size fraction of particles [27]. In this chapter, we describe how exposure experiments with effects of GNP can be performed on the bivalve mollusc, *M. edulis*, and explore how thiol oxidation as a measure of oxidative stress may be routinely performed.

2 Materials

2.1 Animals and Extract Preparation

The blue mussel, *Mytilus edulis* Linnaeus, inhabits temperate Atlantic coasts of Western Europe and North America growing attached to rocky substrates (*see* **Note 1**). Collect animals (4–6 cm) from their substrates at low tide in the intertidal range of unpolluted sites and bring them to the laboratory. After washing with cold tap water, hold them overnight in buckets at 4 °C in clean, natural, aerated seawater. Then place them in groups of 40 in 50 L aquaria for 1 week to acclimate (*see* **Note 2**). Keep the aquaria in a 12 h light/dark cycle at a temperature of 15–16 °C. Feed the animals Phytoplex™ phytoplankton feed (Kent Marine Inc., Acworth, GA, USA) and change the water and feed at 48 h intervals. Handle and treat the control group exactly as groups exposed to GNP.

2.2 Buffers and Reagents

1. Homogenisation buffer: 10 mM Tris–HCl, pH 7.2, 500 mM sucrose, 1 mM ethylenediaminetetraacetic acid (EDTA), 1 mM phenylmethylsulfonyl fluoride, (PMSF). Prepare 1 day before use and store at 4 °C overnight.
2. IAF is prepared in dimethyl sulfoxide (DMSO) as a 20 mM stock.

2.3 Electrophoresis

1. Resolving gel buffer: 1.5 M Tris–HCl, pH 8.8. Store at 4 °C.
2. Stacking gel buffer: 0.5 M Tris–HCl, pH 6.8. Store at 4 °C.
3. Resolving gel: 4 % acrylamide/Bis solution (29.2:0.8). Store in dark at 4 °C.
4. Ammonium persulfate: 10 % aqueous solution.
5. *N*,*N*,*N*,*N'*-tetramethyl-ethylenediamine (TEMED) (Sigma Chemical Co., St. Louis, MO USA). Store at 4 °C.
6. Colloidal coomassie: The colloidal coomassie working solution [28] consists of the following in a total volume of 2 L: 0.02 % coomassie brilliant blue G-250 (0.4 g); 5 % aluminum sulfate hydrate (100 g); 10 % ethanol 200 mL; 2 % orthophosphoric acid (47 mL). Dissolve the aluminum sulfate in milli-Q water. Then add the ethanol and coomassie brilliant blue and mix well. Add the phosphoric acid and make up to 2 L.
7. Sodium dodecyl sulfate polyacrylamide gel electrophoresis (SDS PAGE) running buffer: 0.025 M Tris–HCl, pH 8.3, 0.192 M glycine, 0.1 % SDS.

3 Methods

3.1 Preparation of GNP

1. Prepare citrate-capped GNPs by the method of Grabar et al. [29]. Use nanopure H_2O (18.2 MΩ cm) for all procedures. Take care to first clean all synthetic glassware with *Aqua Regia* (3 HCl:1 HNO_3), and then thoroughly rinse with deionized water.
2. Heat 500 mL of a 1 mM $HAuCl_4$ solution to reflux with vigorous stirring. Rapidly add 50 mL of an aqueous solution of sodium citrate (38.8 mM) until a color change from pale yellow to deep red occurs (usually within 1 min).
3. Stir the nanoparticle dispersion under reflux for 15 min, allow to cool to room temperature, and filter through a 0.45 μm nylon membrane filter (Gelman Sciences Inc, Ann Arbor, MI, USA).
4. Size distribution can be determined by transmission electron microscopy (TEM) in a Jeol JEM-2011 instrument (JEOL Ltd., Akishima, Tokyo, Japan) equipped with a Gatan DualVision 600 CCD. Drop cast a 2.5 μL aliquot of GNP dispersion onto a 300 mesh carbon-coated copper grid, and allow to evaporate under ambient conditions (Fig. 1).

3.2 Animals, GNP Exposure, and Extract Preparation

1. Collect animals (approximate shell length, 5.5 cm) from their substrates at low tide in the intertidal range of unpolluted sites and bring them to the laboratory.
2. After washing with cold water, hold them overnight at 4 °C in seawater. Then place them in groups of 10 in aquaria for 48 h

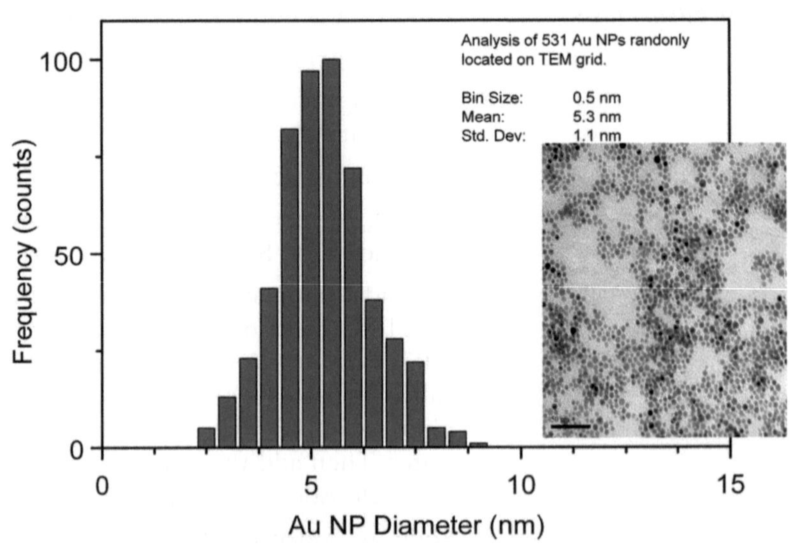

Fig. 1 Analysis of gold nanoparticles by TEM. A typical TEM image is shown as an insert. The average diameter distribution was calculated and found to be approximately 5 nm

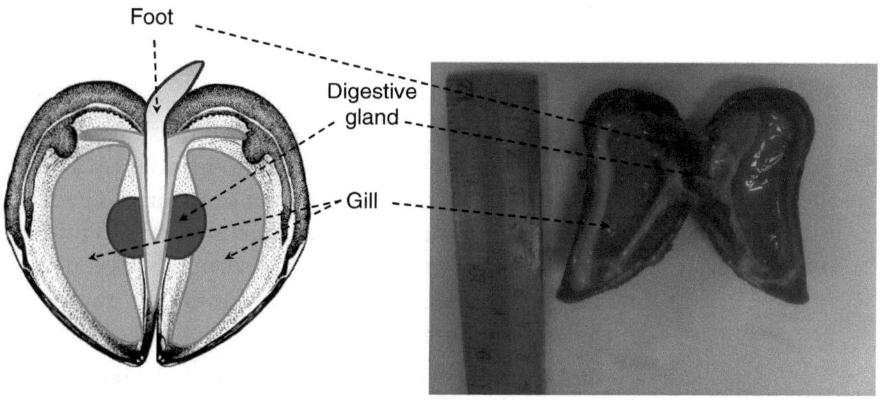

Fig. 2 Dissection of *Mytilus edulis*. The bivalve is forced open with a knife and the two shell-halves hinged back. Gill is evident at the edge of the visceral mass, while digestive gland is a *dark*-colored structure underneath. (**a**) Shows a drawing of this dissection, while (**b**) shows a photograph

to acclimate (*see* **Note 2**). Handle and treat a control group exactly as groups exposed to GNP.

3. Place one group of animals in small aquarium (vol. 4 L) and expose to GNP (e.g., 700 ppb) for 2 h by diluting stock GNP solution in 4 L.
4. Return animals to larger aquaria and allow 24 h to depurate.
5. Dissect digestive gland (Fig. 2; *see* **Note 3**) and pool within treatment or control group. Then, weigh and homogenize in 3 volumes of ice-cold homogenisation buffer. Keep tissues and extracts on ice.
6. Centrifuge at 10,000 × g for 1 h. Collect supernatant.
7. Store in −80 °C as 500 mL aliquots until required.

3.3 Labeling Cysteines in Oxidatively Modified Proteins

1. Thaw protein aliquot.
2. Add IAF stock solution to a final concentration of 200 μM to label protein thiols and incubate at room temperature for 2 h in the dark. Remove excess IAF by passing protein samples through a Zeba desalting column (Pierce, Rockford IL, USA).
3. Apply proteins to 1DE on 12 % SDS PAGE gels [30].
4. Scan gels (excitation λ_{max} of 490–495 nm and emission λ_{max} of 515–520 nm) in a Typhoon9400 scanner (GE Healthcare, Chalfont St Giles, Bucks, UK). Then stain with colloidal coomassie blue G250 [28] (*see* **Notes 4** and **5**). Load equal amounts of protein in all wells (3 replicates per treatment) and repeat at least three times.
5. Calculate the ratio of fluorescence intensity to protein stain for each track and determine the mean ± SD (Fig. 3) (*see* **Note 5**).

4 Notes

1. The genus Mytilus includes a variety of other closely related species such as *M. trossulus* Gould (Baltic), *M. galloprovincialis* Lamarck (Mediterranean and as an invader on Pacific Coast of California), and *M. californianus* Conrad (Pacific Coast of California). Some of these species form hybrids with each other, so there may be hybrid zones between coasts dominated by two species. Animals should be sampled from the middle of the intertidal region of rocky beaches.
2. As filter feeders, mussels bioaccumulate many common pollutants such as heavy metals, organochlorines, and hydrocarbons. It is best to select unstressed animals from unpolluted sites and also to allow 1 week acclimation time for depuration.
3. When dissecting mussels, use a knife to prise apart the two half-shells which are held hermetically sealed together at low tide by the adductor muscle. Dry liquid with some filter papers to better visualize the organs.
4. Prior to staining, wash gel three times (each time for 10 min) in milli-Q water to remove SDS. Shake the colloidal coomassie working solution well before covering the gel. Stain with shaking for 2–12 h. When bands are evident, wash gel three times in milli-Q water.
5. Determining both fluorescence and protein staining intensity for each track allows normalization of fluorescence intensity to protein loading. A decrease in the fluorescence of intensity is indicative of thiol oxidation in response to oxidative stress induced by GNP (Fig. 3).

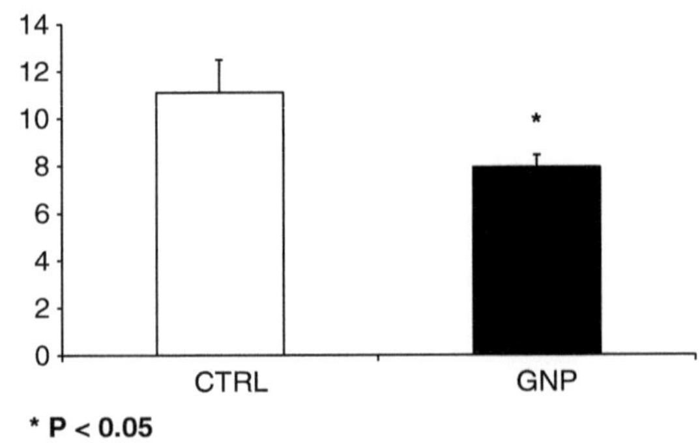

Fig. 3 Thiol oxidation. The ratio of thiol-associated fluorescence to protein staining intensity was calculated. Thiol oxidation resulted in a significant decrease in this ratio (*$p < 0.05$)

References

1. Laurent L, Petit JC (2005) Nanosciences and its convergence with other technologies—new golden age or apocalypse? Hyle 11:45–76
2. Nel A, Xia T, Madler L, Li N (2006) Toxic potential of materials at the nanolevel. Science 311:622–627
3. Moore MN (2006) Do nanoparticles present ecotoxicological risks for the health of the aquatic environment? Environ Int 32:967–976
4. Oberdörster G, Stone V, Donaldson K (2007) Toxicology of nanoparticles: a historical perspective. Nanotoxicology 1:2–25
5. Unfried K, Albrecht C, Klotz LO, Mikecz AV, Grether-Beck S, Schins RPF (2007) Cellular responses to nanoparticles: target structures and mechanisms. Nanotoxicology 1:52–71
6. Tedesco S, Sheehan D (2010) Nanomaterials as emerging environmental threats. Curr Chem Biol 4:151–160
7. Halliwell B, Gutteridge JMC (2008) Free radicals in biology and medicine, 4th edn. Oxford University Press, Oxford, p 704
8. Meister A, Anderson ME (1983) Glutathione. Annu Rev Biochem 52:711–760
9. Farout L, Friguet B (2006) Proteasome function in aging and oxidative stress: implications in protein maintenance failure. Antioxid Redox Signal 8:205–216
10. Xia T, Kovochich M, Brant J, Hotze M, Sempf J, Oberley T, Sioutas C, Yeh JI, Wiesner MR, Nel AE (2006) Comparison of the abilities of ambient and manufactured nanoparticles to induce cellular toxicity according to an oxidative stress paradigm. Nano Lett 6:1794–1807
11. Cedervall T, Lynch I, Lindman S, Berggard T, Thulin E, Nilsson H, Dawson KA, Linse S (2008) Understanding the nanoparticle-protein corona using methods to quantify exchange rates and affinities of proteins for nanoparticles. Proc Nat Acad Sci USA 104:2050–2055
12. Lynch I, Dawson KA (2008) Protein-nanoparticle interactions. Nanotoday 3:40–47
13. Allen RG, Tresini M (2000) Oxidative stress and gene regulation. Free Rad Biol Med 28: 463–499
14. Oktyabrsky ON, Smirnova GV (2007) Redox regulation of cellular functions. Biochemistry (Moscow) 72:132–145
15. Handy RD, Kammer FVD, Lead JR, Hassellöv M, Owen R, Crane M (2008) The ecotoxicology and chemistry of manufactured nanoparticles. Ecotoxicology 17:287–314
16. Handy RD, Valsami-Jones E, Owen R (2008) The ecotoxicology of nanoparticles and nanomaterials: current status, knowledge gaps, challenges, and future needs. Ecotoxicology 17:315–325
17. Mannino S, Scampicchio M (2007) Nanotechnology and food quality control. Vet Res Comm 31:149–151
18. Medley CD, Smith JE, Tang Z, Wu Y, Bamrungsap S, Tan W (2008) Gold nanoparticle-based colorimetric assay for the direct detection of cancerous cells. Anal Chem 80:1067–1072
19. Eisler R (2003) Chrysotherapy: a synoptic review. Inflamm Res 52:487–501
20. Merchant B (1998) Gold, the noble metal and the paradoxes of its toxicology. Biologicals 26:49–59
21. Lovern SB, Owen HA, Klaper R (2008) Electron microscopy of gold nanoparticle intake in the gut of *Daphnia magna*. Nanotoxicology 2:43–48
22. Renault S, Baudrimont M, Mesmer-Dudons N, Gonzalez P, Mornet S, Brisson A (2008) Impacts of gold nanoparticle exposure on two freshwater species: a phytoplanktonic alga (*Scenedesmus subspicatus*) and a benthic bivalve (*Corbicula fluminea*). Gold Bull 41:116–126
23. Tedesco S, Doyle H, Redmond D, Sheehan D (2008) Gold nanoparticles and oxidative stress in Mytilus edulis. Marine Environ Res 66: 131–133
24. Tedesco S, Doyle H, Blasco J, Redmond G, Sheehan D (2010) Oxidative stress and toxicity of gold nanoparticles in *Mytilus edulis*. Aquat Toxicol 100:178–186
25. Tedesco S, Doyle H, Blasco J, Redmond G, Sheehan D (2010) Exposure of the blue mussel, *Mytilus edulis*, to gold nanoparticles and the pro-oxidant menadione. Comp Biochem Physiol Part C 151:167–174
26. Goldberg EG (1986) The mussel watch concept. Environ Monit Assess 7:91–103
27. Baumard P, Budzinski H, Garrigues P, Dizer H, Hansen PD (1999) Polycyclic aromatic hydrocarbons in recent sediments and mussels (*Mytilus edulis*) from the Western Baltic Sea: occurrence, bioavailability and seasonal variations. Marine Environ Res 47:17–47
28. Dyballa N, Metzger S (2009) Fast and sensitive colloidal coomassie G-250 staining for protein in polyacrylamide gels. J Vis Exp 30: e1431. doi: 10.3791/1431
29. Grabar KC, Freeman RG, Hommer MB, Natan MJ (1995) Preparation and characterization of Au colloid monolayers. Anal Chem 67:735–743
30. Laemmli UK (1970) Cleavage of structural proteins during the assembly of the head of bacteriophage T4. Nature 227:680–685

Chapter 13

Oxidative Biomarkers to Assess the Nanoparticle-Induced Oxidative Stress

Rama Narsimha Reddy Anreddy, Narsimha Reddy Yellu, and Krishna R. Devarakonda

Abstract

Nanotechnology involves the creation and manipulation of materials at nanoscale levels to create products that exhibit novel properties. Engineered nanomaterials either metals (like carbon and silver) or metal oxides (like zinc oxide, magnesium oxide, and titanium oxide) induce toxicity and oxidative stress by generating free radicals. Various in vitro and in vivo models are available to estimate the oxidative stress induced by the nanoparticles. In this chapter, we describe the methods for the estimation of oxidative stress markers like reactive oxygen species (ROS), DNA damage estimation, and lipid peroxidation products; total antioxidant capacity (TAC) was mentioned.

Key words Nanoparticles, Oxidative stress, ROS, Reactive oxygen species, Lipid peroxidation, DNA damage, TAC

1 Introduction

Oxidative stress (OS) is proposed to be involved in several human diseases, such as cardiovascular and neurological diseases, cancer, and inflammation-related diseases and aging. Reactive oxygen species (ROS) are physiologically produced in the cell consequently for energy production through molecular oxygen reduction to water during aerobic metabolism [1]. However, excess production or impaired elimination of free radicals leads to increased oxidative stress. In fact, these species may target a wide range of molecules including proteins, nucleic acids, lipids, carbohydrates, and cause marked alterations to a variety of cellular pathways [1].

Nanotechnology involves the creation and manipulation of materials at nanoscale levels to create products that exhibit novel properties. Nanomaterials (NP) have gained enormous attention and are currently being widely used in modern technology. Engineered nanomaterials, or nanoparticles (NP), are defined as materials produced within the range of 1–100 nm in length or

Donald Armstrong and Dhruba J. Bharali (eds.), *Oxidative Stress and Nanotechnology: Methods and Protocols*, Methods in Molecular Biology, vol. 1028, DOI 10.1007/978-1-62703-475-3_13, © Springer Science+Business Media New York 2013

diameter. Nanoparticles are of particular concern due to their extremely high surface area and increased reactivity. Because of these unique physical and chemical properties, NP can potentially impact the health of those exposed to them during industrial manufacturing and production. Although the applications and benefits of these engineered nanomaterials are extensive [2, 3], the human exposure also increases.

Both in vivo and in vitro studies of nanosized particles of various chemistries have been shown to create ROS. ROS production has been found in NP as diverse as carbon nanotubes, C60 fullerenes, quantum dots, and ultrafine particles (UFP), especially under concomitant exposure to light, UV, or transition metals [4, 5]. It has been demonstrated that NP of various sizes and various chemical compositions preferentially mobilize to mitochondria [6]. Since mitochondria are redox-active organelles, there is a likelihood of altering ROS production and thereby overloading or interfering with antioxidant defenses. The exact mechanism by which each of these diverse NP causes ROS is not yet fully understood, but suggested mechanisms include (1) photoexcitation of fullerenes and NP causing intersystem crossing (ISC) to create free electrons; (2) metabolism of NP to create redox-active intermediates, especially if metabolism is via cytochrome P450s; and (3) inflammation responses in vivo which may cause oxyradical release by macrophages [7].

Most of the nanomaterials either metals (like carbon, Fe, silver) or metal oxides (like titanium oxide, magnesium oxide) produce toxicity in vitro and in vivo [8–12]. The induction of oxidative stress by the nanoparticles in vivo [13, 14] and in vitro [12, 15] is also reported and depends on the physicochemical properties of nanoparticles like its size and shape [15].

But the methods used to estimate the oxidative stress induced by these nanomaterials are not specific and are indirect methods like estimation of lipid peroxidation products, LDH release, IL-8, and circulating antioxidants like catalase. In the present study, we have described the specific methods and protocols for the estimation of oxidative biomarkers (like ROS, DNA damage) to assess the nanoparticle-induced OS.

2 Materials

Prepare all solutions using ultrapure water (prepared by purifying deionized water to attain a sensitivity of 18 M Ω cm at 25 °C) and analytical grade reagents. Prepare and store all reagents at cool temperature preferably at −20 °C (unless indicated otherwise). Avoid multiple freeze/thaw cycles. All the kits were procured from Cell Biolabs OxiSelect kits, India (unless indicated otherwise).

2.1 Nanomaterials Suspension

1. Suspend the nanomaterials initially in phosphate buffer saline (PBS) or distilled water. If needed, add any suspending agent like 1 % Tween 80.
2. The final concentrations should be prepared in cell culture medium.
3. To these concentrations add required amount of gentamicin to prevent microbial growth (*see* **Note 1**).

2.2 Cell Culture Medium Components

2.2.1 Preparation of Cell Culture Medium

1. Weigh 10.39 g (one vial) of DMEM (Dulbecco's Modified Eagle Medium) or RPMI 1640 and 2 g of sodium bicarbonate in a sterile conical flask.
2. Dissolve the contents in 900 mL of Milli-Q distilled water by stirring and add the antibiotic solutions (1 mL/L penicillin G sodium, 1 mL/L of streptomycin, and 1 mL/L of amphotericin B) and adjust the pH to 7.2–7.5 with 0.1 N HCl after all the contents are added.
3. Sterilize this solution using 0.22 μm membrane filtration technique in asceptic area.
4. Add 100 mL of sterile fetal bovine serum to 900 mL medium (*see* **Note 2**).
5. Store at cool condition (*see* **Note 3**).

2.2.2 Phosphate Buffer Solution

1. 9.6 % w/v of PBS in distilled water and stored at cool condition after moist heat sterilization (*see* **Note 4**).

2.2.3 Preparation of Trypsin-EDTA

1. 0.125 % or 0.25 % of trypsin and 1 % of EDTA prepared and sterilized using 0.22 μm syringe filtration technique (*see* **Note 5**).

2.3 ROS Estimation Components

1. DCFH-DA stock 10 mM: Prepared by dissolving 5 mg in 1 mL of DMSO. Working stock should be prepared using PBS buffer in amber color tube (*see* **Note 6**).
2. DCF standard: One 100 μL amber tube of a 1 mM solution in DMSO.
3. Hydrogen peroxide: One 100 μL amber tube of an 8.821 M solution.

2.4 Components in 8-OhdG Quantitation Kit

1. 96-Well protein-binding plate: One strip of 96-well plate.
2. Anti-8-OHdG antibody: One 15 μL vial of anti-8-OHdG.
3. Secondary antibody, HRP conjugate: One 50 μL vial.
4. Assay diluent: One 50 mL bottle.
5. 10× wash buffer: One 100 mL bottle.
6. Substrate solution: One 12 mL amber bottle.
7. Stop solution: One 12 mL bottle.

8. 8-OHdG standard: One 100 μL vial of 2 μg/mL 8-OHdG in 1× PBS, 0.1 % BSA.
9. 8-OHdG conjugate: One 20 μL vial of 8-OHdG/BSA conjugate at 1.0 mg/mL in PBS.
10. DNA extraction kit.

2.4.1 Preparation of Reagents

1. 8-OHdG coated plate: Dilute the proper amount of 8-OHdG conjugate (1 mg/mL) to 1 μg/mL in 1× PBS. Add 100 μL of the 1 μg/mL 8-OHdG conjugate to each well and incubate overnight at 4 °C. Remove the 8-OHdG coating solution and wash once with dH_2O. Blot plate on paper towels to remove excess fluid. Add 200 μL of assay diluent to each well and block for 1 h at room temperature. Transfer the plate to 4 °C and remove the assay diluent immediately before use (*see* **Note** 7).
2. 1× Wash buffer: Dilute the 10× wash buffer concentrate to 1× with deionized water. Stir to homogeneity.
3. Anti-8-OHdG antibody and secondary antibody: Dilute immediately before use the anti-8-OHdG antibody 1:500 and secondary antibody 1:1,000 with assay diluent. Do not store diluted solutions.

2.5 Reagents for TBARS Estimation

1. Trichloroacetic acid 15 % (TCA).
2. Thiobarbituric acid (TBA).
3. Malondialdehyde tetraethylacetal (1,1,3,3-tetraethoxypropane).
4. Diethyl maleate, 97 % (DEM).
5. 1-Butanol, spectrophotometric grade.
6. Quick Start Bradford dye reagent, 1×.

2.6 Reagents for Glutathione Estimation

1. 5-5′-Dithiobis(2-nitrobenzoic acid) (DTNB).
2. 5-Sulfosalicylic acid dihydrate (SSA).
3. Diethyl maleate, 97 % (DEM).
4. 1-Methyl-4-vinyl-pyridinium (M4VP).
5. Oxidized glutathione (GSSG).
6. Reduced glutathione (GSH).

2.7 Reagents in TAC Estimation Kit

1. Uric acid standard: One 100 mg tube of powder.
2. Reaction buffer (100×): One 400 μL amber vial.
3. Copper ion reagent (100×): One 1.0 mL vial.
4. Stop solution (10×): One 1.5 mL vial.
5. 1 N NaOH, 1× PBS, and deionized water.

2.7.1 Preparation of Reagents

Reagents may be prepared for either hydrophilic or lipophilic samples. Although many lipophilic samples are soluble upon dilution with

1× PBS, the kit reagents may be prepared in methanol to ensure complete solubility:

1. 1× Reaction buffer: Dilute the reaction buffer 1:100 with 1× PBS (hydrophilic) or with methanol (lipophilic). Mix to homogeneity. Store the 1× reaction buffer at 4 °C up to 3 months.
2. 1× Copper ion reagent: Dilute the copper ion reagent 1:100 with deionized water (hydrophilic) or with methanol (lipophilic). Mix to homogeneity. Store the 1× copper ion reagent at 4 °C up to 3 months.
3. 1× Stop solution: Dilute the stop solution 1:10 with deionized water (hydrophilic) or with methanol (lipophilic). Mix to homogeneity. Store the 1× stop solution at 4 °C up to 3 months.

2.8 Equipment

1. ELISA plate reader.
2. Multichannel pipette.
3. CO_2 incubator.
4. Laminar air flow.
5. Fluorescence multiple plate reader.
6. Cooling centrifuge.
7. Vortex.
8. Sonicator.
9. Filters.
10. UV–visible spectrophotometer.
11. Adjustable pipettes.
12. Shaker.
13. Millipore distillation unit.
14. Tissue homogenizer.

3 Methods

Carry out all procedures at room temperature in an aseptic area unless otherwise mentioned.

3.1 Estimation of ROS

This is a cell-based assay for measuring hydroxyl, peroxyl, or other ROS activity within a cell. The generation of intracellular ROS was determined using dichlorodihydrofluorescein diacetate (DCFH-DA) [16]. The assay employs the cell-permeable fluorogenic probe 2′, 7′-dichlorodihydrofluorescein diacetate (DCFH-DA). In brief, DCFH-DA is diffused into cells and is deacetylated by cellular esterases to nonfluorescent 2′, 7′-dichlorodihydrofluorescein (DCFH), which is rapidly oxidized to highly fluorescent 2′, 7′-dichlorodihydrofluorescein

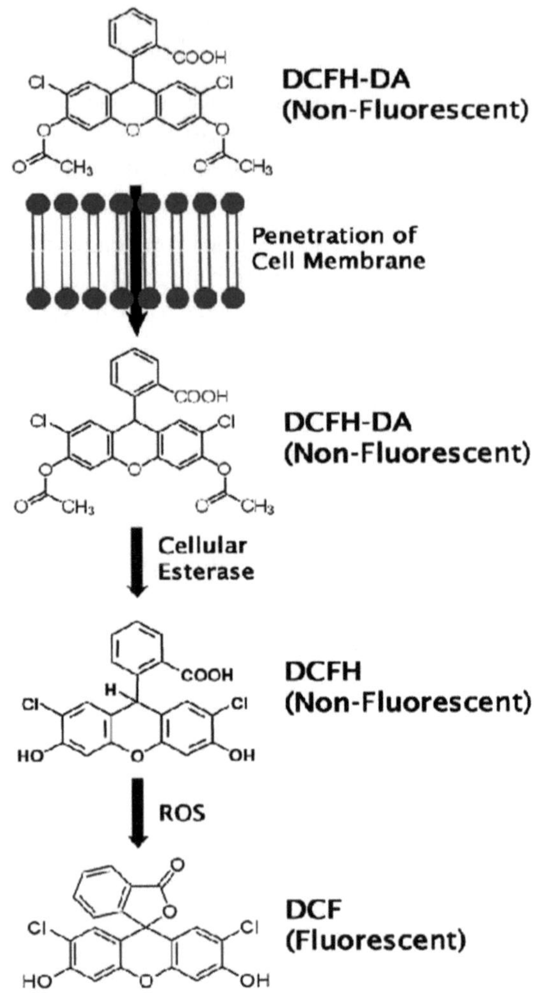

Fig. 1 Mechanism of DCF assay

(DCF) by ROS (Fig. 1). The fluorescence intensity is proportional to the ROS levels within the cell cytosol [17].

1. Seed 0.1 mL of the cells suspension containing 1×10^8 cells to each well in black 96-well plates 24 h before the exposure. Treat the cells and handle them in a light-controlled environment throughout the assay in aseptic laminar flow.
2. Prior to treatment, aspirate the growth media and replace with 300 μL media containing 100 μM DCFH-DA and incubate for 20–30 min under normal cell culture conditions.
3. Centrifuge the plates at $70 \times g$ for 4 min without brake. Then remove the DCFH-DA probe and wash the plate with 200 μL of PBS/media at $70 \times g$ for 4 min.
4. Read zero-time measurement and, then, add 200 μL dosing solutions of nanoparticles.

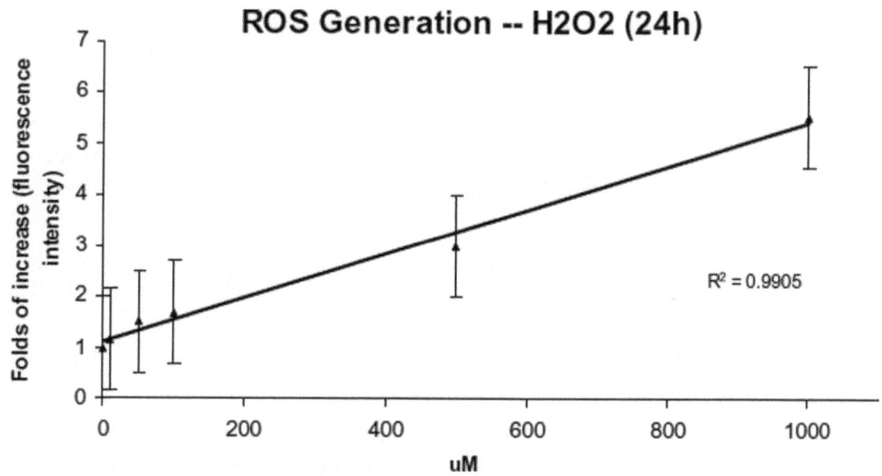

Fig. 2 Standard curve of ROS generation vs. H_2O_2

Fig. 3 DCF fluorescence in H_2O_2-treated HeLa cells after 1 h. Right, 0 μM; *Left side*, 1,000 μM

5. Place back the plates in the incubator and assay after different postexposure interval or at 24 h. Measure the fluorescent values of control and exposed cells at an excitation wavelength of 485 nm and an emission wavelength of 530 nm in the 96-well plate using fluorescent microplate reader.
6. Report the data as a fold of increase when compared to control. Assay the standard curve of H_2O_2 concentrations in conjunction with treated cells (Figs. 2 and 3).

3.2 DNA Damage Quantification

DNA is probably the most biologically significant target of oxidative attack, and it is widely thought that continuous oxidative damage to DNA is a significant marker to assess the oxidative stress induced by the nanoparticles. Among numerous types of oxidative DNA damage, the formation of 8-hydroxydeoxyguanosine (8-OHdG) is a ubiquitous marker of oxidative stress. 8-OHdG, one of the oxidative DNA damage byproducts, is physiologically formed and enhanced by chemical carcinogens. During the repair of damaged DNA in vivo by exonucleases, the resulting 8-OHdG is excreted without further metabolism into urine and is estimated using comet assay [18, 19].

The unknown 8-OHdG samples or 8-OHdG standards are first added to an 8-OHdG/BSA conjugate preabsorbed EIA plate. After a brief incubation, an anti-8-OHdG monoclonal antibody is added, followed by an HRP-conjugated secondary antibody. The 8-OHdG content in unknown samples is determined by comparison with predetermined 8-OHdG standard curve [18, 19].

3.2.1 Urine, Plasma, or Serum Samples

1. Clear urine, plasma, or serum samples can be diluted with assay diluent and used directly in the assay.
2. Samples containing precipitates should be centrifuged at 3,000 × *g* for 10 min, or filtered through 0.45 μm filter, prior to use in the assay.

3.2.2 Cell or Tissue DNA Samples

1. Extract DNA from cell or tissue samples by a desired method or commercial DNA extraction kit.
2. Dissolve extracted DNA in water at 1–5 mg/mL.
3. Convert DNA sample to single-stranded DNA by incubating the sample at 95 °C for 5 min and rapidly chilling on ice.
4. Digest DNA sample to nucleosides by incubating the denatured DNA with 5–20 units of nuclease P1 for 2 h at 37 °C in 20 mM sodium acetate at pH 5.2 and following with treatment of 5–10 units of alkaline phosphatase for 1 h at 37 °C in 100 mM Tris at pH 7.5.
5. The reaction mixture is centrifuged for 5 min at 6,000 × *g* and the supernatant is used for the 8-OHdG ELISA assay.

3.2.3 DNA Damage Quantification Protocols

1. Prepare and mix all reagents thoroughly before use. Each 8-OHdG sample including unknown and standard should be assayed in duplicate. High-content 8-OHdG urine or serum samples should be diluted at least 10–20-fold in assay diluent.
2. Add 50 μL of unknown sample or 8-OHdG standard to the wells of the 8-OHdG conjugate coated plate. Incubate at room temperature for 10 min on an orbital shaker.
3. Add 50 μL of the diluted anti-8-OHdG antibody to each well and incubate at room temperature for 1 h on an orbital shaker.
4. Wash microwell strips three times with 250 μL 1× wash buffer per well with thorough aspiration between each wash. After the last wash, empty wells and tap microwell strips on absorbent pad or paper towel to remove excess 1× wash buffer.
5. Add 100 μL of the diluted secondary antibody-enzyme conjugate to all wells.
6. Incubate at room temperature for 1 h on an orbital shaker.

7. Wash microwell strips three times according to **step 4** above. Proceed immediately to the next step.
8. Warm substrate solution to room temperature. Add 100 μL of substrate solution to each well, including the blank wells. Incubate at room temperature on an orbital shaker. Actual incubation time may vary from 2 to 30 min (*see* **Note 8**).
9. Stop the enzyme reaction by adding 100 μL of stop solution into each well, including the blank wells. Results should be read immediately (color will fade over time).
10. Read absorbance of each microwell on a spectrophotometer using 450 nm as the primary wave length.

3.3 Estimation of Lipid Peroxidation Products

Lipid peroxidation is an indicator of oxidative stress. This protocol describes the estimation for lipid peroxidation products such as malondialdehyde (MDA), following treatment with nanoparticle suspensions.

The thiobarbituric acid-reactive substances (TBARS) assay measures lipid hydroperoxides and aldehydes, such as malondialdehyde (MDA), in the cell culture media and cell lysate. MDA combines with thiobarbituric acid (TBA) in a 1:2 ratio to form a fluorescent adduct that is measured at ex. 530 nm and em. 550 nm. TBARS are expressed as MDA equivalents [20].

3.3.1 Cell Preparation

1. Dilute cells to a density of 7.5×10^5 cells/mL in RPMI 1640 or suitable cell culture media (2 mM l-glutamine, 10 % FBS). Plate 2 mL of diluted cell suspension to each well of a 6-well plate (1.5×10^6 cells/well). All samples and controls are run in triplicate.
2. Incubate plates for 24 h at 5 % CO_2, 37 °C, and 95 % humidity (cells should be approximately 80 % confluent).
3. Replace cell culture media with media containing test nanomaterial or positive control. Treat cells for designated time period.

3.3.2 Cell Media Sample Preparation

1. Collect cell media following treatment with nanomaterials.
2. Add 500 μL of sample media (or appropriate MDA stock for standard curve sample preparation, or blank media) to 400 μL 15 % TCA and 800 μL of 0.67 % TBA/0.01 % BHT in a 5 mL amber vial. Vortex and heat for 20 min in 95 °C water bath. Allow to cool and add 3 mL butanol, gently mix phases, and transfer 200 L of the butanol phase (top) to 96-well plate.

3.3.3 Cell Lysate Sample Preparation

1. Wash 96-well plates with ice-cold PBS.
2. Scrape cells into 1 mL 2.5 % TCA.
3. Centrifuge cells at 13,000 ×*g* for 2 min.
4. Retain pellet for determination of cellular protein by Bradford assay; can be frozen at −20 °C until analysis.

5. Remove 500 μL of lysate supernatant (or appropriate MDA stock for standard curve sample preparation, or blank media) and add to 400 μL 15 % TCA and 800 μL of 0.67 % TBA/0.01 % BHT in a 5 mL amber vial. Vortex and heat for 20 min in 95 °C water bath. Allow to cool and add 3 mL butanol, gently mix phases, and transfer 200 μL of the butanol phase (top) to 96-well plate.
6. Read cell media and cell lysate plate in fluorescence mode, ex. = 530 nm and em. = 550 nm.

3.3.4 Protein Determination (Bradford Assay)

1. Dilute the 2 mg/mL BSA standard to make a standard curve from 0.125 to 1.0 mg/mL in 0.05 N NaOH.
2. Resuspend pellets from **step 4** of subheading 3.3.3 in 0.5 mL of 0.05 N NaOH.
3. Add 5 μL of standard, resuspended protein pellet or 0.05 N NaOH blank to each well of a microtiter plate in duplicate.
4. Add 250 μL of 1× dye reagent to each well of the plate. Incubate at room temperature for at least 5 min, but not longer than 1 h.
5. Read on a microtiter plate at 595 nm [21].

3.3.5 Calculations

1. TBARS assay concentrations are determined by comparison to an MDA standard curve following linear regression analysis ($y = x$(slope) + y int), expressed as MDA equivalents and normalized to total protein.
2. Protein concentration is determined from the BSA standard curve following linear regression analysis ($y = x$(slope) + y int). Total protein is determined from the equation: total protein = (mg/mL protein × 0.5 mL).
3. Total lysate or media MDA equivalents normalized to total protein = [(MDA, ng/mL) × 1 mL/total protein, mg] = ng MDA/mg protein.
4. Mean, SD, and %CV should be calculated for each positive control and sample (*see* **Notes 9** and **10**).

3.4 Estimation of Glutathione

This protocol describes the analysis of cell suspension (Hep G2) for reduced and oxidized glutathione, following treatment with nanoparticle formulations. A shift from reduced to oxidized glutathione is indicative of oxidative stress, while a decrease in the overall reduced glutathione pool is indicative of conjugative metabolism or impaired synthesis [22]. This protocol utilizes a colorimetric method for glutathione determination.

Reduced glutathione (GSH) reduces with 5-5′-dithiobis(2-nitrobenzoic acid) (DTNB) to form the colored product 2-nitro-5-thiobenzoic acid, which is measured kinetically at 415 nm, and oxidized glutathione (GSSG). GSSG is then reduced by glutathione reductase to form reduced glutathione GSH, which is again measured by the preceding method. Preincubation of the sample with the thiol-masking agent 1-methyl-4-vinyl-pyridinium prevents measurement of GSH, resulting in measurement of GSSG only [23].

3.4.1 Cell Preparation (Or as Recommended by Supplier)

1. Plate 2 mL of diluted cells (1.5×10^6 cells/well) to each well of a 6-well plate. Test samples and positive controls are run in triplicate. Incubate plates for 24 h at 5 % CO_2, 37 °C, and 95 % humidity (cells should be approximately 80 % confluent).
2. Replace cell culture media with media containing test nanomaterials or positive control. Treat cells for designated time period. Remove media and wash well with 1 mL of room temperature PBS.
3. Remove PBS, add 100 μL ice-cold 5 % 5-sulfosalicylic acid dihydrate (SSA) to the plate, and scrape cells. Transfer lysed cells to 0.6 mL Eppendorf tubes and incubate for 10 min on ice. Centrifuge at $8{,}000 \times g$ for 5 min. Prepare supernatants and retain pellet for determination of cellular protein by Bradford assay. Pellet can be frozen at −20 °C until analysis.
4. Total GSH assay: Dilute 5 μL of supernatant 1:2 with 5 % SSA dilution, further dilute to 1:2 with 400 mM sodium carbonate, and then further dilute 1:8 with phosphate-EDTA dilution buffer (total dilution 1:32). Transfer supernatant to 96-well plate.
5. GSSG assay: Dilute 10 μL of supernatant 1:2 with 10 μL of M4VP-masking reagent in Eppendorf tube and then further dilute to 1:2 with 400 mM sodium carbonate (total dilution 1:4). Incubate samples for 2 min at room temperature. GSSG QC standard is treated in the same manner as the samples; total dilution will be 1:4. Transfer samples to 96-well plate.

3.4.2 Glutathione Assay Protocol

1. Add 170 μL of reaction buffer to each standard and sample well of a 96-well plate. Incubate for 10 min at RT.
2. Add 20 μL of each GSH standard, total GSH QC standard, GSSG QC standard, sample, DEM positive control, or blank to each well. Mix and incubate for 10 min at RT.
3. Add 10 μL of DTNB substrate solution to each well. Vortex.
4. Read 415 nm absorbance kinetically, 5 cycles at 300 s intervals on a microplate reader.

3.4.3 Calculations

1. Assay total GSH and GSSG concentrations are determined from their respective GSH standard curves following linear regression analysis ($y = x$(slope) + y int) and then normalized to total protein (*see* **Note 11**).

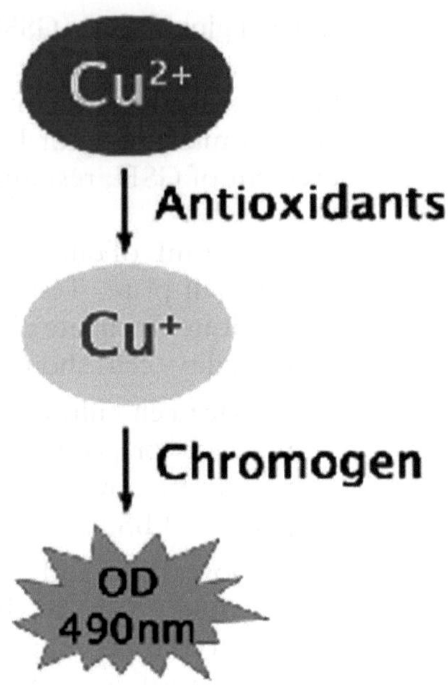

Fig. 4 TAC activity assay principle

3.5 Total Antioxidant Capacity Assay

Samples are compared to a known concentration of uric acid standard within a 96-well microtiter plate format. Samples and standards are diluted with a reaction reagent and, upon the addition of copper, the reaction proceeds for a few minutes. The reaction is stopped and read with a standard 96-well spectrophotometric microplate reader at 490 nm (Fig. 4). Antioxidant capacity is determined by comparison with the uric acid standards [24, 25].

Samples should be stored at −80 °C prior to performing the assay. Samples should be prepared at the discretion of the user. The following recommendations are only guidelines and may be altered to optimize or complement the user's experimental design (*see* **Note 12**):

1. Tissue lysate: Sonicate or homogenize tissue sample on cold PBS and centrifuge at 10,000 × *g* for 10 min at 4 °C. Aliquot the supernatant for storage at −80 °C, protein determination, and subsequent TAC assay.
2. Cell culture: Wash cells three times with cold PBS prior to lysis. Lyse cells with sonication or homogenation in cold PBS and centrifuge at 10,000 × *g* for 10 min at 4 °C. Aliquot the supernatant for storage at −80 °C, protein determination, and subsequent TAC assay.
3. Plasma: Collect blood with heparin and centrifuge at 4 °C for 10 min. Remove the plasma and aliquot samples for testing.

3.5.1 Preparation of Uric Acid Standard Curve

1. Prepare fresh uric acid standards by weighing out the uric acid powder for a 10 mg/mL solution in 1 N NaOH. This 10 mg/mL is equivalent to a concentration of 60 mM. Use the 60 mM uric acid solution to prepare a 2 mM solution of uric acid (e.g., add 100 μL of the 60 mM uric acid standard to 2.900 mL of deionized water).
2. Prepare a series of the remaining uric acid standards.

3.5.2 Total Antioxidant Capacity Assay Protocols

Each uric acid standard and sample should be assayed in duplicate or triplicate. A freshly prepared standard curve should be used each time the assay is performed:

1. Add 20 μL of the diluted uric acid standards or samples to the 96-well microtiter plate.
2. Add 180 μL of the 1× reaction buffer to each well using either a multichannel pipette or a plate-reader liquid-handling system. Mix thoroughly.
3. Obtain an initial absorbance by reading the plate at 490 nm.
4. To initiate the reaction, add 50 μL of the 1× copper ion reagent into each well. Incubate for 5 min on an orbital shaker.
5. Add 50 μL of 1× stop solution to each well to terminate the reaction.
6. Read the plate again at 490 nm.

3.5.3 Calculation of Results

1. Calculate the net absorbance by subtracting the initial absorbance readings for samples and standards (**step 3**) from the final readings taken for each (**step 6**).
2. Calculate the antioxidant capacity of unknown samples by comparing the net OD 490 nm values of samples to the uric acid standard curve.

4 Notes

1. If the nanomaterials are thermostable, sterilize the powdered nanoparticles at 360 °C for 1 h using hot air oven technique. Then prepare the nanoparticles suspension with sterile distilled water and is ready to use.
2. Check the prepared cell culture medium for the sterility by keeping 5–10 mL of the medium in an incubator at 37 °C for 24 h.
3. Before using the cell culture medium should be at room temperature.
4. PBS solution is used as washing buffer to remove the dead cells/debris in a tissue culture flask.

5. Do not add excess amount (or do not incubate for longer periods) of trypsin solution to the monolayer of the cell culture during trypsinization as it causes cell membrane damage.
6. We can use methanol also as solvent for DCFH-DA. Due to light-induced autoxidation, both stock and working DCFH-DA solutions must be protected from light.
7. 8-OHdG coated plate is not stable. We recommend using it within 24 h after coating.
8. Watch plate carefully; if color changes rapidly, the reaction may need to be stopped sooner to prevent saturation.
9. The positive control and sample replicate coefficient of variations should be within 50 %. The assay is acceptable if this condition is met. Otherwise, the assay should be repeated until acceptance criteria are met.
10. If statistical assays determine that the total protein normalized control and treated fluorescence are significantly different from one another, then the fold change in fluorescence can be considered meaningful. This result would indicate that sample treatment significantly affected cellular lipid peroxidation.
11. The protein normalized total GSH concentration of the DEM positive control should be at least 40 % lower than the media control at 6 h. The positive control and sample replicate coefficient of variations should be within 50 %. The assay is acceptable if these conditions are met. Otherwise, the assay should be repeated until acceptance criteria are met.
12. EDTA can interfere with the TAC assay and should not be present in any sample.

References

1. Griendling KK, FitzGerald GA (2003) Oxidative stress and cardiovascular injury: part I: basic mechanisms and in vivo monitoring of ROS. Circulation 108:1912–1916
2. Hoet PH, Bruske-Hohlfeld I, Salata OV (2004) Nanoparticles—known and unknown health risks. J Nanobiotechnol 1:12–14
3. Thomas K, Sayre P (2005) Research strategies for safety evaluation of nanomaterials, part I: evaluating the human health implications of exposure to nanoscale materials. Toxicol Sci 2:316–321
4. Brown DM, Stone V, Findlay P et al (2000) Increased inflammation and intracellular calcium caused by ultrafine carbon black is independent of transition metals or other soluble components. Occupat Environ Med 57:685–691
5. Derfus AM, Chan WCW, Bhatia SN et al (2004) Probing the cytotoxicity of semiconductor quantum dots. Nano Lett 4(1):11–18
6. Foley S, Crowley C, Smaihi M et al (2002) Cellular localisation of a water-soluble fullerene derivative. Biochem Biophys Res Commun 294(1):116–119
7. Oberdorster G, Maynard A, Yang H et al (2005) Principles for characterizing the potential human health effects from exposure to nanomaterials: elements of a screening strategy 1. Part Fibre Toxicol 2(1):8–14
8. Reddy ARN, Rao MV, Krishan DR et al (2012) Pulmonary toxicity assessment of multi-wall carbon nanotubes in rats following intratracheal instillation. Environ Toxicol 27(4): 211–219

9. Anreddy RNR, Devarakonda RK, Vurimindi H et al (2011) In vitro toxicity of multi wall carbon nanoparticles on Hep G 32 liver cell lines. Lat Am J Pharm 30(1):177–80
10. Lam C, James TJ, McClusky R, Hunter LR (2003) Pulmonary toxicity of carbon nanotubes in mice 7 and 90 days after intratracheal instillation. Toxicology 72(S-1):44–48
11. Hussain SM, Hess KL, Gearhart JM, Geiss KT (2005) In vitro toxicity of nanoparticles in BRL 3A rat liver cells. Toxicol In Vitro 19:975–983
12. Anreddy RNR, Yellu NR, Devarakonda RK et al (2010) Multi wall carbon nanoparticles induce oxidative stress and cytotoxicity in human embryonic kidney (HEK293) cells. Toxicology 272:11–16
13. Reddy ARN, Rao MV, Krishan DR et al (2011) Evaluation of oxidative stress and antioxidant status in rats following exposure of carbon nanotubes. Regul Toxicol Pharmacol 59: 251–257
14. Reddy ARN, Rao MV, Krishan DR et al (2011) Induction of oxidative stress and cytotoxicity by carbon nanomaterials is dependent on physical properties. Toxicol Ind Health 27:3–10
15. Oberdorster E (2004) Manufactured nanomaterials (fullerenes, C60) induce oxidative stress in the brain of juvenile largemouth bass. Environ Health Perspect 112:1058–1062
16. Reddy ARN, Rao MV, Krishan DR et al (2011) Induction of oxidative stress and cytotoxicity by carbon nanomaterials is dependent on physical properties. Toxicol Ind Health 27:3–10
17. Wang H, Joseph JA (1999) Quantifying cellular oxidative stress by dichlorofluorescein assay using microplate reader. Free Radic Biol Med 5–6:612–616
18. Keston AS, Brandt R (1965) The fluorometric analysis of ultramicro quantities of hydrogen peroxide. Anal Biochem 11:1–5
19. Patel PR, Bevan RJ, Mistry N et al (2007) Evidence of oligonucleotides containing 8-hydroxy-2′-deoxyguanosine in human urine. J Free Radic Biol Med 42:552–558
20. Shen J, Deininger P, Hunt JD, Zhao H (2007) 8-Hydroxy-2′-deoxyguanosine (8-OH-dG) as a potential survival biomarker in patients with non-small cell lung cancer. Cancer 109(3): 574–580
21. Dubuisson ML, de Wergifosse B, Trouet A, Baguet F et al (2000) Antioxidative properties of natural coelenterazine and synthetic methyl coelenterazine in rat hepatocytes subjected to tert-butyl hydroperoxide-induced oxidative stress. Biochem Pharmacol 60:471–478
22. Bradford MM (1976) A rapid and sensitive method for the quantitation of microgram quantities of protein utilizing the principle of protein-dye binding. Anal Biochem 72:248–254
23. Deleve LD, Kaplowitz N (1991) Glutathione metabolism and its role in hepatotoxicity. Pharmacol Ther 52:287–305
24. Shaik IH, Mehvar R (2006) Rapid determination of reduced and oxidized glutathione levels using a new thiol-masking reagent and the enzymatic recycling method: application to the rat liver and bile samples. Anal Bioanal Chem 385:105–113
25. Cerutti P, Trump B (1991) Inflammation and oxidative stress in carcinogenesis. Proc Cancer Cell 3:1–7
26. Trachootham D, Lu W, Ogasawara MA et al (2008) Redox regulation of cell survival. Antioxid Redox Signal 10:1343–1374

Chapter 14

Vesicular Antioxidants: Role in Age-Related Cerebral Oxidative Injury

Sibani Sarkar, Ardhendu Kumar Mandal, and Nirmalendu Das

Abstract

Oxidative stress, due to the generation of reactive oxygen species, is a major factor in cerebral ischemic damage and changes the activities of antioxidant enzymes and substantially influences the aging process. Free chemical antioxidant is almost ineffective to treat brain ischemia as blood–brain barrier exists in between blood and brain interstitial fluid, limiting component to pass from the circulation into cerebral region. Different compounds have been tested in vivo in different vesiculated forms to prevent cerebral ischemia. Nanoparticle-encapsulated drug treatment resulted in a significant protection of the antioxidant enzymes in both young and old rats. Nanocapsulated drug treatment causes a substantial protection against cerebral ischemia–reperfusion-induced oxidative damage to all parts of brain specifically hippocampal regions of all age groups of rat brain.

Key words Oxidative stress, Antioxidant, Blood–brain barrier, Nanocapsulated drugs, Aging, Hippocampal region

1 Introduction

When parts of the brain suffer a deprivation of oxygen, such as stroke, damage to neuron occurs that is exuberated by the subsequent reperfusion of the affected part of the brain by oxygenated blood. So, maintaining a continuous oxygen supply to the brain is important for its survival. Massive oxidative stress damages the neurons under these conditions, including cell death by necrosis and apoptosis. Studies have shown that elevation in ROS during cerebral ischemia reperfusion results in irreversible neuronal damages [1]. These effects play an important role in the process of aging. Oxidative stress is a major factor in cerebral ischemic damage because the brain contains large amount of polyunsaturated fatty acids and consumes huge quantities of oxygen. The reactive oxygen and nitrogen species can lead to lipid peroxidation, protein oxidation, and DNA damage in brain after ischemia/reperfusion injury [2]. Moreover, oxidative damage due to reactive oxygen

Donald Armstrong and Dhruba J. Bharali (eds.), *Oxidative Stress and Nanotechnology: Methods and Protocols*, Methods in Molecular Biology, vol. 1028, DOI 10.1007/978-1-62703-475-3_14,

radicals has complex interactions with excitotoxicity, apoptosis, and inflammation [3]. The multiple pathophysiological mechanisms (excitotoxicity, oxygen free radicals, per infarct depolarization, inflammation, and apoptosis) are involved in post-ischemic neuronal progressive cell death in the penumbra [4, 5]. Mitochondria are the major target of free radical attack that leads to human aging. The rate of mitochondrial generation of the superoxide anion and declination of mitochondrial function are increased gradually with aging. In the initial phase of free radical-mediated neurodegenerative diseases, neuronal cells trigger their effective protection mechanism to counter the oxidative damage. Thus, the survival of neurons depends on the equilibrium between free radical attack and their antioxidant defense mechanism.

In aging of the brain, increase of oxidative potential and loss of proper antioxidant defense appear to be involved. By introducing biological antioxidant attempts will be made to protect those cells from oxygen free radical attack in those diseases. Hence, there is a need for better and more potent compounds, especially on dietary origin, to boost antioxidant defense. Increased production of reactive oxygen species during ischemia and reperfusion injury leads to a rapid consumption and depletion of endogenous scavenging antioxidants. In this context, it is necessary to introduce exogenous antioxidant as a drug as well as scavenger to counter cerebral ischemia and reperfusion oxidative attack especially in aged individual. Some exogenous antioxidant compounds such as quercetin [6], Ginkgo biloba, and CDP-choline [7] have already been reported to combat cerebral ischemia–reperfusion injury in young and aged rat brain.

1.1 Presence of Blood–Brain Barrier and Obstacles of Drug Entry into the Brain

Drug delivery to the brain is a challenge, because this tissue benefits from a very efficient protective barrier. Blood–brain barrier (BBB) protects the brain from foreign substances and also restricts the entry of many potentially therapeutic agents. The BBB is the major barrier for the passage of active molecules from the blood compartment to the brain. It is located at the level of the brain capillaries, where there is a convergence of different cell types: endothelial cells, pericytes, astrocytes and microglias (perivascular macrophages).

The integrity of the BBB in cerebrovascular disease due to hypertension is variable, because the underlying cerebral ischemia varies with respect to the mechanism, the severity, and the duration [8]. After ischemia, drastic reductions in cerebral blood flow in the core of the lesions typically result in rapid cell death within minutes. Activation of cytokines such as tumor necrosis factor (TNF) and interleukin-1 (IL-1) and the upregulation of cell adhesion molecules can also be observed during this disease. Lymphocytes can then penetrate the BBB, releasing proteases, particularly metalloproteinases, which induce the opening of BBB. By virtue of this opening, migration and adhesion of neutrophils, monocytes, and macrophages to the site of injury occur [9]. These mediators have been shown to alter the structure and function of the BBB [10]. The cytokines, predominantly TNF-α and IL-1 [11], and metalloproteinases contribute to open the BBB [12].

The blood–brain barrier imposes lots of limitations in the process of drug delivery toward the brain. The most challenging part of CNS treatment is drug delivery through the BBB rather than designing a new treatment. Most of the drugs that have been designed for treatment of brain disorders such as antibiotics and anticancer agents do not pass the BBB [13].

1.2 Chemotherapy Treatment of Different Vesicular Antioxidants to Combat Brain Ischemia Through Blood–Brain Barrier

In modern days central nervous system diseases represent the most dreadful and painstaking in medical science due to the presence of blood–brain barrier (BBB). Different vesicles such as liposomes, niosomes, microsphere, and nanoparticles are used for transporting the drug into the brain through BBB. Different vesicles also can be utilized to maintain drug levels in a therapeutically desirable range and increase half-lives, solubility, stability, and permeability of drugs into the brain. Importantly, they can be structurally adapted to deliver a variety of drugs, improve delivery efficiency, and reduce side effects by targeted delivery.

Liposome- or polylactide nanocapsules-mediated drug delivery to brain cells is potent among other delivery devices not only for their nontoxic nature, biodegradability, and lack of immunogenic properties but also for helping the entrapped drug to reach the target organ effectively. Liposomes >100 nm are known as effective drug carrier to the CNS, especially during cerebral ischemia–reperfusion [14].

Furthermore, owing to the presence of mannosyl receptors on the surface of brain cells, mannose-coated liposomes are effective in the site-specific drug delivery to brain cells. Again, the mannose receptors which are transmembrane glycoproteins are expressed on brain cells, particularly on the macrophages and microglia [15], and facilitate endocytosis of any mannosylated moieties binding to them.

Nanocapsule has been well accepted in present age as a potent drug delivery vesicle because of its nontoxic, biodegradability, non-immunogenic, and sustained drug-releasing ability in the biological system. The stability of the nanoparticle in the bioenvironment offers the possibility of using these vesicles as drug carriers. The use of nanoparticles as drug carriers has advantages over existing approaches to circumvent the BBB. Loaded with both hydrophobic and hydrophilic drugs, this colloidal carrier has the ability to cross the BBB to enhance drug bioavailability via particle uptake mechanisms, and therefore, it appears to have significant promise in delivering therapeutic molecules [16].

2 Materials

1. Phosphatidylethanolamine (PE), polylactide(PLA), dichloromethane, isopropyl myristate, and Tween 80 were purchased from Sigma Chemicals (St. Louis, MO, USA).
2. Urethane, ferricytochrome C, xanthine, and xanthine oxidase were purchased from Sigma Chemicals (St. Louis, MO, USA). Quercetin, Ginkgo biloba, and CDP-choline were also purchased from Sigma Chemicals.
3. All other reagents were of analytical grade.

3 Methods

3.1 Preparation of Drug-Incorporated Polylactide Nanoparticles

Nanoparticle is prepared according to the method of Seyler et al. [17]:

1. 70 mg of PLA is dissolved in 10 ml of dichloromethane (DCM).
2. Dissolve 100 mg phosphatidyl ethanolamine (PE) in 250 μl of isopropyl myristate. 4 mg of quercetin is to be added to the DCM phase.
3. Allow the organic solution to run slowly into 50 ml of 0.025 M phosphate-buffered saline (PBS) (pH 7.2) containing 0.4 % of the nonionic surfactant Tween 80 under moderate magnetic stirring for 3 h, and the traces left, if any, are removed by passing gaseous nitrogen for 5 min (*see* **Note 1–3**).
4. Centrifuge (ultra) the suspension then at 105,000 × g (35,000 rpm) in Sorvall RC-5B Plus using the rotor Sorvall T-865 for 1 h (*see* **Note 4**).
5. Wash the pellet of nanoparticles with PBS and then collect and resuspend in PBS (*see* **Note 5**).

3.2 Study Design in Animal Model

Male Wistar rats of two age groups (2 months and 20 months) weighing 160–180 g and 415–440 g, respectively, are used. The animals are acclimatized to conditions in the laboratory (16–18 °C, 60–80 % relative humidity, 12 h light/dark cycle) for 10 days before the commencement of the treatment during which they receive food (commercial pelleted rat chow, purchase from Hindustan Unilever Limited, Maharashtra, India) and drinking water ad libitum. Rats from each category (2 months old and 20 months old) randomly be subdivided into different groups of ten animals each. Two groups from each age group are given saline 5 ml/kg body weight (one group for normal but sham-operated and other group as untreated ischemia–reperfused control).

The remaining other experimental groups of rats are used for treatment with free drug and different vesiculated drug, respectively. Void vesicles do not show any therapeutic effect. As per the body weight of animals, inject nanocapsulated drug into the tail vein 3 h

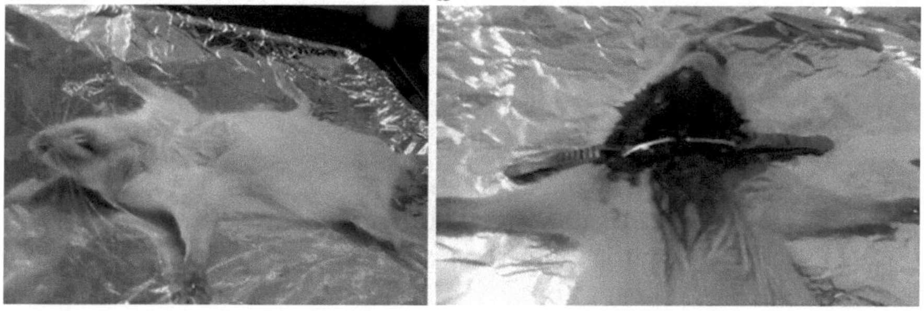

Fig. 1 Animal undergoing bilateral common artery occlusion (BCCAO) procedure. (**a**) control animal (**b**) clipping of carotid artery after BCCAO procedure

before cerebral ischemia. In control experiments the same dose of free drug in physiological saline is injected into tail vein of rats 3 h before ischemic insult. All the rats are used in the study in proper care in compliance with the Animal Ethics Committee. Minimum possible number of animal experiments are designed with reduced number of rats and shorter time period and be handled with professional care to reduce probable sufferings of animals.

3.3 Induction of Global Transient Cerebral Ischemia

1. Animals from both young and aged groups (excluding normal group) are anesthetized by an IP injection of urethane (35 mg/kg) and are made ischemic by BCCAO for 30 min (Fig. 1) (*see* **Note 6**).
2. Blood flow restores after 30 min and rats are killed by decapitation [18]. Maintain body temperature of the rats by keeping the animals under filament lamp.
3. Rat brain is isolated immediately and surface rinsed with ice-cold saline. Remove the subarachnoid membrane and vessels gently and the brain is dissected into several coronal sections [19].
4. A fraction is used to isolate mitochondria. The rest part of the brain is homogenized in phosphate-buffered saline with a Teflon-coated homogenizer (*see* **Note 7**).
5. Sham-operated animals are subjected to the same surgical procedure without BCCA.

3.4 Isolation of Mitochondria from Rat Brain

1. Dissected rat brain from experimental as well as sham-operated animals is homogenized (10 % w/v) in buffer (IM) containing 20 mM Hepes–KOH, 10 mM KCl,1.5 mM MgCl2, 1 mM EDTA, 1 mM EGTA, 1 mM DTT, and 0.1 mM PMSF [20].
2. The homogenate is centrifuged at 750 × g for 5 min in Sorvall high-speed cold centrifuge at 4 °C to remove large cell debris and nuclei. Collect the supernatant and centrifuge at 8,000 × g at 4 °C.
3. The pellet is resuspended in 2 ml 12 % (v/v) Percoll in IM loaded on a discontinuous density gradient consisting of 3 ml

26 % (v/v) Percoll and 4 ml (v/v) Percoll in IM and centrifuged at 31,000 × *g* in Sorvall SS-34 rotor.

4. Three major bands are obtained and the purified mitochondria are collected from the bottom band containing high-density mitochondria.
5. Finally wash the mitochondria with IM by centrifuging twice for 5 min at 10,000 × g and resuspend it in 200 μl IM at a mitochondrial protein concentration of approximately 12 mg/ml.

3.5 Lipid Peroxidation Assay in Brain Homogenate

1. For the determination of conjugated diene, lipids in one part of the brain homogenate (10 ml) from young and aged [19] rats are extracted in chloroform–methanol mixture (2:1, v/v).
2. Evaporate the extracted lipid solvent mixture to dryness under nitrogen atmosphere at 25 °C and redissolve in 10 ml cyclohexane. Lipid in cyclohexane solvent is assayed at 234 nm and the results are expressed as μ mol of lipohydroperoxide/mg brain protein by using an ε_m of $2.52 \times 10^4\ M^{-1}\ cm^{-1}$ [21].
3. Estimate proteins as per Lowry et al. [22].

3.6 Preparation of Cytosolic Fraction of Brain Homogenate for the Determination of Enzymes in Molecular and Enzymatic Antioxidant Level

1. Re-homogenize remaining part of the brain homogenate in 0.25 M sucrose with a Teflon-coated homogenizer and centrifuge at 10,000 × g for 15 min.
2. A portion of supernatant is further centrifuged at 105,000 × g for 60 min.
3. The supernatant from the second centrifugation contains the cytosolic fraction of brain cells [23].

3.7 The Estimation of Antioxidant Enzymes Activities from the Cytosolic Fraction of Brain Homogenate

The catalase activity in young and aged rat brain homogenate is measured spectrophotometrically by calculating the rate of degradation of H_2O_2 [24]:

1. Estimate superoxide dismutase (EC1.15.1.1) (SOD) activity in cytosolic fraction of brain homogenate of all the groups of rats (young and aged) by following the methods of Beyer and Fridovich [25].
2. Glutathione level in first part of brain homogenate is determined [26] by using tetrachloroacetic acid with EDTA as protein precipitating reagent.
3. Measure the glutathione peroxidase (GSH-Px) activity from cytosolic fraction of brain spectrophotometrically [27].
4. Glutathione reductase (GR) activity, glutathione S-transferase (GST), and glucose-6-phosphate dehydrogenase (G6PDH) are determined [28–30].

3.8 Measurement of ROS Level in Rat Brain Mitochondria

Intracellular reactive oxygen species (ROS) level is measured in brain mitochondria[31]:

1. Rat brain mitochondrial suspension (0.4 mg/ml) has been loaded with the cell-permeant probe CM-H_2DCFDA (5-(and-6)-chloromethyl-2′,7′-dichlorodihydrofluorescein diacetate acetyl ester) (2 μM) for 15 min at 30 °C in dark, and fluorescence is monitored.
2. Using H_2DCFDA, an uncharged, cell-permeable fluorescent probe readily diffuses through the membrane bilayer and gets hydrolyzed by intracellular esterase to yield H_2DCF, which is trapped inside the cell.
3. Then it is oxidized from the nonfluorescent form to a highly fluorescent compound dichlorofluorescein by hydrogen peroxide (H_2O_2) or other low-molecular-weight peroxides produced in the cells.
4. Thus, the fluorescence intensity is proportional to the amount to the H_2O_2 produced by the mitochondria.
5. Measure fluorescence through a spectrofluorometer (LS-3B, PerkinElmer, USA) by using 499 nm as excitation and 520 nm as emission wavelengths.

3.9 Quantization of Cerebral Edema

Cerebral edema, the entry of plasma water into swollen brain of young and aged rats, is expressed as a percentage of water content in whole brain of normal, sham-operated as well as other experimental groups.

1. Brain samples obtained after experiment have been immediately weighed on an electronic balance (Sartorious VP61S) to obtain the wet weight.
2. Brain samples are then dried at 100 °C for 24 h to obtain the dry weight and the tissue water content calculated as

$$\left(W_{\text{wet}} - W_{\text{dry}}\right) \times \frac{100}{W_{\text{wet}}},$$

where W_{wet} is the initial weight of brain tissue and W_{dry} is the dry weight of the same tissue [32].

3.10 Histological Examination of Hippocampal Slices of Experimental Animals by Eosin–Hematoxylin Staining

1. Hippocampus region from brains of experimental animals is to be removed quickly and further fixed overnight in 4 % paraformaldehyde at 4 °C.
2. Postfixed hippocampi are embedded in paraffin, and 5 μm coronal sections are cut on a microtome.
3. The sections have been stained with hematoxylin–eosin and examined under a light microscope [33].

3.11 Electron Microscope Study of the Mitochondria Isolated from Rat Brain

Electron microscope study is done on mitochondria isolated from rat brain [7]:

1. For thin-section transmission electron microscopy, isolated mitochondria are fixed with 6 % glutaraldehyde in 0.125 M phosphate buffer (pH 7.2) for 14–16 h and then with 1 % osmium tetroxide in Kellenberger buffer for 16–20 h at room temperature.
2. The fixed tissues is washed for 2 h in 0.5 % uranyl acetate in the same buffer, dehydrated with increasing concentrations of ethanol, and embedded in Spur medium at 70 °C for 48 h.
3. Sections are cut with a DuPont diamond knife in an LKB ultramicrotome and are stained with uranyl acetate and lead citrate and are examined under a JEOL 100CX transmission electron microscope at 60 kV.

4 Notes

1. During nanoparticle preparation the organic solution will be added to the phosphate-buffered saline (PBS) slowly and dropwise; otherwise, clumping of the substance will take place.
2. Stirring will be done slowly by a magnetic stirrer for 3 h or more until evaporation will occur properly.
3. If any traces of organic material are left, nitrogen gas will be passed to the solution until and unless all the organic material will be removed from the solution.
4. Ultracentrifugation and lyophilization will be done after the preparation of nanoparticle for its preservation.
5. After the preparation of nanoparticle, it will be used within 2 or 3 days. Otherwise size of the nanoparticle will be deferred and clumping will take place.
6. During BCCAO procedure proper clipping of carotid artery will be done properly.
7. During the preparation of mitochondria, density gradient centrifugation will be done cautiously; otherwise pure isolation of mitochondria will not be possible.

References

1. Bora KS, Sharma A (2011) Evaluation of antioxidant and cerebroprotective effect of Medicago sativa Linn. against ischemia and reperfusion insult. Evid Based Complement Alternat Med 2011:792167
2. Mergenthaler P, Dirnagl U, Meisel A (2004) Pathophysiology of stroke: lessons from animal models. Metab Brain Dis 19:151–167
3. Warner DS, Sheng H, Batinic-Haberle I (2004) Oxidants, antioxidants and the ischemic brain. J Exp Biol 207:3221–3231
4. Dirnagl U, Iadecola C, Moskowitz MA (1999) Pathobiology of ischaemic stroke: an integrated view. Trends Neurosci 22:391–397
5. Lipton P (1999) Ischemic cell death in brain neurons. Physiol Rev 79:1431–1568

6. Sarkar S, Das N (2006) Mannosylated liposomal flavonoid in combating age-related ischemia-reperfusion induced oxidative damage in rat brain. Mech Ageing Dev 127: 391–397
7. Ghosh S, Das N, Mandal AK, Dungdung SR, Sarkar S (2010) Mannosylated liposomal cytidine 5′ diphosphocholine prevent age related global moderate cerebral ischemia reperfusion induced mitochondrial cytochrome c release in aged rat brain. Neuroscience 171:1287–1299
8. McIntosh TK, Saatman KE, Raghupathi R, Graham DI, Smith DH, Lee VM, Trojanowski JQ (1998) The Dorothy Russell Memorial Lecture. The molecular and cellular sequelae of experimental traumatic brain injury: pathogenetic mechanisms. Neuropathol Appl Neurobiol 24:251–267
9. Huang SH, Jong AY (2001) Cellular mechanisms of microbial proteins contributing to invasion of the blood-brain barrier. Cell Microbiol 3:277–287
10. Wolka AM, Huber JD, Davis TP (2003) Pain and the blood-brain barrier: obstacles to drug delivery. Adv Drug Deliv Rev 55:987–1006
11. Petty MA, Lo EH (2002) Junctional complexes of the blood-brain barrier: permeability changes in neuroinflammation. Prog Neurobiol 68:311–323
12. Bolton SJ, Anthony DC, Perry VH (1998) Loss of the tight junction proteins occludin and zonula occludens-1 from cerebral vascular endothelium during neutrophil-induced blood-brain barrier breakdown in vivo. Neuroscience 86:1245–1257
13. Hawkins BT, Davis TP (2005) The blood-brain barrier/neurovascular unit in health and disease. Pharmacol Rev 57:173–185
14. Fresta M, Puglisi G (1997) Survival rate improvement in a rat ischemia model by long circulating liposomes containing cytidine-5I-diphosphate choline. Life Sci 61:1227–1235
15. Kohler-Staub D, Kohler HP (1989) Microbial degradation of beta-chlorinated four-carbon aliphatic acids. J Bacteriol 171:1428–1434
16. Lockman PR, Mumper RJ, Khan MA, Allen DD (2002) Nanoparticle technology for drug delivery across the blood-brain barrier. Drug Dev Ind Pharm 28:1–13
17. Seyler I, Morin C, Barratt G, Appel M, Devissaguet JP, Puisieux F (1995) Induction of macrophage No-synthase by an immunomodulator entrapped within polymeric nanocapsules. Eur J Pharm Biopharm 41:49–54
18. Sinha J, Das N, Basu MK (2001) Liposomal antioxidants in combating ischemia-reperfusion injury in rat brain. Biomed Pharmacother 55:264–271
19. Das S, Mandal AK, Ghosh A, Panda S, Das N, Sarkar S (2008) Nanoparticulated quercetin in combating age related cerebral oxidative injury. Curr Aging Sci 1:169–174
20. Navarro A, Boveris A, Bandez MJ, Sanchez-Pino MJ, Gomez C, Muntane G, Ferrer I (2009) Human brain cortex: mitochondrial oxidative damage and adaptive response in Parkinson disease and in dementia with Lewy bodies. Free Radic Biol Med 46:1574–1580
21. Recknagel RO, Glende EA (1984) Spectrophotometric detection of lipid conjugated dienes. In: Parker L (ed) Methods in enzymology. Academic, New York, pp 331–337
22. Lowry OH, Rosebrough NJ, Farr AL, Randall RJ (1951) Protein measurement with the Folin phenol reagent. J Biol Chem 193:265–275
23. Sarkar A, Bishayee A, Chatterjee M (1995) Beta-carotene prevents lipid peroxidation and red blood cell membrane protein damage in experimental hepatocarcinogenesis. Cancer Biochem Biophys 15:111–125
24. Aebi H (1984) Catalase in vitro. Methods Enzymol 105:121–126
25. Beyer WF Jr, Fridovich I (1987) Assaying for superoxide dismutase activity: some large consequences of minor changes in conditions. Anal Biochem 161:559–566
26. Davila JC, Davis PJ, Acosta D (1991) Changes in glutathione and cellular energy as potential mechanisms of papaverine-induced hepatotoxicity in vitro. Toxicol Appl Pharmacol 108:28–36
27. Levander OA, DeLoach DP, Morris VC, Moser PB (1983) Platelet glutathione peroxidase activity as an index of selenium status in rats. J Nutr 113:55–63
28. Carlberg I, Mannervik B (1985) Glutathione reductase. Methods Enzymol 113:484–490
29. Warholm M, Guthenberg C, von Bahr C, Mannervik B (1985) Glutathione transferases from human liver. Methods Enzymol 113:499–504
30. Lohr GW, Waller HD (1974) Glucose-6-phosphate dehydrogenase. In: Bergmeyer HU (ed) Methods of enzymatic analysis, 2nd edn. Verlag Chemie, Weinheim, pp 636–643
31. Batandier C, Fontaine E, Keriel C, Leverve XM (2002) Determination of mitochondrial reactive oxygen species: methodological aspects. J Cell Mol Med 6:175–187
32. Gong Y, Hua Y, Keep RF, Hoff JT, Xi G (2004) Intracerebral hemorrhage: effects of aging on brain edema and neurological deficits. Stroke 35:2571–2575

33. Chowdhury AR, Mandal S, Goswami A, Ghosh M, Mandal L, Chakraborty D, Ganguly A, Tripathi G, Mukhopadhyay S, Bandyopadhyay S, Majumder HK (2003) Dihydrobetulinic acid induces apoptosis in *Leishmania donovani* by targeting DNA topoisomerase I and II: implications in antileishmanial therapy. Mol Med 9:26–36

Chapter 15

Methods for Detection of Oxidative Stress and Genotoxicity of Engineered Nanoparticles

Ashutosh Kumar, Vyom Sharma, and Alok Dhawan

Abstract

The distinctive characteristics of engineered nanoparticles (ENPs) such as higher surface-to-volume ratio find immense applications in personal care products, food packaging, drug delivery systems, therapeutics & biosensors and others. The exponential increase in the ENP containing consumer products in the last 5 years has also increased their inadvertent release in the environment and a debate towards their adverse effects to the human and environment health. A variety of ENPs with different size, shape, and surface properties have been shown to induce genotoxicity, cytotoxicity, and oxidative stress in different cellular models. Here we describe the techniques and protocols used in the assessment of the genotoxicity (single-cell gel electrophoresis (comet) assay, cytokinesis block micronucleus assay) and oxidative stress parameters (reactive oxygen species, lipid peroxidation, and glutathione depletion) induced by the ENPs in the cells.

Key words ENPs, Oxidative stress, Genotoxicity, Comet assay, CBMN, ROS, LPO, GSH

1 Introduction

Engineered nanoparticles (ENPs) are the constructs that have unique physical and chemical properties associated with their small size (1–100 nm). The development of technology in controlling the synthesis of materials at nanoscale has lead to increased and diverse production of ENP-based consumer products in fields ranging from personal care products, electronics, optics, energy, cellular imaging and others. The size of the ENPs is in the range of biomolecules such as antibodies, membrane receptors, nucleic acids, and proteins [1]. These properties together with the high surface-to-volume ratio as well as the possibility of manipulation in particle properties have made them a valuable tool for diagnostics and therapeutics. Studies have also revealed that properties such as size, large surface area, solubility, geometry, and presence of atoms on the surface make the ENPs unique but at the same time are responsible for their potential hazard to human and environment [2–4].

Donald Armstrong and Dhruba J. Bharali (eds.), *Oxidative Stress and Nanotechnology: Methods and Protocols*, Methods in Molecular Biology, vol. 1028, DOI 10.1007/978-1-62703-475-3_15, © Springer Science+Business Media New York 2013

Zinc oxide (ZnO) NPs are the most extensively used particles in consumer products such as cosmetics, food packaging, drug delivery systems, therapeutics, and biosensors. Several studies have reported that the size, shape, and concentrations of the ZnO NPs affect the toxicity of the particles. With regard to the size of the particles, Lin et al. have shown that the ZnO NPs of size 70 nm are more toxic in human lung epithelial cells than their bulk (470 nm) counterparts [5]. In context to the shape-dependent toxicity, Heng et al. have shown that in human bronchial epithelial cells (BEAS-2B), ZnO NPs of spherical shape (LD_{50}, 13.3 μg/mL) are more toxic as compared to the sheet (LD_{50}, 16 μg/mL)-shaped particles [6]. Regarding ENPs dose-dependent cytotoxicity, higher concentrations of ZnO NPs resulted in higher levels of cell death [7, 8]. Surface modifications have also been reported to play a crucial role in the cytotoxicity. Studies with ZnO NPs demonstrated that their toxicity could be attributable to the surface molecules surrounding the metallic core. Therefore, it can be inferred that the properties as well as the toxicity of the same particles vary with size, geometry, concentration, surface composition, as well as the experimental conditions [9].

There are still many mysteries in the field of nanotoxicology that await resolution. Studies undertaken to identify the molecular basis for ZnO NP-induced toxicity have shown that reactive oxygen species (ROS) play a key role [2]. It has also been reported that ROS generation induces genotoxicity and oxidative stress in the cells that finally lead to the cell death. However, it is hypothesized that ROS alone could not be responsible for the generation of oxidative stress and the induction of apoptosis cascade in the cell. The possibility for the involvement of metabolic and respiratory pathways also cannot be ruled out. Nevertheless, the precise mechanisms responsible for generating the oxidative damage and the responsible molecules involved in it are still unclear.

Genotoxic damage is the most extensively studied end point, because it has direct and long-lasting effects on the organism. Additionally, assessment of DNA damage has been considered as an essential preliminary test for the safety of products for human use. Single-cell gel electrophoresis, cytokinesis block micronucleus (CBMN) assay, and chromosomal aberrations are some of the tests that are frequently used for genotoxicity assessment. For oxidative stress assessment, lipid peroxidation (LPO), ROS, glutathione depletion (GSH), SOD, and catalase are the major markers. NPs have a tendency to produce ROS in aqueous suspension as well as after interactions with cellular organelles. ROS is known to induce LPO which can cause several types of damage including oxidation of sulfhydryl groups, reduction of disulfides, protein–protein crosslinks, peptide fragmentation, and modification of prosthetic groups. This can also lead to a decrease in the cellular antioxidant levels.

This chapter deals with the methodologies used in the assessment of the genotoxicity [single-cell gel electrophoresis (comet) assay, cytokinesis block micronucleus (CBMN) assay] and oxidative stress (ROS, LPO and GSH) induced by the engineered nanoparticles (ENPs) in the cells.

2 Materials

1. Phosphate-buffered saline (PBS; Ca^{2+}-, Mg^{2+}-free) and fetal bovine serum (FBS) were purchased from Life Technologies (India) Pvt. Ltd. New Delhi, India.
2. Low-melting-point agarose (LMPA), ethidium bromide (EtBr), 2, 7-dichlorofluorescein diacetate, cytochalasin B, diethyl maleate (DEM; D97703-1006), 97 %, ethyl methanesulfonate (CAS No: 62-50-0), butylated hydroxytoluene (BHT, Cat No. B1378), and bovine serum albumin (BSA) were purchased from Sigma-Aldrich Co. (St. Louis, MO, USA).
3. Sodium chloride, normal melting agarose (NMA), May Grunwald stain (powder), Triton X-100, ethylenediaminetetraacetic acid disodium salt [$(EDTA)Na_2$], trypsin–EDTA, L-glutamine, Tris buffer, antibiotic/antimycotic solution (10,000 U/mL penicillin, 10 mg/mL streptomycin, and amphotericin-B) were purchased from Hi-Media Pvt. Ltd., Mumbai, India.
4. Giemsa stain was purchased from BDH Pvt. Ltd. (England, UK). Sodium hydroxide, xylene, DPX mountant, dimethyl sulfoxide (DMSO), and methanol were purchased from Merck Ltd. (New Delhi, India).
5. Ethanol was purchased from Changshu Yangyuan Chemical (China).
6. 5, 5′-dithiobis-2-nitrobenzoic acid (DTNB), glacial acetic acid, glutathione (reduced), monosodium and disodium hydrogen phosphate, potassium chloride, thiobarbituric acid (TBA), and trichloroacetic acid (TCA) were purchased from Sisco Research Laboratories Pvt. Ltd. (Mumbai, India).
7. All other chemicals were purchased locally and were of analytical reagent grade.
8. Tissue culture flasks (25 cm^2, 75 cm^2) and black bottom 96-well culture plate were purchased from NUNC, India. Syringe filters (0.22 μm) and filter holders were purchased from Millipore Co. Pvt. Ltd. (Mumbai, India).
9. End-frosted conventional glass slides (75 mm × 25 mm, with 19 mm frosted end) and coverslips (No. 1, 24 × 60 mm) were purchased from Blue Label (Mumbai, India).

10. Screw cap bottles (100–2,000 mL) were purchased from Schott Duran (Mainz, Germany).
11. Cytofunnel was purchased from Thermo Scientific Pvt. Ltd (India).

3 Methods

3.1 Genotoxicity

3.1.1 Single-Cell Gel Electrophoresis (Comet) Assay

The comet assay is a simple, rapid, and sensitive technique used to detect the single- and double-stranded DNA damage in individual cells (in vitro and in vivo). The methodology was first described by Singh et al. and has since become the most widely used screening test for the quantification of alkali-labile sites, oxidative DNA damage, DNA–DNA or DNA–protein cross-linking, and abasic sites DNA damage [10]. A schematic of the single-cell gel electrophoresis assay is given in Fig. 1.

Preparation of Culture Medium

1. Dissolve the powder media as per manufacturer's protocol in sterile Milli-Q water. This should be supplemented with sodium bicarbonate (2 g/L), L-glutamine (300 mg/L), and antibiotic/antimycotic solution (10 mL/L) to avoid contamination.
2. Maintain the pH to 7.2 and make up the volume up to 1 L.
3. Sterilize the media using 0.22 μm membrane filters, kept at room temperature overnight to check contamination and then store at 4 °C.

Phosphate-Buffered Saline (PBS; Ca^{2+}-, Mg^{2+}-Free)

1. Mix 9.6 g of Dulbecco's PBS in 990 mL distilled water (dH_2O).
2. Adjust the pH to 7.4, make the volume of the solution up to 1,000 mL, and store at room temperature. The calcium- and magnesium-free PBS inhibits the endonuclease activities of the cells.

Preparation of Trypsin–EDTA (0.125 %)

1. Prepare a solution of trypsin–EDTA in PBS at a concentration of 0.125 %.
2. This is achieved by diluting the stock solution (0.25 %) of trypsin in PBS or incomplete media.

Lysing Solution

1. Add 146.1 g NaCl, 37.2 g EDTA, 1.2 g trizma base in ~700 mL dH_2O and stir it to mix properly.
2. Add ~8 g NaOH in it and allow the mixture to dissolve (about 20 min). Adjust the pH to 10.0 using HCl or NaOH.
3. Make up the volume up to 1 L and store at room temperature.
4. Prepare the final lysing solution (fresh) by adding 1 % Triton X-100.
5. Refrigerate the final lysing solution for at least 30 min prior to use.

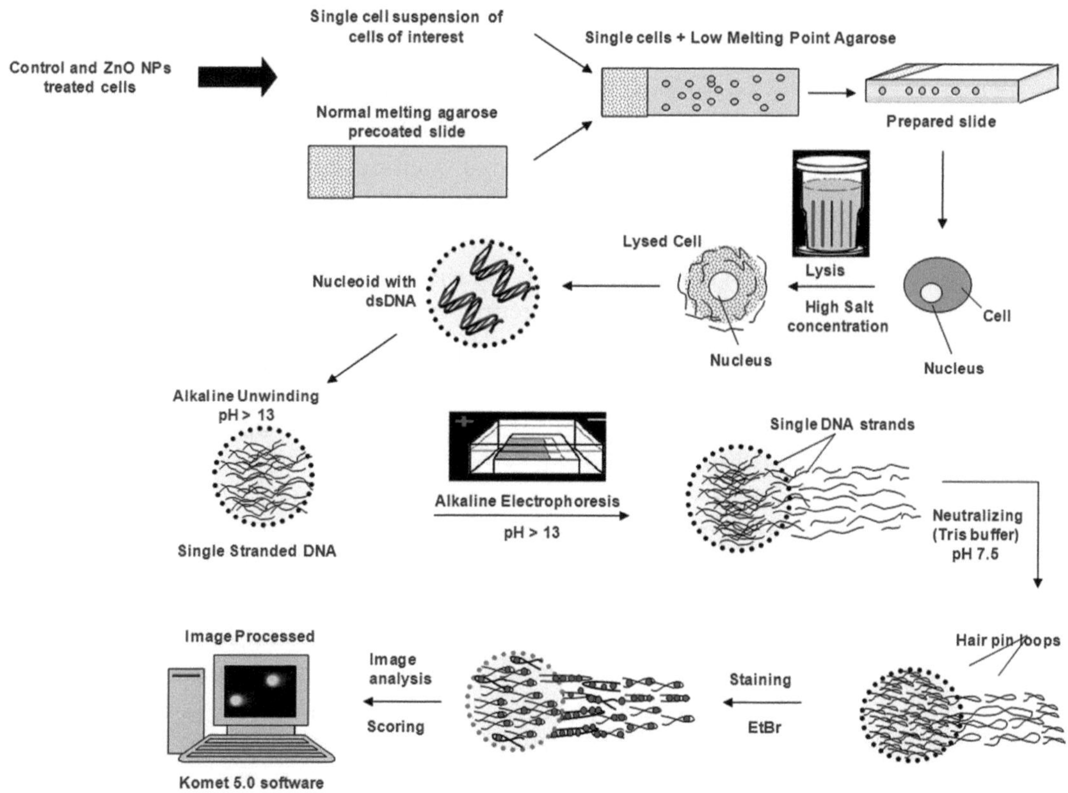

Fig. 1 Schematic showing the process of single-cell gel electrophoresis (comet assay)

Electrophoresis Buffer

1. Prepare the stock solutions of 10 N NaOH by dissolving 200 g in 500 mL dH_2O.
2. Also, prepare a stock solution of 200 mM EDTA by adding 14.89 g of EDTA in 200 mL dH_2O, pH 10.
3. The stock solutions can be stored at room temperature for 2 weeks.
4. To prepare 1× electrophoresis buffer, add 30 mL of NaOH stock solution and 5.0 mL of EDTA stock solution, and make up the volume to 1,000 mL.
5. Electrophoresis buffer should be prepared fresh before each experiment.

Neutralization Buffer

1. Add 48.5 g of Trizma base in 800 mL dH_2O. Mix it well and adjust the pH to 7.5 with concentrated (>10 M) HCl.
2. Make up the volume to 1,000 mL with dH_2O.
3. The solution can be stored at room temperature up to 2 weeks.

Staining Solution

1. Ethidium bromide (EtBr; 10× Stock—20 mg/mL): add 10 mg of EtBr in 50 mL dH_2O and store at room temperature.
2. For working solution, mix 1 mL of stock in 9 mL dH_2O.

3.1.2 Preparation of Base Slides

1. Dip the end-frosted slides in methanol and burn them over a blue flame to remove dusts and other contaminants. Slides should be labelled using graphite pencil and please use the number format instead of alphabets.
2. Prepare 1.0 % normal melting agarose (NMA) by adding 500 mg NMA in 50 mL Milli-Q water. Microwave or heat until near boiling and dissolve the agarose properly.
3. While NMA is hot, dip the slides up to one-third of the frosted area and remove gently.
4. Wipe the underside of slide to remove agarose and lay the slide on a flat surface to dry.
5. Store the slides at room temperature until used; avoid high humidity or heat.

3.1.3 Cell Preparation and Treatment

1. Seed the cells (2.5×10^4 cells/well) in a 48-well plate or proportionally higher or lower well plates depending on the study design.
2. Plate the cells in such a manner that control, positive control (EMS; 1 μM) and the treatments run in triplicate.
3. After 24 h of seeding (more than 80 % confluency), treat the cells with the NPs and EMS (positive control) for designated time point.
4. After the treatment, remove media from the wells and add trypsin–EDTA (0.005 %) to detach the cells from the surface.
5. Add equal amount of complete medium to quench the activity of trypsin.
6. Centrifuge the cells at 250 × *g* for 5 min, remove the supernatant, and resuspend the pellet in PBS.

3.1.4 Slide Preparation and Cell Lysis

1. Resuspend the pellet (~10,000 cells) in 100 μL of 1× PBS and add 100 μL of 1 % low-melting-point agarose (LMPA).
2. Pour the cell suspension on NMA-coated slide and place a coverslip over it for even spreading.
3. Place the slides on an ice pack (~0 °C) to solidify the LMPA layer.
4. Gently slide off the coverslip and add a third agarose layer (90 μL LMPA) to the slide.
5. Replace the coverslip and return to the icepack until the agarose layer solidifies (~5–10 min).
6. Remove coverslip and slowly dip the slide into cold, freshly prepared lysing solution in a coupling jar.
7. Protect from light and refrigerate for a minimum of 3 h for cell lysis.

3.1.5 Electrophoresis of Microgel Slides

1. After the lysis, gently remove the slides from the lysing solution. Place the slides side by side on the horizontal electrophoretic chamber near one end; slide them close together as much possible.
2. Fill the tank with freshly prepared electrophoresis buffer until it completely covers the slides (avoid bubbles over the agarose).
3. Leave the slides in the alkaline buffer for 20 min to allow for unwinding of the DNA and the expression of alkali-labile sites.
4. Turn on the power supply to 24 V (~0.74 V/cm) and adjust the current to 300 mA by adjusting the buffer level.
5. Depending on the purpose of the study and on the extent of migration in control samples, electrophorese the slides for ~30 min.

3.1.6 Neutralization and Staining of Microgel Slides

1. Turn off the power and gently lift the slides from the buffer and place on a drain tray.
2. Dropwise coat the slides with neutralization buffer and let it sit for at least 5 min. Drain the slides and repeat the process for at least three more times.
3. Stain the slides with 80 μL, 1 × EtBr. Leave the slides for 5 min, and then dip them in chilled distilled water to remove excess stain.
4. Put a coverslip over it and store the slides in a moist slide box at 4 °C.

3.1.7 Scoring of the Microgel Slides

1. Score the slide after 2 h under yellow/dimmed light to prevent the DNA damage from bright white light.
2. For visualization of DNA damage, see the EtBr-stained DNA using a 40× objective on a fluorescent microscope.
3. Count approximately 50–100 randomly selected cells per sample using Komet 5 image analysis software Andor Technology (Belfast, UK).
4. DNA damage in the treated cells is compared to that of the control samples by the length of DNA migration, the percentage of migrated DNA, and the Olive tail moment. This gives a measurement of qualitative and quantitative DNA damage in the cells.

3.1.8 Acceptance Criteria

- The extent of the DNA damage in the EMS (positive control)-treated cells should be at least 70 % more than the negative control after 3 h exposure.
- The positive control and sample replicate coefficient of variations should be within 50 %.

3.2 Cytokinesis Block Micronucleus (CBMN) Assay

Micronucleus is a chromatin-containing structure in the cytoplasm surrounded by a membrane without any detectable link to the cell nucleus. They are formed during the anaphase stage of cell division from the chromosomal fragments or whole chromosomes that are left behind when the nucleus divides. The CBMN method is routinely used for the measurement of chromosome breakage, impairment in DNA repair, chromosome loss, nondisjunction, necrosis, and apoptosis.

3.2.1 Preparation of Culture Medium

Dissolve the powder media as per manufacturer's protocol in sterile Milli-Q water. This should be supplemented with sodium bicarbonate (2 g/L), L-glutamine (300 mg/L), and antibiotic/antimycotic solution (10 mL/L) to avoid contamination. Maintain the pH to 7.2 and make up the volume up to 1 L. The media is then sterilized using 0.22 μm membrane filters, kept at room temperature overnight to check contamination, and then stored at 4 °C.

Phosphate-Buffered Saline (PBS; Ca^{2+}-, Mg^{2+}-Free)

Mix 9.6 g Dulbecco's PBS in 990 mL distilled water (dH_2O), and adjust the pH to 7.4. Make the volume of the solution up to 1,000 mL and store at room temperature.

Preparation of Trypsin–EDTA (0.125 %)

A solution of trypsin–EDTA in PBS is prepared at a concentration of 0.125 %. This is achieved by diluting the stock solution (0.25 %) of trypsin in PBS or incomplete media.

3.2.2 Cytochalasin B

Prepare 0.5 mg/mL stock of cytochalasin B in DMSO. Store the aliquots of 500 μL in −20 °C till use. Use the final concentration of cytochalasin B at 3 μg/mL in culture medium.

3.2.3 Hypotonic Solution (Potassium Chloride, KCl)

Prepare 0.56 % KCl in dH_2O and store at room temperature.

3.2.4 Carnoy's Fixative

Prepare a solution of methanol and glacial acetic acid in the ratio 3:1 v/v freshly and always use chilled.

3.2.5 Staining Solution

Giemsa Stock Solution

Dissolve 500 mg of Giemsa in a solution, consisting of 20 mL methanol and 30 mL glycerol. Keep the solution for maturation for 2–3 days.

Sorenson's Buffer

Weigh 0.445 g of Na_2HPO_4 and 0.34 g of potassium dihydrogen phosphate. Dissolve them in 200 mL of distilled water and adjust the pH to 6.8 using HCl or NaOH.

Working Staining Solution

Prepare 10 % Giemsa in Sorenson's buffer (10 mL of Giemsa stock in 90 mL of Sorenson's buffer) and store at room temperature.

CBMN Assay

1. Seed the cells (2.0×10^5 cells/well) in a six-well plate.

2. Plate the cells in such a manner that control, positive control, and the treatments run in triplicate.
3. After 24 h (~80 % confluency) treat the cells with the ENPs and EMS (positive control; 6 mM) for designated time point.
4. After the treatment, remove the culture media from the wells and wash the cells with PBS.
5. Grow the cells further in complete medium containing cytochalasin B (final concentration, 3 μg/mL) to inhibit the cytokinesis, till the doubling time of the cells.
6. Remove the culture media from the wells after incubation and add trypsin–EDTA (0.005 %) to detach the cells from the surface.
7. Add equal amount of culture medium (with FBS) to quench the activity of trypsin.
8. Centrifuge the cells at 250×g for 10 min and aspirate the supernatant carefully.
9. Fix the cells in chilled Carnoy's fixative (5 mL) and incubate at 4 °C for 15 min.
10. Centrifuge the cells again at 250×g for 10 min and aspirate the supernatant carefully.
11. Mix the cells finally in 600 μL of chilled Carnoy's fixative.
12. Distribute the cell suspension in a Shandon Double Cytofunnel and prepare the dots on the slides by centrifugation at 880×g for 10 min using a cytospin (Thermo Shandon, Hampshire, UK).
13. Fix the slides in chilled methanol (90 %) for 5 min, air-dry and store until staining.
14. Stain the slides with 10 % Giemsa (working solution) for 10 min.
15. Air-dry the slides and dip in xylene and mount with Dibutyl Phathalate Xylene (DPX).
16. Examine the presence of micronucleus in the binucleate cells using a bright field microscope.
17. Count a minimum of 1,000 binucleate cells with well-preserved cytoplasm from each slide of the sample and determine the frequency of micronucleated binucleate cells.

3.2.6 Acceptance Criteria

1. The extent of the micronucleus frequency in the EMS (positive control)-treated cells should be at least 1.5-fold greater than the negative control after 3 h of exposure.
2. The positive control and sample replicate coefficient of variations should be within 50 %.

3.3 Measurement of Reactive Oxygen Species

Oxidative stress in the cells is a cellular state arouse due to the imbalance in the normal redox state or the over production of free radicals which can cause toxic effects in cells. The generation of the free radicals may further damage all components of the cell, including proteins, lipids, and DNA. The level of ROS, malondialdehyde, and the glutathione in the cells usually define the stress condition.

The level of intracellular ROS generation can be estimated using 2, 7-dichlorofluorescein diacetate (DCF-DA) dye [11]. DCF-DA is non-fluorescent, but the deacetylation of the DCF-DA in the presence of esterase enzyme converts it into DCFH, which is oxidized by free radicals present in the cells and forms a fluorescent DCF. This can be estimated at an excitation/emission wavelength of 485/528 nm, respectively.

3.3.1 Preparation of Culture Medium

1. Dissolve the powder media as per manufacturer's protocol in sterile Milli-Q water.
2. This should be supplemented with sodium bicarbonate (2 g/L), L-glutamine (300 mg/L), and antibiotic/antimycotic solution (10 mL/L) to avoid contamination.
3. Maintain the pH to 7.2 and make up the volume up to 1 L.
4. Sterilize the media using 0.22 μm membrane filters, kept at room temperature overnight to check contamination, and then store at 4 °C.

Phosphate-Buffered Saline (PBS; Ca^{2+}-, Mg^{2+}-Free)

Mix 9.6 g of Dulbecco's PBS in 990 mL distilled water (dH_2O) and adjust the pH to 7.4. Make the volume of the solution up to 1,000 mL and store at room temperature.

Preparation of Trypsin–EDTA (0.125 %)

A solution of trypsin–EDTA in PBS is prepared at a concentration of 0.125 %. This is achieved by diluting the stock solution (0.25 %) of trypsin in PBS or incomplete media.

Preparation of DCF-DA

DCF-DA stock (10 mM): Dissolve 5 mg of DCF-DA powder in 1 mL of DMSO.

DCF-DA working stock (20 μM): Dissolve 20 μL of 10 mM stock to 10 mL PBS buffer. The preparation of stock and working suspension should be performed in dark. The working solution should be prepared immediately prior to use.

3.3.2 Preparation of Diethyl Maleate (DEM) Positive Control

Prepare 5 mM of the DEM solution in the culture medium.

Measurement

1. Seed the cells (1.0×10^4 cells/well) in a 96-well black bottom culture plate.
2. After 24 h, treat the cells with the ENPs and DEM (positive control) for different time points. For positive control, cells are treated with DEM for 2 h.
3. Aspirate the treatment media and wash the cells twice with PBS.
4. Add culture medium containing DCF-DA dye (20 μM)/PBS to each well and incubate it for 30 min at 37 °C.

5. Decant the medium containing DCF-DA after incubation and add 200 μL of PBS to each well.
6. The fluorescence intensity was measured in a Synergy HT multi-well plate reader, Bio-Tek (USA), or equivalent using software at an excitation/emission wavelengths of 485/528 nm, respectively.
7. Assess the autofluorescence and interference of ENPs with the DCF-DA dye by performing a parallel experiment in similar manner without the cells.
8. Calculate the percent ROS generation using the following formula, after correcting for background fluorescence:

$$\text{Percent ROS generation} = \frac{(\text{F485}/528_{\text{sample}} - \text{F485}/528_{\text{sample blank}})}{(\text{F485}/528_{\text{control}} - \text{F485}/528_{\text{control blank}})} \times 100.$$

9. Mean, SD and SE should be calculated for each positive control and unknown sample.
10. Compare the mean values of percent ROS generation in treated cells among various concentrations with the control and analyze the results using analysis of variance (ANOVA).

3.3.3 Acceptance Criteria

1. The fluorescence intensity of the DEM positive control should be at least 140 % of media control after 2 h of exposure.
2. The positive control and sample replicate coefficient of variations should be within 50 %.

3.4 Measurement of Lipid Peroxidation

Lipid peroxidation (LPO) is a chain reaction process which is generated due to the oxidation of unsaturated fatty acids present in the cell membrane. It is an indicator of the cellular oxidative stress. The thiobarbituric acid reactive substances (TBARS) assay measures lipid hydroperoxides and aldehydes, such as malondialdehyde (MDA), in the cell culture media and cell lysate. MDA forms a stable pink-colored adduct with thiobarbituric acid (TBA) in 1:2 ratios, and its quantification at 532 nm corresponds to the extent of LPO in the cell.

3.4.1 Preparation of Culture Medium

Dissolve the powder media as per manufacturer's protocol in sterile Milli-Q water. This should be supplemented with sodium bicarbonate (2 g/L), L-glutamine (300 mg/L) and antibiotic/antimycotic solution (10 mL/L) to avoid contamination. Maintain the pH to 7.2 and make up the volume up to 1 L. The media is then sterilized using 0.22 μm membrane filters, kept at room temperature overnight to check contamination, and then stored at 4 °C.

Phosphate-Buffered Saline (PBS; Ca^{2+}-, Mg^{2+}-Free)

Mix 9.6 g of Dulbecco's PBS in 990 mL distilled water (dH_2O) and adjust the pH to 7.4. Make the volume of the solution up to 1,000 mL, and store at room temperature.

Preparation of Trypsin–EDTA (0.125 %)

A solution of trypsin–EDTA in PBS is prepared at a concentration of 0.125 %. This is achieved by diluting the stock solution (0.25 %) of trypsin in PBS or incomplete media.

Preparation of 15 % TCA (w/v)

Add 7.5 g TCA to 50 mL Milli-Q water. This can be prepared in advance and stored at 4 °C.

Preparation of 2.5 % TCA (w/v)

Add 1.25 g TCA to 50 mL Milli-Q water. This can be prepared in advance and stored at 4 °C.

Preparation of Diethyl Maleate Positive Control

Prepare 5 mM of the DEM solution in the culture medium.

Preparation of 0.67 % TBA (w/v) /0.01 % Butylated Hydroxytoluene (w/v)

Add 0.335 g TBA and 0.005 g BHT to 50 mL Milli-Q water. The solution should be prepared freshly and stored in ice.

Preparation of Solution for MDA Standard Curve

Prepare the MDA dilutions of known concentrations 0.007, 0.015, 0.031, 0.063, 0.125, 0.25, 0.5, 1, 2, and 4 nmol/mL.

Measurement

1. Seed the cells (2.0×10^5 cells/well) in a six-well plate.
2. Plate the cells in such a manner that control, positive control, and the treatment run in triplicate.
3. After 24 h (~80 % confluency) treat the cells with the ENPs and DEM (positive control) for designated time point.
4. After treatment, collect the media.
5. Add 500 μL of sample media to 400 μL 15 % TCA and 800 μL of 0.67 % TBA/0.01 % BHT in a 5 mL amber vial.
6. Vortex and heat the solution for 20 min at 95 °C in a water bath.
7. Allow the solution to cool, then add 3 mL butanol, gently mix phases, and transfer 200 μL of the butanol phase (top) to a 96-well plate for taking reading.
8. For cell lysate sample preparation, process the cells with following protocol.
9. After the completion of treatment, scrap the cells out in 1 mL chilled PBS.
10. Sonicate the cells (5 pulse, 20 w) on ice.
11. Centrifuge the cells at $13{,}000 \times g$ for 2 min.
12. Store the pellet for the estimation of protein. The sample can be stored at −20 °C until analysis.
13. Remove 500 μL of lysate supernatant and add to 400 μL 15 % TCA and 800 μL of 0.67 % TBA/0.01 % BHT in a 5 mL amber vial.

14. Vortex and heat the solution for 20 min in 95 °C water bath.
15. Allow the solution to cool and add 3 mL butanol, mix it gently, and transfer 200 μL of the butanol phase (top) to a 96-well plate for taking reading.
16. Measure the absorbance of cell media and cell lysate in a microtiter at 532 nm.
17. Estimation of protein by Bradford assay [15].
18. Prepare the protein standard curve using the 1 mg/mL stock of bovine serum albumin (BSA).
19. Add 10 μL of known concentration of the BSA (ranging from 0.001 to100 μg/mL) to 990 μL of Bradford reagent.
20. Mix the solution well and incubate for 15 min.
21. Resuspend the pellet in 500 μL of 0.05 N NaOH solutions.
22. Add 10 μL of the sample to 990 μL of Bradford reagent.
23. Mix the solution well and incubate for 15 min.
24. Dispense 200 μL of the dye reagent mixture to a 96-well (transparent bottom) plate in triplicate.
25. Measure the absorbance in a microtiter plate at 595 nm.
26. Calculate the protein concentration of sample using the BSA standard curve.
27. MDA level in the test samples can be calculated by comparison with MDA standard curve and expressed in equivalent MDA after normalization with total protein.

3.4.2 Acceptance Criteria

1. The level of the MDA in the DEM (positive control)-treated cells should be more than two fold than the negative control after 3 h exposure.
2. The positive control and sample replicate coefficient of variations should be within 50 %.

3.5 Measurement of Glutathione

Glutathione (GSH) is a ubiquitous tripeptide which serves several vital functions including detoxifying electrophiles, scavenging free radicals, and providing a reservoir for cysteine. The sulfhydryl group of GSH reduces 5,5′-dithiobis (2-nitrobenzoic acid) (DTNB) and forms the yellow color product 5-thio-2-nitrobenzoic acid (TNB) [12]. Measurement of the absorbance of TNB at 412 nm provides an accurate estimation of GSH in the sample.

3.5.1 Preparation of Culture Medium

Dissolve the powder media as per manufacturer's protocol in sterile Milli-Q water. This should be supplemented with sodium bicarbonate (2 g/L), L-glutamine (300 mg/L), and antibiotic/antimycotic solution (10 mL/L) to avoid contamination. Maintain the pH to 7.2 and make up the volume up to 1 L. The media is then sterilized using 0.22 μm membrane filters, kept at room temperature overnight to check contamination, and then stored at 4 °C.

3.5.2 Phosphate-Buffered Saline (PBS; Ca^{2+}-, Mg^{2+}-Free)

Mix the 9.6 g of Dulbecco's PBS in 990 mL of distilled water (dH_2O), adjust the pH to 7.4, make the volume of the solution up to 1,000 mL, and store at room temperature.

3.5.3 Preparation of Trypsin–EDTA (0.125 %)

A solution of trypsin–EDTA in PBS is prepared at a concentration of 0.125 %. This is achieved by diluting the stock solution (0.25 %) of trypsin in PBS or incomplete media.

3.5.4 Preparation of 5 % TCA (w/v)

Add 2.5 g TCA to 50 mL Milli-Q water. This can be prepared in advance and stored at 4 °C.

3.5.5 Preparation of Diethyl Maleate Positive Control

Prepare 5 mM of the DEM solution in the culture medium.

3.5.6 Preparation of Phosphate–EDTA Dilution Buffer

100 mM Na_3PO_4 and 1 mM EDTA, pH 7.4: Dissolve 8.2 g of Na_3PO_4 and 208 mg of EDTA to 500 mL Milli-Q water. Adjust pH to 7.4.

3.5.7 Preparation of DTNB Substrate (0.01 %)

Dissolve 5 mg DTNB in 50 mL of phosphate–EDTA dilution buffer.

3.5.8 Preparation of 0.02 % Glutathione

Dissolve 4 mg of GSH in 20 mL phosphate–EDTA dilution buffer.

3.5.9 GSH Standard Curve

Prepare the GSH dilutions of known concentrations ranging from 0.005 to 100 μg/mL.

3.5.10 Preparation of 5-Sulfosalicylic Acid Dehydrates (SSA; 5 %)

Dissolve 500 mg SSA in 10 mL Milli-Q water.

3.5.11 Measurements

1. Seed the cells (2.0×10^5 cells/well) in a six-well plate.
2. Plate the cells in such a manner that control, positive control, and the treatments run in triplicate.
3. After 24 h (~80 % confluency) treat the cells with the ENPs and DEM (positive control) for designated time point.
4. After the treatment, remove media from the wells and wash the cells with 1 mL of room temperature PBS.
5. Scrap the cells out in 500 μL of 5 % SSA.
6. Add 500 μL of 5 % TCA and incubate at room temperature for 30 min.
7. Centrifuge the cells at $13,000 \times g$ for 2 min.

8. Take 500 μL of lysate supernatant and add to 2.5 mL of DTNB solution.
9. Vortex the solution and incubate it at room temperature for 20 min.
10. Dispense 200 μL of the solution to a 96-well (transparent bottom) plate in triplicate.
11. Measure the absorbance in a microtiter plate reader at 412 nm.

3.5.12 Estimation of Protein by Bradford Assay

1. Prepare the protein standard curve using the 1 mg/mL stock of bovine serum albumin (BSA).
2. Add 10 μL of known concentration of the BSA (ranging from 0.001 to 100 μg/mL) to 990 μL of Bradford reagent [15].
3. Mix the solution well and incubate for 15 min.
4. Add 10 μL of the test sample to 990 μL of Bradford reagent.
5. Mix the solution well and incubate for 15 min.
6. Dispense 200 μL of the dye reagent mixture to a 96-well (transparent bottom) plate in triplicate.
7. Measure the absorbance in a microtiter plate reader at 595 nm.
8. Calculate the concentration of sample using the BSA standard curve.
9. GSH level in the test samples can be calculated by comparing the readings with GSH standard reading and expressed in equivalent GSH in μmol/mg protein after normalization with total protein.

3.5.13 Acceptance Criteria

1. The level of the GSH in the DEM (positive control)-treated cells should be at least 40 % less than the negative control after 6 h of exposure.
2. The positive control and sample replicate coefficient of variations should be within 50 %.

Acknowledgments

The authors gratefully acknowledge the funding from CSIR, New Delhi, under its network project (NWP34, NWP35), Supra Institutional Project (SIP-008) and OLP 009. The funding from the Department of Science and Technology, Government of India, under the nano mission project DST-NSTI grant (SR/S5/NM-01/2007) and UK India Education and Research Initiative (UKIERI) standard award to Institute of Life Sciences, Ahmedabad University, Ahmedabad, India (IND/CONT/E/11-12/217) the Department ofBiotechnology, under the New INDIGO programme (NanoLINEN project); and NanoValid (contract: 263147-EU-FP7).

References

1. Sanvicens N, Marco MP (2008) Multifunctional nanoparticles—properties and prospects for their use in human medicine. Trends Biotechnol 26:425–433
2. Kumar A, Pandey AK, Singh SS, Shanker R, Dhawan A (2011) Engineered ZnO and TiO(2) nanoparticles induce oxidative stress and DNA damage leading to reduced viability of *Escherichia coli*. Free Radic Biol Med 51:1872–1881
3. Sharma V, Singh SK, Anderson D, Tobin DJ, Dhawan A (2011) Zinc oxide nanoparticle induced genotoxicity in primary human epidermal keratinocytes. J Nanosci Nanotechnol 11:3782–3788
4. Shukla RK, Kumar A, Gurbani D, Pandey AK, Singh S, Dhawan A (2011) TiO(2) nanoparticles induce oxidative DNA damage and apoptosis in human liver cells. Nanotoxicology. doi:10.3109/17435390.17432011.17629747
5. Lin W, Xu Y, Huang CC, Ma Y, Shannon KB, Chen DR, Huang YW (2009) Toxicity of nano- and micro-sized ZnO particles in human lung epithelial cells. J Nanopart Res 11:25–39
6. Heng BC, Zhao X, Tan EC, Khamis N, Assodani A, Xiong S, Ruedl C, Ng KW, Loo JS (2011) Evaluation of the cytotoxic and inflammatory potential of differentially shaped zinc oxide nanoparticles. Arch Toxicol 85:1517–1528
7. Sharma V, Shukla RK, Saxena N, Parmar D, Das M, Dhawan A (2009) DNA damaging potential of zinc oxide nanoparticles in human epidermal cells. Toxicol Lett 185:211–218
8. Shukla RK, Kumar A, Pandey AK, Singh SS, Dhawan A (2011) Titanium dioxide nanoparticles induce oxidative stress-mediated apoptosis in human keratinocyte cells. J Biomed Nanotechnol 7:100–101
9. Dhawan A, Sharma V (2010) Toxicity assessment of nanomaterials: methods and challenges. Anal Bioanal Chem 398:589–605
10. Singh NP, McCoy MT, Tice RR, Schneider EL (1988) A simple technique for quantitation of low levels of DNA damage in individual cells. Exp Cell Res 175:184–191
11. Wan C, Cardus L, McGreevy B, Lewis V, Johnson J, Robertson WO (1993) Content audit of POISINDEX. Vet Hum Toxicol 35:168–169
12. Ellman GL (1959) Tissue sulfhydryl groups. Arch Biochem Biophys 82:70–77
13. Sharma V, Anderson D, Dhawan A (2011) Zinc oxide nanoparticles induce oxidative stress and genotoxicity in human liver cells (HepG2). J Biomed Nanotechnol 7:98–99
14. Kumar A, Pandey AK, Singh SS, Shanker R, Dhawan A (2011) Cellular uptake and mutagenic potential of metal oxide nanoparticles in bacterial cells. Chemosphere 83:1124–1132
15. Bradford MM (1976) A rapid and sensitive method for the quantitation of microgram quantities of protein utilizing the principle of protein-dye binding. Anal Biochem 72:248–254

Chapter 16

Size-Dependent Study of Pulmonary Responses to Nano-sized Iron and Copper Oxide Nanoparticles

Rajiv Kumar and Dattatri K. Nagesha

Abstract

The application of nanotechnology in various fields has resulted in a tremendous increase in the synthesis of variety of engineered nanoparticles (NPs). These applications are possible only due to the small size and large surface area of the NPs which imparts them unique properties. Inorganic oxide NPs as iron and copper oxide NPs are widely used in several biomedical and synthetic applications. The beneficial aspects of these NPs are concurrently associated with several drastic and deleterious effects as well. Size of the NPs plays a critical role in systemic clearance from the body. Initial studies have confirmed inflammatory responses in mice associated with non-biodegradable oxide NPs. The associated oxidative stress varied from mild effects to reactive oxygen species generation which can potentiate DNA damage or even induced carcinogenesis. Copper oxide NPs, in particular, induced acute toxicity and inflict neutrophil infiltration. This chapter focuses on the applicability of various in vivo techniques for studying the effect of these NPs, especially on the pulmonary system. These in vivo techniques would certainly provide a better understanding and insight into the mechanistic pathways by which these NPs interact with various organ systems in human body.

Key words Nanoparticles, Toxicity, Pulmonary inflammation, Intratracheal instillation, Bronchoalveolar lavage (BAL) fluid, Macrophages, Immunohistochemistry

1 Introduction

The potential of medical applications of nanotechnology has laid some early inroads in understanding many complex diseases. Engineered nanomaterials have gained a tremendous interest among scientists from interdisciplinary backgrounds [1]. Until now, a large number of new nanotechnology-based concepts for therapeutic and diagnostic medicines have emerged, and their feasibility has been demonstrated successfully [2]. The two key parameters which make the NPs as the most desired and successful candidate in medical applications are the sub-nanometer size and the large surface area. Rapid development of engineered nanomaterials has resulted in vast exposure of human being via several

Donald Armstrong and Dhruba J. Bharali (eds.), *Oxidative Stress and Nanotechnology: Methods and Protocols*, Methods in Molecular Biology, vol. 1028, DOI 10.1007/978-1-62703-475-3_16, © Springer Science+Business Media New York 2013

routes which include inhalation of the airborne nanomaterials, dermal penetration, oral ingestion through food or water, and sometimes direct systemic injection [3, 4]. The same "novel" properties of the NPs result in a high reactivity associated with large surface area and enhanced efficiency to diffuse across biological membranes and tissues due to smaller size [5]. This in turn raises a humming questionable scenario for the biocompatibility and the risks associated with these nanomaterials and presents new challenges in understanding potential health risks.

Metal oxide NPs, like iron and copper oxide, represent a broad class of engineered nanomaterials with wide applications not only in the medical field but also in photonics, catalysis, cosmetics, etc. Copper oxide (CuO) NPs are used in antimicrobial preparations, heat transfer fluids, semiconductors, or intrauterine contraceptive devices [6, 7]. Iron oxide NPs are used in catalysis, as magnetic resonance imaging (MRI) contrast agents, targeted drug delivery agent, and in hyperthermia [8–10]. The application of these NPs has raised concerns regarding the genotoxicity, associated stress, and inflammatory responses because of the strong size-dependent physicochemical properties. Interaction of NPs with tissues/cells can inflict pathogenic mechanism which involves inflammation-driven effects like fibrosis, oxidative stress, and DNA damage [11, 12]. Pulmonary system is one of the most susceptible organ systems for the accumulation of these NPs as a result of their incidental/accidental inhalation from the environment and intentional administration via inhalation or systemic delivery. There are several positive attributes associated with the potential use of NPs in diagnosis, therapeutics, and prophylaxis of various lung diseases; however, the accumulation of these NPs should be carefully evaluated in terms of risks and benefits. Iron and copper oxide NPs are classified as inorganic NPs which are nonbiodegradable and due to their small size can exert toxic effects on the lungs which include inflammation and fibrosis. [13, 14]. Depending upon the size (hydrodynamic diameter), iron oxide NPs have been classified into three main groups: oral (300 nm to 3.5 μm), standard (50–150 nm), and ultrasmall (<50 nm). The NPs in the range of 10–100 nm are exploited for most in vivo applications. The fate of these NPs is strongly dependent on the size as well as charge and route of entry which determines the biodistribution, metabolism, and toxicity profile of the accumulated NPs [15–17].

The primary effect associated with the accumulation of iron/copper oxide in pulmonary tissues is inflammation. Inflammation in a tissue usually represents complex collective responses induced by the external stimuli, which depending on the chronicity can be very harmful. The extent and outcome of the inflammation can be different and depends to a large extent on the nature of the stimulus responsible. Complexity of the inflammation arises because of the combined inflictive cellular and humoral interactions with in intact vascular system. In vitro studies provide only an indicative assessment of the inflammation and thus lack at times the information on the severity, chronicity, and persistence of the inflammation. Both iron and copper ions are known to generate oxidative stress; thus, it is highly possible that iron/copper oxide NPs can be associated with cancer and several mechanisms for iron-induced carcinogenesis and have been known to generate ROS that can potentiate direct damage to DNA, proteins, and lipid peroxidation. Oxidative stress generated in cells exposed to NPs may stimulate inflammatory responses, oxidize lipids, or even lead to cell death [18, 19]. CuO NPs are known to induce acute toxicity with severe inflammatory changes in neutrophil infiltration and edema in rats instilled intratracheally [20]. It has also been established that CuO NPs are more toxic than their Cu ions counterpart because of the inability of the Cu ions to penetrate the cell membrane [21]. Iron oxide NPs have shown to activate oxidative stress by generation of hydroxyl radicals via the Fenton's reaction inside the cells [22]. It has also been established that larger-sized iron oxide NPs were more toxic as compared to smaller sized as assessed by in vitro FDA and WST-1 assay [23]. However, little is known about the mechanisms involved in the potential adverse effects of these NPs, and more emphasis should be given on the further evaluation of these NPs for development as potential clinical agents. In this chapter, we focus on the different in vitro and in vivo assays involved for studying the interaction of different NPs with the pulmonary system of the animals.

2 Materials

2.1 Nanoparticles Formulations and Characterization

1. Aqueous suspension of iron oxide NPs (a) 5 nm size and (b) 10 nm size, both at a concentration of 5 mg/ml (Sigma-Aldrich). Copper oxide NPs of (a) 25 nm size and (b) 40 nm size in powdered form (US Research Nanomaterials Inc.) (*see* **Note 1**):
2. HPLC grade water (Sigma).
3. Phosphate-buffered saline (1×), pH 7.2 (Invitrogen).
4. Four-sided quartz cuvettes (Starna Cells), 3 ml volume.
5. UV–visible spectrophotometer (Shimadzu UV3600).
6. Zetasizer (90Plus-ZetaPALS, Brookhaven Instruments, NY).

2.2 In Vivo Animal Exposure

2.2.1 Intratracheal Instillation

1. Animals: male C57Bl/6 mice, 4–6 weeks old (Charles River Laboratory). The animals can be divided into three groups, with six animals per group for nanoparticles formulation of iron oxide (Group 1, control; Group 2, treated with 5 nm iron oxide

NPs; and Group 3, 25 nm iron oxide NPs) and copper oxide (Group 1, control; Group 2, treated with 25 nm copper oxide NPs; and Group 3, 40 nm copper oxide NPs).

2. 0.9 % NaCl in sterile water: Prepare the saline solution by adding 900 mg of sodium chloride salt in 100 ml of sterile water.
3. Bronchoalveolar lavage (BAL fluid) (*see* Subheading 2.3).
4. NPs suspensions: Different NPs suspension can be prepared by adding a specific amount (as per experimental needs) of powdered NPs to 0.9 % saline containing 10 % (v/v) BAL fluid. The mixture should be sonicated for 15 min with occasionally cooling over ice.
5. Anesthetizing agent: xylazine and ketamine (Sigma).
6. Sterile water (Invitrogen).
7. Tuberculin Monoject syringes (1 ml, 27 G, Cat. # 305620, Becton Dickinson, USA) and SGE glass syringe (Cat # 006050, 250 μl volume, SGE Analytical Sciences).
8. Heating plates for maintaining body temperature of the animals (Fisher Sci.).
9. BD Insyte catheter with needle, 24 G (Cat. # 381212, Becton Dickinson, USA).
10. Sonicator (Fisher Sci.).

2.2.2 Inhalation

1. Inhalation/exposure chamber.
2. High-frequency sonicator (Fisher Scientific, Model #550).
3. Collision nebulizer (6-Jet, BGI Inc., MA).
4. Air source (HEPA filtered).
5. Static discharge device (Simco Inc., PA).
6. Exposure chamber fitted with Scanning Mobility Particle Sizer (SMPS) consisting of a condensation particle counter (Cat # 3010; TSI Inc., St. Paul, MN) and an electrostatic classifier with a "long" differential mobility analyzer (Cat # 3071; TSI Inc., St. Paul, MN).
7. Nanoparticles suspensions (*see* Subheading 2.1).

2.3 Bronchoalveolar Lavage Fluid Isolation

1. Dissection kit (World Precision Instruments).
2. Syringes (1 cc), lavage tube 23G, and hemocytometer (Becton Dickinson, USA).
3. Sterile saline solution (*see* Subheading 2.2).
4. Falcon tubes (15 ml) (Fisher Sci.) on ice.
5. RPMI media (Invitrogen, Cat #12633-012).
6. Trypan blue solution (0.4 %) (Sigma, Cat # T8154) in 0.81 % sodium chloride and 0.06 % potassium phosphate, dibasic.

7. Pen–strep (Invitrogen, Cat # 15140).
8. Centrifuge (Eppendorf, Model 5804R).

2.4 Macrophage Cell Culture and Cytokine Measurements

1. Cell pellets obtained from BAL fluid isolation (*see* Subheading 3.3).
2. RPMI 1640 cell culture medium with 10 % FBS (Invitrogen, Cat # 12633-012).
3. 24-well cell culture plates (Fisher Sci., Cat # 07-200-84).
4. Lipopolysaccharide (5 μg/ml) (Sigma, Cat # L3024) in RPMI medium.
5. ELISA kits for cytokine measurements using Quantikine ELISA kits from R&D Systems for INF-γ (Cat # MIF00), IL-13 (Cat # D1300), macrophage inflammatory protein-2 (MIP-2) (Cat # MM200), tumor necrosis factor-α (TNF-α) (Cat # MTA00B), and IL-1β/IL-1F2 (Cat # MLB00C) (*see* **Note 2**).
6. Cell culture CO_2 incubator (Sanyo, Model MCO-18AIC) and centrifuge (Eppendorf, Model 5804R).

2.5 LDH Assay

1. RPMI cell culture medium (Invitrogen) with 1 % FBS, phenol red-free.
2. Lyophilized catalyst solution (Biovision Cat # K311-400-1).
3. Dye solution (Biovision Cat # K311-400-2).
4. 96-well culture plates (Fisher Sci., Cat # 07-200-87).
5. Spectrophotometer/BioTek microplate reader.
6. Sterile water (Invitrogen).
7. Triton-X 100 (Sigma Cat # T8787).

2.6 BCA Protein Assay

The protein assay is carried out by using Pierce BCA protein assay kit (Catalogue # 23225).

1. BCA reagent A which contains sodium carbonate, sodium bicarbonate, bicinchoninic acid, and sodium tartrate in 0.1 M sodium hydroxide.
2. BCA reagent B containing 4 % cupric acid.
3. Bovine serum albumin (BSA) 2 mg/ml in 0.9 % saline and 0.05 % sodium azide.
4. Eppendorf tubes (0.5 ml), 96-well cell culture plates, and 37 °C water bath.
5. Spectrophotometer (Shimadzu UV3600)/plate reader (BioTek).

2.7 Comet Analysis (See Note 3)

1. Phosphate-buffered saline (Ca^{+2}- and Mg^{+2}-free) (Invitrogen).
2. Low-gelling-temperature agarose (Sigma, Cat # A-4018, Type VII).

3. Staining solution: propidium bromide (10 μg/ml) (Sigma, Cat # P-4170).
4. Disodium EDTA (0.5 M): Prepare by adding 55.8 g of Na_2 EDTA and 6.4 g NaOH to 270 ml of sterile water. Dissolve completely and adjust the pH to 8 by adding NaOH.
5. Proteinase K (EMD chemicals, Cat # 24568-2).
6. 5 M NaOH stock solution.
7. Alkaline lysis buffer for single-strand break detection: 1.2 M NaCl, 100 mM Na_2 EDTA, 0.1 % sodium lauroyl sarcosinate, and 0.26 M NaOH (pH > 13), and buffer must be freshly prepared and kept at 4 °C.
8. Alkaline rinse and electrophoresis buffer: 0.03 M NaOH and 2 mM Na_2 EDTA (pH 12.3).
9. Neutral lysis solution for double-strand break detection: 2 % sarkosyl, 0.5 M Na_2EDTA, and 0.5 mg/ml proteinase K (pH 8.0), and equilibrate at 4 °C.
10. Neutral rinse and electrophoresis solution: 90 mM Tris-HCl buffer, 90 mM boric acid, and 2 mM Na_2EDTA (pH 8.5).
11. Falcon tubes (5 ml), water bath, hemocytometer/cell counter, glass slides, and covered glass slide chamber (Fisher Sci.).
12. Horizontal gel electrophoresis apparatus with power supply (Fisher Sci), epifluorescence microscope with green excitation filters.

2.8 Micronuclei Assay

1. Sterile water (Invitrogen).
2. Hoechst 33342 (Sigma, Cat. # B2261). Hoechst staining solution is composed of 1.8 μM Hoechst in water and should be freshly prepared and protected from light.
3. Hypotonic buffer: 75.5 μM KCl and 1.15 μM $CaCl_2$.
4. Hypotonic buffer with methanol: 1:1 v/v solution of hypotonic buffer with methanol.
5. Methanol fixing solution: 1:1 v/v methanol/acetone solution stored at −20 °C.
6. Fluorescence microscope (Zeiss).

2.9 Histological Analysis of Lung Tissues

2.9.1 Fixation and Dehydration

1. Paraformaldehyde (4 %) or buffered formalin bottles (10 %) (Fisher Sci.).
2. RNAse-free phosphate-buffered saline (Invitrogen).
3. Tissue samples (*see* **Note 4**).
4. Histological grade ethanol (different v/v % solution in water), xylene (Sigma, Cat # 95672), and paraffin (Emsdiasum).

2.9.2 Hematoxylin and Eosin Staining (H&E Staining)

1. Eosin stock solution: Prepare a working solution of eosin by mixing 1 ml of 1 % aq. eosin (Sigma, Cat # 2853), 0.1 ml 1 % aq. phloxine B (Sigma, Cat # P4030), 7.8 ml 95 % ethanol, and 0.4 ml glacial acetic acid.
2. Hematoxylin stock solution (Sigma, Cat # HHS-32).
3. Lithium carbonate (1.36 % w/v) prepared by mixing 4.7 g of lithium carbonate (Sigma, Cat # 13010) in 350 ml of distilled water.
4. Hydrochloric acid solution (0.25 %): Prepare a working solution by adding 2.785 L of 95 % ethanol, 950 ml of distilled water, and 9 ml hydrochloric acid.
5. Cytoseal mounting medium (Emsdiasum, Cat # 18006).

2.9.3 Immunohistochemistry

1. Phosphate-buffered saline (Invitrogen) 1×, pH 7.0, and distilled water.
2. Sodium citrate buffer (0.01 M, pH 6.0): Antigen retrieval solution can be prepared by mixing 9 ml of 0.1 M citric acid (Sigma, Cat # C2404) solution, 41 ml of 0.1 M sodium citrate (Sigma, Cat # W302600), and 450 ml of distilled water. Adjust the pH to 6.0.
3. Hydrogen peroxide (3 %) (Sigma, Cat # 16911).
4. Blocking buffer (Dulbecco's phosphate-buffered saline, 1× + 10 % serum).
5. Antigen-specific primary antibodies, secondary HRP-conjugated anti-rabbit antibodies (*see* **Note 5**).
6. DAB (3,3′-diaminobenzidine tetrahydrochloride) substrate (Pierce, Cat # 34001): Dissolve DAB in 50 mM Tris-HCl, pH 7.2 at 1 mg/ml.

3 Methods

3.1 Nanomaterials Formulations and Characterization

1. For basic characterization of the NPs, disperse a small amount of NPs powder (~1 mg) in 5 ml of distilled water and sonicate the solution for 10 min.
2. Take 3 ml of NPs suspension in the quartz cuvette and study the optical signature by UV–visible spectrophotometer. Record the absorbance of the suspension. Copper NPs show a characteristic surface plasmon peak at 560 nm (*see* **Note 6**).
3. Measure the size of the NPs suspension by taking 3 ml of suspension in disposable cuvette and reading the particle size using the Zetasizer (90Plus instrument). Also, on the same instrument, measure the zeta potential (charge) of the suspension by placing the instrument electrode into the cuvette with approximately 1.5 ml of NPs suspension (*see* **Note 7**).

3.2 In Vivo Animal Exposure to Animals

3.2.1 Instillation of Nanoparticles

For intratracheal instillation, some important factors must be considered which include the method of intubation, the specific vehicle containing the test material, the volume of the liquid used, method of anesthesia, and dose of the test material administered.

1. Anesthetize the animals with ketamine/xylazine mixture (ketamine (100 mg/kg) and xylazine (10 mg/kg)) at 10 ml/kg intraperitoneal injection with a 27 gauge ½ in. attached needle on a 0.5 ml tuberculin syringe.
2. Keep the sedated mice at 37 °C using a temperature controlled heating plate.
3. Position the mice at an angle of approximately 40° against an angled restraining stand (*see* **Note 8**).
4. Press the tongue towards the lower jaw using a small spatula.
5. Intubate the trachea using a 24G BD Insyte catheter with a shortened needle (*see* **Note 9**).
6. Take 50 μl of NPs suspension in a 250 μl SGE glass syringe and instill the suspension through the intubated catheter followed by injecting 150 μl of air (*see* **Note 10**).
7. Remove the intubation catheter and keep the animal in vertical position with head up for 5–10 min.
8. Keep the animal on 37 °C heating plate until the animal recovers from anesthesia.
9. The deposition and distribution of the instilled NPs can be verified by installing Evan's blue dye or radioactive gold seeds.

3.2.2 Inhalation of Nanoparticles

The inhalation exposure system usually comprised of a fluidized bed aerosol generator, an animal exposure chamber, and assorted aerosol monitoring and control devices for generating and characterizing the NPs aerosol.

3.2.3 Exposure System

The exposure system can vary from laboratory to laboratory. This protocol represents a generalized system applicable for exposing the animals.

1. The exposure chamber can be of varying sizes to accommodate several animals placed in mesh cage for free flow of particles around mice and minimize crowding.
2. A vacuum pump is connected to chamber to maintain airflow and water vapor equipped with a high-efficiency particulate air (HEPA) filter.

3.2.4 Aerosol Generation and Characterization

1. Nanoparticles suspension at a specific concentration is prepared by adding powdered NPs in water as described in Subheading 2.1.
2. Sonicate the NPs suspension with a high-frequency probe for 10 min.

3. Add the sonicated solution to the reservoir of the 6-jet collision nebulizer operating at 20 psi from HEPA-filtered air source.
4. The nebulized droplets are dried in the heated brass pipes, connecting to a static discharge device before entering the chamber, to completely evaporate the water.
5. The generated aerosol is characterized using the control devices like mobility particle sizer and mobility analyzer.

3.2.5 Inhalation Exposure of Mice

1. Animals in a group of six (as described in Subheading 2.2) are exposed to the NPs aerosol for 4 h duration or for 4 h duration per day for 10 days by placing the animals in the exposure chamber, depending upon the type of experiment performed (acute or subacute experiment).
2. Mice exposed with nebulized water act as control group.

3.3 BAL Fluid Isolation

1. Sacrifice the animals after 24 h of NPs instillation/inhalation by intraperitoneal injection of sodium phenobarbitone solution (200 mg/ml) at a conc. of 150–200 mg/kg of body weight.
2. Using scissors, make a small incision around the neck area, and dissect out the tissue to expose trachea.
3. Carefully make a small incision in the trachea enough to allow passage of a 23G lavage tube into trachea.
4. Cut 1–1.5 in. of lavage tube and carefully place a 23G needle inside the tube.
5. Inject 0.9 ml of sterile saline using a 1 ml syringe through the inserted needle. Care must be taken to avoid moving the tubing by holding it with a forcep.
6. Aspirate the saline from the lungs by using the same syringe while holding the trachea with forceps.
7. Remove the syringe from the needle and collect the recovered lavage fluid in a 15 ml Falcon tube placed on ice. Repeat the process four times.
8. Remove 1 ml of the fluid and place it in a 1.5 ml tube on ice. Centrifuge rest of the lavage at 400 × g for 10 min to obtain a cell pellet.
9. From the initial fraction of 1 ml of fluid, prepare 1:10 dilution of lavage fluid with RPMI cell culture medium with 10 % trypan blue. Load 10 μl of suspension in hemocytometer and count the cells concentration per ml of fluid with percentage viability.
10. Take the centrifuged lavage, remove the supernatant, and store at −80 °C for further protein and LDH assay.
11. Resuspend the cell pellet in 1 ml PBS (with Ca^{+2} and Mg^{+2}) for determination of cell storage and composition for comet analysis.

3.4 Macrophage Cell Culture and Cytokine Measurements

1. Take the cell plate obtained after BAL Fluid isolation.
2. Wash and resuspend the cells in RPMI cell culture medium with 10 % FBS. Adjust the concentration of the cells to 1×10^6 cells per ml of RPMI.
3. Plate the cells in a 24-well cell culture plate by adding 0.5 ml aliquots in each well.
4. Incubate the plate for 1 h in humidified incubator at 37 °C, 5 % CO_2 to allow adherence.
5. Wash the adhered cells twice with 0.5 ml of serum-free RPMI medium and allow to equilibrate in 0.5 ml of serum-free RPMI for 30 min.
6. For macrophage activation add appropriate amount of LPS (1 μl of 5 μg/ml) per well and incubate the plate in incubator at 37 °C for 18 h.
7. Collect the cell supernatant and store at −80 °C until further analysis.
8. Measure different cytokines (INF-γ, IL-13, MIP-2, TNF-α, and IL-1β) in the cell-free culture supernatants by sandwich ELISA kits as per manufacturer's instructions.

Figure 1 shows confocal fluorescence image of the macrophages extracted from BAL fluid. The cells are treated with rhodamine dye for 2 h and imaged with 488 nm laser.

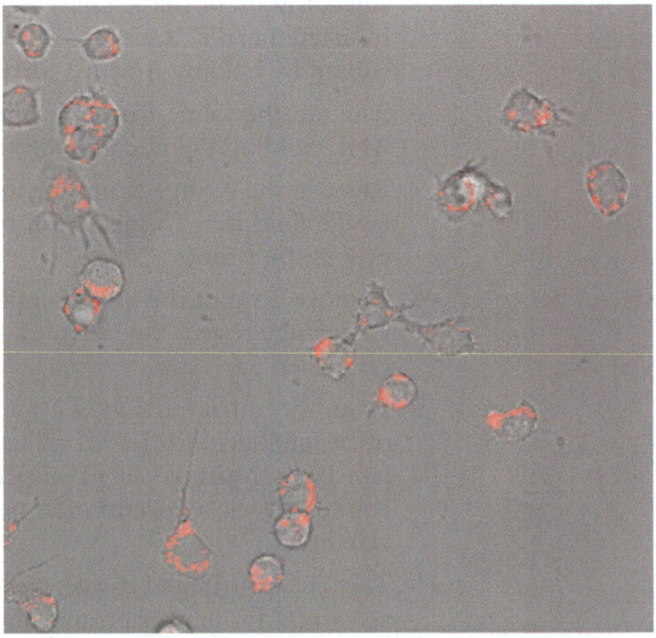

Fig. 1 Confocal fluorescence image of the macrophages extracted from BAL fluid. The cells are treated with rhodamine dye for 2 h and imaged with 488 nm laser

3.5 LDH Assay

1. Reconstitute catalyst by adding 1 ml of sterile water, and vortex occasionally for 10 min to dissolve completely. The reconstituted solution remains stable for 2 weeks at 4 °C.
2. Prepare reaction mixture by adding 250 μl of reconstituted catalyst solution to 11.25 ml of dye solution (for one 96-well plate).
3. Collect cells and wash with cell culture medium with 1 % FBS. In a 96-well culture plate prepare samples in triplicate as:

   Background sample: Add 200 μl of cell culture media per well.

   Low control: Add approx $1–2 \times 10^4$ cells per well in 200 μl of cell culture media.

   High control: Add approx $1–2 \times 10^4$ cells/well in 200 μl of cell culture media with 1 % Triton-X 100.

   Test sample: Add approx $1–2 \times 10^4$ cells/well in 200 μl of cell culture media with test substance.
4. Incubate the cells for appropriate time in cell culture incubator. After that centrifuge the cells at 250 × g for 10 min.
5. Take 100 μl of the supernatant and add to a new 96-well transfer plate followed by addition of 100 μl of reaction mixture to each well.
6. Incubate the plate for 30 min at room temperature in dark and measure the absorbance of samples at 490 nm using a spectrophotometer/plate reader.
7. Calculate the percentage toxicity by the formula:

$$\text{Percent cytotoxicity} = \frac{\text{Test sample} - \text{low control}}{\text{High control} - \text{low control}} \times 100.$$

3.6 BCA Protein Assay

1. Equilibrate all reagents and samples (resuspended lavage fluid) and mix thoroughly.
2. Prepare fresh working reagent for all the standards and samples by mixing reagent A and reagent B in 50:1 ratio (*see* **Note 11**).
3. In a 96-well cell culture plate, add 200 μl of working reagent to each standard or sample per well. All standard and samples must be plated in replicates.
4. Add 25 μl of standard or sample to each well and mix gently by vortexing for 30 s.
5. Incubate the 96-well plate at 37 °C for 30 min or 60 °C for 30 min.
6. Cool the plate at room temperature and measure the absorbance at 562 nm using the spectrophotometer or plate reader.
7. Correct the background absorbance by subtracting the absorbance measurement of the blank standard replicates from the

absorbance measurement of all other individual standard and unknown sample replicates.

8. Prepare a standard curve by plotting the average corrected absorbance for each BSA concentration in μg/ml. Use the standard curve to determine the protein concentration of each unknown sample.

3.7 Comet Assay

1. Prepare 1 % agarose gel by dissolving the powdered agarose in sterile water at 100 °C in water bath for several minutes.
2. Place the dissolved agarose at 40 °C water bath (*see* **Note 12**).
3. Coat the slides with agarose by gently dipping the slides in molten 1 % agarose. Wipe one side of slide clean.
4. Allow slides to air-dry to form thin films on slides. Coated slides can be stored with desiccant.
5. Trypsinize the cells to obtain single cell suspension and adjust the cell counts to 2×10^4 cells/ml by adding PBS. PBS minimizes cell aggregation and DNA repair.
6. Take 0.4 ml of cell suspension in Falcon tube and add 1.2 ml of 1 % agarose gel at 40 °C. Mix well and rapidly transfer 1.2 ml cell suspension onto the agarose coated slides and allow it to gel.
7. Carefully submerge the coated slides horizontally in alkaline lysis solution at 4 °C and incubate overnight. In case of neutral lysis, incubate the slides at 37 °C.
8. After overnight lysis, remove slides carefully and rinse with alkaline rinse solution by dipping in for 20 min at room temperature. For neutral lysis use the respective neutral lysis buffer for 30 min. Repeat the rinsing two more times.
9. Carry out the electrophoresis in respective alkaline electrophoresis buffer or neutral buffer for 25 min at 0.6 V/cm.
10. Remove slides from electrophoresis chamber, rinse the slides, and neutralize in 400 ml sterile water.
11. Stain the slides with 2.5 μg/ml of propidium bromide in distilled water for 20 min. Wash off the excess dye solution by keeping the slides in 400 ml sterile water.
12. Analyze cells using the epifluorescence microscope and acquire minimum of 50 images.
13. Using the comet image analyses software, analyze individual images for total intensity (DNA content), tail length, percent DNA in tail, and tail moment. The use of five damage classes introduced by Collins has been widely adopted for estimating the DNA damage [24].

3.8 Micronulei Assay

1. Seed cells at an approximate concentration of 500 cells per well from the preculture in a 96-well cell culture plate.

2. Incubate the cells in 100 μl of high-glucose DMEM cell culture media for 24 h (media without antibiotic) (*see* **Note 13**).
3. After the incubation time period, replace the culture medium with the equivalent volume of hypotonic buffer.
4. After a specific time period, the hypotonic solution is further replaced with hypotonic buffer with methanol.
5. Fix the cells by wetting the cells with methanol fixing solution for 5 min.
6. Treat the fixed cells with an aq. solution of Hoechst for 1 h.
7. Wash the excess dye with sterile water.
8. Dry the stained cells in air and count the number of micronuclei using a fluorescence microscope equipped with an excitation source at 365 nm.
9. Record the number of micronuclei per 500 cells.

3.9 Histological Analysis of Lung Tissues

3.9.1 Fixation and Dehydration

1. Dissect out the tissue and immediately place it in ice-cold PBS.
2. Wash the tissue thoroughly to remove any blood and thereafter place the tissue in fixative vial for an hour.
3. Cut the tissue into small sizes (approx 2 mm × 2 mm) for the proper fixation.
4. Place the tissue sections back in fixative and leave it overnight in dark for fixation at 4 °C.
5. Wash the tissue sections with PBS for 1 min followed by leaving the sections in PBS for another 30 min.
6. Dehydrate the sections by placing in an increasing concentration of ethanol solution as 30 % ethanol in water (2 h), 50 % ethanol (2 h), 70 % ethanol (4 h), 95 % ethanol (overnight), 100 % ethanol (1 h at −20 °C), and finally in 100 % ethanol (1 h, at room temperature, 3 times). The samples should always be covered by solutions and should not be dried at any stage.
7. Transfer the sections to xylene for 30–40 min (2 times).
8. Put the sample in paraffin for 40 min (3 times).
9. Finally embed the samples and section them using a microtome (3–10 μm) and store the sections at 4 °C.

3.9.2 Hematoxylin and Eosin Staining (H&E Staining)

1. Place the paraffinized sections in xylene for 5 min to remove the paraffin (2 times). The sections should be cooled down to room temperature before dipping into xylene to avoid any loss of sections.
2. Rehydrate the sections by keeping it in a decreasing concentration of ethanol solution as 100 % ethanol (3 min, 2 times), 95 % ethanol (3 min, 2 times), and 70 % ethanol (3 min).
3. Rinse the sections in distilled water for 5 min.

4. Immerse the sections in hematoxylin solution for 6 min followed by rinsing in running tap water for 20 min (*see* **Note 14**).
5. Wash the sections with HCl solution for 1 s and rinse under tap water for 5 min.
6. Immerse the sections in lithium carbonate solution for 3 s and again rinse in running tap water for 5 min.
7. Counterstain the sections with eosin solution for 3 min followed by dehydration of the sections in 95 % ethanol (3 min, 2 times) and 100 % ethanol (3 min, 2 times).
8. Place the sections in xylene solution for 5 min and mount the sections with Cytoseal mounting media.
9. Place the coverslip and analyze the sections under the microscope for histopathological signatures.

3.9.3 Immunohistochemistry

1. Dry paraffinized slides over night at 58 °C.
2. Deparaffinize and hydrate the samples as described in the previous section as described previously.
3. Take the rehydrated slides and dip in 3 % H_2O_2 for 13 min followed by washing with PBS for 15 min.
4. Immerse the washed slides in a dish containing antigen retrieval solution (sodium citrate buffer).
5. Put the covered dish in a pressure cooker and add ¾ cup water to cooker and cook for about 10–15 min and thereafter let it cool at room temperature for 30 min.
6. Wash slides with water for 10 min followed by washing with PBS for another 10 min.
7. Block the slides with blocking solution for 1 h.
8. Make proper dilution of the primary antibody (as per recommendation in data sheet) and apply primary antibody solution the each sections and incubate overnight at 4 °C.
9. Wash slides with PBS at room temperature for 10 min.
10. Add the secondary HRP-conjugated anti-rabbit antibody to each section after making proper dilutions in blocking buffer (as per manufacturer's instructions) and incubate for 1 h at room temperature.
11. Wash slides with PBS again for 10 min.
12. Prepare DAB substrate solution and add 2–3 drops of the mixture on sections at room temperature for 2–5 min. Suitable stain is usually developed within 2–5 min after substrate addition.
13. Rinse the sections with water and counterstain with hematoxylin for 5–10 min.

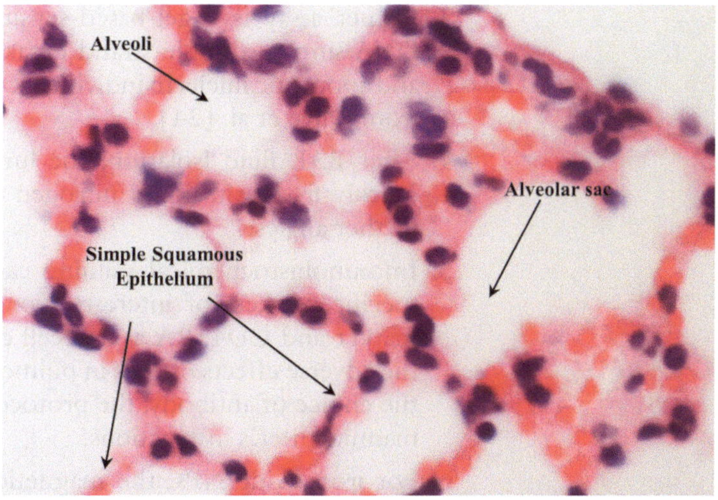

Fig. 2 Histology of the lung sections with hematoxylin and eosin staining (H&E staining)

14. Rinse again with water and dip slides in 0.1 % HCl (3 times) followed by washing with water.
15. Dip slides in 0.1 % NH_4OH solution (3 times) and rinse again with water.
16. Dehydrate the slides again by dipping it in 70 % ethanol (5 times), 80 % ethanol (5 times), 90 % ethanol (5 times), 100 % ethanol (5 times), and xylene (2 × 5 times).
17. Mount coverslip on the slides after adding Cytoseal mounting medium.

Figure 2 shows the histology of the lung sections with hematoxylin and eosin staining (H&E staining).

4 Notes

1. Depending upon the conditions and requirements of the experiments, different sized NPs can be used. Here, we selected different sizes of NPs which are commercially available. However, to study the in vivo interactions based on the charge, shape, and size of NPs, surface-modified NPs can be synthesized from reported literatures.
2. There can be varied choice of cytokines which can be selected for studying the effect of NPs in pulmonary system. Here, we have selected some of the common cytokines which are expressed as a result of oxidative stress and inflammation caused in pulmonary system. Several other cytokines which are expressed include macrophage/monocyte chemoattractant protein-1 (MCP-1) and IL-6.

3. Comet assay is a gel-based-sensitive method for analysis of DNA single-/double-strand breaks, cross-links, base damage, and apoptotic nuclei in individual cells. This protocol is adapted from Olive et al. [24].
4. After BAL fluid isolation, the lungs were dissected out and quickly frozen in liquid nitrogen and stored at −80 °C until further analysis.
5. Immunohistochemical staining can be selected for variety of specific markers of interests. For example, CD3 as a T-cell marker and CD45RA as a B cell can be selected for studying the adverse effects of NPs in pulmonary system. Depending on the choice of antibody, the protocol should be followed as per manufacturer's instructions.
6. For iron oxide NPs, the magnetic properties can be studied using a SQUID magnetometer.
7. The concentration of the NPs in the suspension plays a crucial role in making accurate measurements. If the sample seems to be too concentrated, then dilute the samples accordingly. Also, make sure there is no sedimentation or precipitation of the NPs from the suspension.
8. For illumination a directed light source can be used as described by Brain et al. [25].
9. In an alternative approach, insert a cannula in the mouth of animal and place it between the vocal cords and into the lumen of the trachea. A ball-tipped needle can be guided to the epiglottis, which avoids the use of a laryngoscope.
10. Prepare the NPs suspension by adding a specific amount of NPs to 0.9 % NaCl Milli-Q water containing 10 % BAL fluid (v/v) isolated from the C57Bl mice as described in Subheading 3.3. The NPs suspension should be vigorously mixed or briefly sonicated immediately before instillation.
11. The BCA microassay has a linear has a range of 20–200 μg/ml of protein using a 1:1 sample to reagent ratio and a higher linear range of 125–2,000 μg/ml can be obtained using a 1:20 sample to reagent ratio. A standard curve must be prepared using BSA protein using the same protocol.
12. If the cell samples contain RBC or hemoglobin, add 2 % DMSO to agarose or to alkaline lysis solution to avoid damage by iron-catalyzed reactive oxygen species, which can cause strand breaks.
13. The incubation time for the micronucleus assay depends on the system tested as well as the test substances. Thus, it is recommended to have several incubation time points.
14. Hematoxylin must be protected from the light and filtered before use to eliminate oxidized golden sediments.

Acknowledgement

This work was supported by Electronics Materials Research Institute (eMRI), Department of Physics, Northeastern University.

References

1. Dvir T, Timko BP, Kohane DS, Langer R (2011) Nanotechnological strategies for engineering complex tissues. Nat Nanotech 6:13–22
2. Dobrovolskaia MA, McNeil SE (2007) Immunological properties of engineered nanomaterials. Nat Nanotech 2:469–478
3. Teow Y, Asharani PV, Hande MP, Valiyaveettil S (2011) Health impact and safety of engineered nanomaterials. Chem Commun 47:7025–7038
4. Maynard AD, Aitken RJ, Butz T, Colvin V, Donaldson K, Oberdörster G, Philbert MA, Ryan J, Seaton A, Stone V, Tinkle SS, Tran L, Walker NJ, Warheit DB (2006) Safe handling of nanotechnology. Nature 444:267–269
5. Cho K, Wang X, Nie S, Chen Z, Shin DM (2008) Therapeutic nanoparticles for drug delivery in cancer. Clin Cancer Res 14:1310–1316
6. Ren G, Hu D, Cheng EW, Vargas-Reus MA, Reip P, Allaker RP (2009) Characterisation of copper oxide nanoparticles for antimicrobial applications. Int J Antimicrob Agents 33: 587–590
7. Eastman JA, Choi SUS, Li S, Yu W, Thompson LJ (2001) Anomalously increased effective thermal conductivities of ethylene glycol-based nanofluids containing copper nanoparticles. Appl Phys Lett 78:718–720
8. Guo L, Huang Q, Li X, Yang S (2001) Iron nanoparticles: synthesis and applications in surface enhanced Raman scattering and electrocatalysis. Phys Chem Chem Phys 3:1661–1665
9. Gupta AK, Gupta M (2005) Synthesis and surface engineering of iron oxide nanoparticles for biomedical applications. Biomaterials 26: 3995–4021
10. Ito A, Shinkai M, Honda H, Kobayashi T (2005) Medical application of functionalized magnetic nanoparticles. J Biosci Bioeng 100:1–11
11. Karlsson HL, Gustafsson J, Cronholma P, Möller L (2009) Size-dependent toxicity of metal oxide particles—a comparison between nano- and micrometer size. Toxicol Lett 188:112–118
12. Oberdorster G, Stone V, Donaldson K (2007) Toxicology of nanoparticles: a historical perspective. Nanotoxicology 1:2–25
13. Weissleder R, Stark DD, Engelstad BL, Bacon BR, Compton CC, White DL, Jacobs P, Lewis J (1989) Superparamagnetic iron oxide: pharmacokinetics and toxicity. AJR Am J Roentgenol 152:167–173
14. Yokohira M, Hashimoto N, Yamakawa K, Suzuki S, Saoo K, Kuno T, Imaida K (2009) Lung carcinogenic bioassay of CuO and TiO_2 nanoparticles with intratracheal instillation using F344 rats. J Toxicol Pathol 22: 71–78
15. Nel A, Xia T, Mädler L, Li N (2006) Toxic potential of materials at the nanolevel. Science 311:622–627
16. Singh N, Jenkins GJS, Asadi R, Doak SH (2010) Potential toxicity of superparamagnetic iron oxide nanoparticles (SPION). Nano Rev 1:5358–5372
17. Warheit DB, Webb TR, Reed KL, Frerichs S, Sayes CM (2007) Pulmonary toxicity study in rats with three forms of ultrafine-TiO_2 particles: differential responses related to surface properties. Toxicology 230:90–104
18. Braydich-Stolle L, Hussain S, Schlager JJ, Hofmann MC (2005) In vitro cytotoxicity of nanoparticles in mammalian germline stem cells. Toxicol Sci 88:412–419
19. Xia T, Kovochich M, Brant J, Hotze M, Sempf J, Oberley T, Sioutas C, Yeh JI, Wiesner MR, Nel AE (2006) Comparison of the abilities of ambient and manufactured nanoparticles to induce cellular toxicity according to an oxidative stress paradigm. Nano Lett 6:1794–1807
20. Cho W-S, Duffin R, Poland CA, Howie SEM, MacNee W, Bradley M, Megson IL, Donaldson K (2010) Metal oxide nanoparticles induce unique inflammatory footprints in the lung: important implications for nanoparticle testing. Environ Health Perspect 118: 1699–1706
21. Fahmy B, Cormier SA (2009) Copper oxide nanoparticles induce oxidative stress and cytotoxicity in airway epithelial cells. Toxicol In Vitro 23:1365–1371
22. Guo B, Zebda R, Drake SJ, Sayes CM (2009) Synergistic effect of co-exposure to carbon black and Fe_2O_3 nanoparticles on oxidative

stress in cultured lung epithelial cells. Part Fibre Toxicol 6:4–16

23. Ying E, Hwang H-M (2010) In vitro evaluation of the cytotoxicity of iron oxide nanoparticles with different coatings and different sizes in A3 human T lymphocytes. Sci Total Environ 408:4475–4481

24. Olive PL, Banáth JP (2006) The comet assay: a method to measure DNA damage in individual cells. Nat Protoc 1:23–29

25. Brian JD, Knudson DE, Sorokin SP, Davis MA (1976) Pulmonary distribution of particles given by intratracheal instillation or by aerosol inhalation. Environ Res 11:13–33

Chapter 17

In Vitro Techniques to Investigate the Oxidative Effects of Quantum Dots

Ridhima Juneja and Indrajit Roy

Abstract

Quantum dots are semiconductor nanoparticles with unique, size-tunable optical properties. Owing to these properties, these nanoparticles are fast emerging as versatile and multimodal agents in diagnostic imaging and drug delivery applications. In the past decade, their interaction with various biological models, ranging from isolated cells and tissues to small animals, has been extensively investigated. However, along with the various beneficial effects, the presence of heavy metals in most quantum dots has led the scientific community to view the biomedical applications of quantum dots in a more cautious manner. One potential deleterious effect of quantum dots is triggering or exacerbating the process of oxidative stress, which can interfere with or destroy key biomolecular components or processes in the body. This can lead to premature ageing or diseases such as dementia and cancer. In this chapter, we describe various in vitro techniques which are fundamental in investigating the oxidative effects of quantum dots following their internalization within cells in culture.

Key words Quantum dots, Oxidative stress, Fluorescence, Confocal imaging, Cell viability assay

1 Introduction

When the dimensions of semiconductor materials are reduced to the nanoscale, they are known as quantum dots (QDs) as they are in a quantum-confined state [1, 2]. This quantum confinement renders unique optical properties to these QDs, which include bright, highly photostable photoluminescence, broad absorption spectra, and narrow, size-tunable emission spectra [3–6]. QDs are composed of a semiconductor core, which generally belongs to the following categories, group II–VI (e.g., CdS, CdSe, CdTe), IV–VI (e.g., PbSe, PbS), III–V (e.g., InP, InAs,), and IV (e.g., Si) [6, 7]. These cores are protected by a shell of wide bandgap semiconductor, such as ZnS, followed by a further polymeric coating that enables water solubility for biological application.

The unique photophysical properties of QDs have led to their widespread use in biological and medical fields. QDs have quickly

Donald Armstrong and Dhruba J. Bharali (eds.), *Oxidative Stress and Nanotechnology: Methods and Protocols*, Methods in Molecular Biology, vol. 1028, DOI 10.1007/978-1-62703-475-3_17, © Springer Science+Business Media New York 2013

filled in the role of organic dyes in biological analysis, on the account of some of their properties, one of the most immediately obvious being brightness (owing to the high extinction coefficient combined with a comparable quantum yield to fluorescent dyes) as well as their stability (allowing much less photobleaching). Another application that takes advantage of the extraordinary photostability of quantum dot probes is the real-time tracking of molecules and cells over extended periods of time. Researchers were able to observe quantum dots in lymph nodes of mice for more than 4 months. Semiconductor quantum dots have also been employed for in vitro imaging of targeted cells. The ability to image single-cell migration in real time is expected to be important to several research areas such as embryogenesis, cancer metastasis, stem-cell therapeutics, and lymphocyte immunology. Attempts have been made to use quantum dots for tumor targeting under in vivo conditions. QD probes provide unique functionality of traceable drug delivery, as biodistribution of carriers and intracellular uptake can be monitored via fluorescence.

Although the QDs offer potentially invaluable societal benefits such as drug targeting and in vivo biomedical imaging, a number of significant barriers have hindered the clinical translation of this technology. Among them, their toxicity is an urgent concern. QD surface plays a key role in evaluating the extent of their toxicity [8]. If the QD surface is well protected by polymers or proteins, it exerts little harm on cells or mammals. Otherwise, it manifests noticeable toxicity, mainly those containing heavy metals such as cadmium and lead. In addition, the erosion of the shell may cause undesirable reactions in vivo. Cytotoxicity of QDs has been observed in a large number of in vitro studies, affecting cell growth and viability [9–12]. Various mechanisms proposed to explain the toxicity of QDs include release of toxic metals from the core of nanoparticles and generation of reactive oxygen species (ROS). QDs could generate ROS by electron transfer to molecular oxygen. When this generated ROS overcomes the cellular antioxidant defense system, cells undergo oxidative stress. Results of this stress include modification and damage of cellular components, such as lipids, proteins, and DNA [13, 14]. Oxidation of unsaturated fatty acids results in the generation of lipid peroxides that start a chain reaction leading to oxidation of nearby unsaturated fatty acids that can culminate in the disruption of the plasma membrane and membranes of other cellular organelles. Cells are able to respond to even small changes of intracellular redox status, which is sufficient to inhibit proliferation and induce differentiation [15]. The type of ROS formed strongly depends on the composition of the QDs [16]. A number of fluorescent probes and other techniques have been used to reveal ROS in live cells. This chapter provides a description of various in vitro protocols routinely used to investigating the oxidative effects of QDs on treated cells in culture.

2 Materials

2.1 Characterization Techniques of Commercial Quantum Dots

1. Commercial quantum dots (a) Qdot-655-ITK-amino(PEG), or QDa, at a concentration of 8 μM in water and (b) Qdot-655-ITK-carboxyl, or QDc, at a concentration of 8 μM in water (Invitrogen) (*see* **Notes 1** and **2**).
2. Phosphate-buffered saline (PBS, 1×, pH 7.2). Prepare this buffer solution by dissolving 8 g NaCl, 0.2 g KCl, 1.44 g Na_2HPO_4, and 0.24 g K_2HPO_4 in 800 ml of distilled water. Adjust the pH to 7.2 by adding 0.1 M HCl solution. Then bring up the total volume of the solution to 1 L by adding water.
3. Standard plastic and glass/quartz cuvettes (maximum capacity 3 ml).
4. ZetaPlus DLS Instrument (Brookhaven Instruments, Inc., Holtsville, NY).
5. Shimadzu UV-3600 or any other standard spectrophotometer.
6. Fluorolog-3.11 (Jobin Yvon, Longjumeau, France) or any other standard spectrofluorimeter.

2.2 Confocal Microscopic Analysis of Uptake of QDs in Treated Cells

1. Phosphate-buffered saline (PBS, 1×, pH 7.2).
2. DMEM medium (Invitrogen).
3. Fetal bovine serum (FBS), heat treated (Invitrogen).
4. Penicillin–streptomycin, containing 5,000 U of penicillin (base) and 5,000 μg of streptomycin (base)/ml water, utilizing penicillin G (sodium salt) and streptomycin sulfate in 0.85 % saline (Invitrogen).
5. Complete growth medium. Make this by adding 50 ml of FBS and 5 ml of penicillin–streptomycin to 500 ml of DMEM media.
6. Cell line MiaPaCa-2 (available from ATCC, Manassas, VA), maintained in complete growth medium (*see* **Note 3**).
7. Sterile cell culture dishes in 35 mm format.
8. Laminar flow tissue culture hood.
9. Cell culture incubator (37 °C, 5 % CO_2).
10. Nikon Eclipse TE2000 confocal microscope equipped with Nuance GNIR imaging system (Cambridge Research & Instrumentation Inc., Cambridge, MA) or any other standard confocal microscope.

2.3 Quantitative Estimation of Cellular Uptake of the QDs Using Spectrofluorimetry

1. Cells, PBS, media, and media supplements, as described in Subheading 2.2.
2. Sterile cell culture dishes in 12-well format.
3. Standard glass/quartz cuvettes.
4. Sterile cell scraper.

5. 1× Cell culture lysis reagent. Prepare this by dissolving 1 % (mass/volume) of surfactant Triton-X 100 (Sigma-Aldrich) in 1× PBS, pH 7.2.
6. Sterile microcentrifuge tubes, maximum capacity 1.5 ml.
7. Cell culture incubator (37 °C, 5 % CO_2).
8. Any regular microcentrifuge.
9. Fluorolog-3.11 (Jobin Yvon, Longjumeau, France) or any other standard spectrofluorimeter.

2.4 Cell Viability (MTS) Assay to Probe Effects of QDs on Treated Cells

1. Sterile cell culture plates in 96-well format.
2. *N*-acetyl-L-cysteine, or NAC (Sigma-Aldrich).
3. MTS assay (CellTiter 96® AQueous One Solution Cell Proliferation Assay) reagent (Promega).
4. Spectrophotometric plate reader, e.g., Opsys MR Autoreader, Dynex Technologies Inc., Chantilly, VA, USA.

2.5 Flow Cytometry to Probe Effects of QDs on Treated Cells

1. Sterile cell culture dishes in 35 mm format.
2. 15 ml centrifuge tubes.
3. Regular laboratory centrifuge.
4. Trypsin–EDTA. This contains the following components in 1 L water: 400 mg potassium chloride (KCl), 60 mg potassium phosphate monobasic (KH_2PO_4), 350 mg sodium bicarbonate ($NaHCO_3$), 8,000 mg sodium chloride (NaCl), 90 mg sodium phosphate dibasic (Na_2HPO_4–$7H_2O$), 1,000 mg D-glucose (dextrose), 10 mg phenol red, 380 mg sodium EDTA (Na_2 EDTA), and 2,500 mg trypsin (Invitrogen).
5. Fluorescein isothiocyanate (FITC)-conjugated annexin-V (Sigma-Aldrich).
6. *N*-acetyl-L-cysteine (NAC).
7. FACSCalibur flow cytometer (Becton Dickinson, USA).

2.6 Intracellular ROS Measurement

1. Sterile cell culture plates in 96-well format.
2. 20,70-dichlorofluorescin diacetate (DCFH-DA) (Beyotime Institute of Biotechnology, Jiangsu, China).
3. Serum-free medium. Make this by adding 5 ml of FBS and 1 ml of penicillin–streptomycin in 500 ml DMEM.
4. Trypsin–EDTA (Invitrogen).
5. Fluorescence microplate reader (Safire2, Tecan, Switzerland).

2.7 Glutathione (GSH) Assay

1. Sterile cell culture dishes in 35 mm format.
2. Sterile cell scraper.
3. Eppendorf tubes (0.6 ml).
4. Crushed ice.

5. Diethyl maleate (Sigma-Aldrich).
6. 5-5#-Dithiobis(2-nitrobenzoic acid) (DTNB) (Sigma-Aldrich).
7. 1-Methyl-4-vinyl-pyridinium (M4VP) (Sigma-Aldrich).
8. 5-Sulfosalicylic acid (SSA) (Sigma-Aldrich).
9. Sodium carbonate (Sigma-Aldrich).
10. Glutathione reductase (Sigma-Aldrich).
11. NADPH (Sigma-Aldrich).
12. Regular laboratory centrifuge.
13. Spectrophotometric plate reader, e.g., Opsys MR Autoreader, Dynex Technologies Inc., Chantilly, VA, USA.

2.8 Lipid Peroxidation Assay

1. Sterile cell culture dishes in 35 mm format.
2. 15 ml centrifuge tubes.
3. Sterile cell scraper.
4. Regular laboratory centrifuge.
5. Malondialdehyde, or MDA (Sigma-Aldrich).
6. 1-Butanol.
7. Trichloroacetic acid, or TCA (Sigma-Aldrich).
8. Thiobarbituric acid, or TBA (Sigma-Aldrich).
9. Spectrophotometric plate reader, e.g., Opsys MR Autoreader, Dynex Technologies Inc., Chantilly, VA, USA.

3 Methods (*See* Note 3)

3.1 Characterization Techniques (See Note 4)

1. Dilute QDs by adding 50 μl of QDs (8 μM in water) to 3 ml of PBS in standard plastic cuvettes (total capacity 3 ml). Observe their hydrodynamic size using a ZetaPlus DLS Instrument (*see* **Notes 5** and **6**).
2. Dilute QDs by adding 25 μl of QDs (8 μM in water) to 1.5 ml of PBS in standard plastic cuvettes (total capacity 3 ml). Dip the electrode in the cuvette (provided with the instrument), and measure the surface charge (zeta potential) using a ZetaPlus DLS Instrument (*see* **Note 6**).
3. Dilute QDs by adding 50 μl of QDs (8 μM in water) to 3 ml of PBS in standard glass/quartz cuvettes (total capacity 3 ml). Record absorption spectra using a standard spectro-photometer.
4. Dilute QDs by adding 50 μl of QDs (8 μM in water) to 3 ml of PBS in standard glass/quartz cuvettes (total capacity 3 ml). Record the emission spectra using a standard spectrofluorimeter, following excitation at an appropriate wavelength as determined from absorption measurements (*see* **Note 7**).

3.2 Confocal Microscopic Analysis of Uptake of QDs in Treated Cells In Vitro

1. Take 6 mm × 35 mm tissue culture dishes, mark them A1, A2, B1, B2, C1, and C2. Seed 200,000 cells in 2 ml of complete growth medium in each dish, and allow them to attach overnight.
2. Next day, aspirate the overnight growth medium, and add 2 ml of fresh growth medium in each dish.
3. Add 100 μl aqueous solutions of QDa (8 μM in water) each to dishes A1 and A2, and QDc (8 μM in water) each to dishes B1 and B2. Add 100 μl of 1× PBS solution each to dishes C1 and C2. Mix by gentle swirling. Return dishes to the cell culture incubator (37 °C, 5 % CO_2).
4. After 1 h of incubation, take out dishes A1, B1, and C1, aspirate the existing medium in the dishes, rinse twice with 1× PBS, add 2 ml of 1× PBS in each dish.
5. Immediately image the plates on a Nikon Eclipse TE2000 confocal microscope equipped with Nuance GNIR imaging system, which is capable of multispectral (wavelength-resolved) imaging in the range of 500–950 nm. Use custom-designed filter cube containing corresponding dichroic and emission filters acquired from Omega Optical to cut off excitation light and obtain high-contrast fluorescence images of the QDs in cells.
6. After 4 h of incubation from **step 3**, take out dishes A2, B2, and C2, and repeat **steps 4** and **5** with these dishes (*see* **Note 8**).

3.3 Quantitative Estimation of Cellular Uptake of the QDs Using Spectrofluorimetry

1. Take 2 × 12-well tissue culture plates, and mark them 1 and 2. Seed 100,000 cells in 1 ml of complete growth medium in 9 (out of 12) wells in each plate (total 18 wells), and allow them to attach overnight.
2. Next day, aspirate the overnight growth medium, and add 1 ml of fresh growth medium in each well.
3. Add 50 μl aqueous solutions of QDa (8 μM in water) to three wells each in Plate 1 and three wells each in Plate 2. Similarly, add 50 μl aqueous solutions of QDc (8 μM in water) to three wells each in Plate 1 and three wells each in Plate 2. Finally, add 50 μl of 1× PBS to three wells each in Plate 1 and three wells each in Plate 2. Mix by gentle swirling. Return plates to the cell culture incubator (37 °C, 5 % CO_2).
4. After 1 h of incubation, take out Plate 1. Aspirate the existing medium in all the wells, and rinse twice with 1× PBS. Immediately, lyse the cells by adding 200 μl of 1× cell culture lysis reagent to each well, followed by incubation for 30 min in with gentle shaking.
5. Analyze the cell lysates for QD uptake by quantifying the fluorescence recovered from the lysates. For this, scrape the cells (in the wells containing the lysis buffer) with a sterile scraper,

add further 1 ml of PBS/well and mix well with the lysed cells in lysis buffer, and transfer the contents of each well in separate microcentrifuge tubes (maximum capacity 1.5 ml).

6. Separate the cell debris from the lysate (containing the QDs) by centrifugation (704´*g*, 5 min) (*see* **Note 9**). Collect the supernatant (centrifugate) for fluorescence analysis.
7. Examine the amount of the fluorescent material in the centrifugate in a spectrofluorimeter, as described before (in Subheading 3.1, **step 4**).
8. After 4 h of incubation of cells from **step 3**, take out Plate 2. Repeat **steps 4** through 7 with the cells in this plate.

3.4 Cell Viability (MTS) Assay to Probe Effects of QDs on Treated Cells

1. Take 2×96-well tissue culture plates, and mark them 1 and 2. Seed 3,000 cells in 100 μl of complete growth medium in 18 (out of 96) wells in each plate (total 36 wells), and allow them to attach overnight.
2. Next day, aspirate the overnight growth medium, and add 100 μl of fresh growth medium in each well.
3. Immediately, to nine wells each in both plates, add 1 μl of NAC solution (100 μM in PBS) to the existing media, mix by gentle swirling, and return to incubator for 2 h (*see* **Note 10**).
4. After incubation, aspirate the existing medium from each well, and add 100 μl of fresh growth medium in each well.
5. Immediately, add 5 μl aqueous solutions of QDa (8 μM) to six wells each in Plate 1 (three wells each treated and non-treated with NAC), and six wells each in Plate 2 (three wells each treated and non-treated with NAC). Similarly, add 5 μl aqueous solutions of QDc (8 μM) to six wells each in Plate 1 (three wells each treated and non-treated with NAC), and six wells each in Plate 2 (three wells each treated and non-treated with NAC). Finally, add 5 μl of 1× PBS to six wells each in Plate 1 (three wells each treated and non-treated with NAC), and six wells each in Plate 2 (three wells each treated and non-treated with NAC). Mix by gentle swirling. Return both plates to the cell culture incubator (37 °C, 5 % CO_2).
6. After 1 h of incubation, take out Plate 1. Aspirate the existing medium in all the wells, and replace with 200 μl of fresh complete growth medium per well. Return plates to the incubator (37 °C, 5 % CO_2).
7. After 3 days of further incubation of Plate 1, analyze the cells for viability by a modified MTS assay protocol (the following experimental steps).
8. Aspirate the media and replace with fresh serum-free media (200 μl/well).

9. Add 40 μl of the MTS assay reagent to each well, mix and return to the incubator for 1 h. Dim the lights when performing this experimental step.
10. Transfer 100 μl of the liquid mixture from each well to the individual wells in a new 96-well plate, and read the plate using a spectrophotometric plate reader (Opsys MR Autoreader, Dynex Technologies Inc., Chantilly, VA, USA) at wavelength of 490 nm. Run the experiment in triplicates to access the statistical significance of the data (*see* **Note 11**).
11. Similarly, after 4 h of incubation from **step 5**, take out Plate 2. Repeat **steps 6** through **10** with the cells in this plate.

3.5 Flow Cytometry to Probe Effects of QDs on Treated Cells (See Note 12)

1. Take 12 mm × 35 mm tissue culture dishes, and mark them A1, A2, A3, A4, B1, B2, B3, B4, C1, C2, C3, and C4. Seed 200,000 cells in 2 ml of complete growth medium in each dish, and allow them to attach overnight.
2. Next day, aspirate the overnight growth medium, and add 2 ml of fresh growth medium in each dish. To dishes A3, A4, B3, B4, C3, and C4, further add 20 μl of NAC solution (100 μM in PBS), mix by gentle swirling, and return all dishes to incubator for 2 h.
3. After incubation, aspirate the existing medium from each dish, and add 2 ml of fresh growth medium in each dish.
4. Immediately, add 100 μl aqueous solutions of QDa (8 μM in water) each to dishes A1, A2, A3, and A4 and 100 μl of QDc (8 μM in water) each to dishes B1, B2, B3, and B4. Also, add 100 μl of 1× PBS solution each to dishes C1, C2, C3, and C4. Mix by gentle swirling. Return dishes to the cell culture incubator (37 °C, 5 % CO_2).
5. After 1 h of incubation, take out dishes A1, A3, B1, B3, C1, and C3. Aspirate existing medium, and rinse each dish twice with PBS.
6. Harvest cells by adding 100 μl of trypsin–EDTA to each dish, followed by brief incubation (for 2–3 min) in the incubator.
7. Add 2 ml of PBS to each dish, mix well with trypsinized cells, and pipette out the mixture in sterile 15 ml centrifuge tubes (separate tube for each dish).
8. Centrifuge the tubes at 313´*g* for 5 min to separate the cells, and wash the cell pellets twice with 5 ml of PBS. Redisperse the cell pellets by adding 3 ml of PBS to each tube.
9. Label the trypsinized cells with 5 g/ml FITC-conjugated annexin-V for 15 min.
10. Measure the fluorescence of annexin-V-positive cells with FACSCalibur flow cytometer (Becton Dickinson, USA) on the FL-1 detector (530 ± 15 nm, band pass) (*see* **Note 13**).

11. After 4 h of incubation from **step 4**, take out dishes A2, A4, B2, B4, C2 and C4. Repeat **steps 5** through **10** for these dishes.

3.6 Intracellular ROS Measurement (See Note 14)

1. Take 2 × 96-well tissue culture plates, and mark them 1 and 2. Seed 3,000 cells in 100 μl of complete growth medium in 9 (out of 96) wells in each plate (total 18 wells), and allow them to attach overnight.
2. Next day, aspirate the overnight growth medium, and add 100 μl of fresh growth medium in each well.
3. Immediately, add 5 μl aqueous solutions of QDa (8 μM in water) to three wells each in Plates 1 and 2. Similarly, add 5 μl aqueous solutions of QDc (8 μM in water) to three wells each in Plates 1 and 2. Finally, add 5 μl of 1× PBS to three wells each in Plates 1 and 2. Mix by gentle swirling. Return both plates to the cell culture incubator (37 °C, 5 % CO_2).
4. After 1 h of incubation, take out Plate 1. Aspirate the existing medium in all the wells, and add in each well 100 μl of serum-free medium (media containing 0.5 % FBS) mixed with DCFH-DA at a final concentration 10 μM. Return plates to the incubator (37 °C, 5 % CO_2).
5. After incubation for 30 min, determine DCF fluorescence at 485 nm excitation and 520 nm emission using a fluorescence microplate reader (Safire2, Tecan, Switzerland) (*see* **Note 13**).
6. Similarly, after 4 h of incubation from **step 3**, take out Plate 2. Repeat **steps 4** and **5** in this plate.

3.7 Glutathione (GSH) Assay (See Note 15)

1. Take 8 mm × 35 mm tissue culture dishes, and mark them A1, A2, B1, B2, C1, C2, D1, and D2. Seed 200,000 cells in 2 ml of complete growth medium in each dish, and allow them to attach overnight.
2. Next day, aspirate the overnight growth medium, and add 2 ml of fresh complete growth medium in each dish. Immediately, add 100 μl aqueous solutions of QDa (8 μM in water) each to dishes A1 and A2 and 100 μl of QDc (8 μM in water) each to dishes B1 and B2. Also, add 100 μl of 1× PBS solution each to dishes C1 and C2, and 100 μl of diethyl maleate solution (0.1 mM in PBS) each to dishes D1 and D2. Mix by gentle swirling. Return dishes to the cell culture incubator (37 °C, 5 % CO_2).
3. After 1 h of incubation, take out dishes A1, B1, C1, and D1. Aspirate the existing medium, wash each dish twice with 2 ml 1× PBS, and add 100 μl of ice-cold 5 % 5-sulfosalicylic acid (SSA) in each dish. Then scrape the cells with a sterile scraper.

4. From each dish, transfer the scraped cells with the 5 % SSA in separate 0.6 ml Eppendorf tubes, incubate the tubes on ice for 10 min, and centrifuge them at 8,000 × *g* for 5 min.
5. Separate the supernatant from the cell pellet. Use the supernatant for glutathione (GSH) determinations in the next **steps** (**6** through **12**).
6. For determination of total glutathione (GSH + GSSG) from each tube, take 5 μl of supernatant from **step 5** in fresh 0.6 ml Eppendorf tubes, and add to it first 5 μl of 5 % SSA, then 10 μl of sodium carbonate solution (400 mM in PBS), and finally 140 μl of PBS.
7. For determination of total GSSG from each tube, take 5 μl of supernatant from **step 5** in fresh 0.6 ml Eppendorf tubes, and add to it first 5 μl of 0.5 mM M4VP-masking reagent in 5 % SSA, followed by 10 μl of sodium carbonate solution (400 mM in PBS).
8. Incubate the assay samples (four tubes each from **steps 6** and 7) for 5 min at room temperature.
9. Transfer 20 μl from each tube to individual wells in a 96-well microtiter plate. Also, transfer 20 μl each of six preformed GSSG standards (containing 5, 4, 3, 2, 1, and 0.5 ng of GSSG/1 μl of PBS) to individual wells in the same 96-well microtiter plate.
10. To each well, add 170 μl of reaction mixture, containing 1.9 U/ml glutathione reductase and 0.4 mM NADPH in PBS, and mix the contents by gentle swirling.
11. To each well, further add 10 μl of DTNB substrate solution (4.5 mM in PBS), mix the contents by gentle swirling, and incubate for 5 min at room temperature.
12. Read the absorbance from each well at 415 nm on a microplate spectrophotometer every 5 min for 25 min at room temperature. Determine GSH concentration by subtracting GSSG concentration from total glutathione concentrations.
13. After 4 h of incubation from **step 2**, take out dishes A2, B2, C2, and D2, and repeat **steps 3** through **12**.

3.8 Lipid Peroxidation Assay (See Note 16)

1. Follow **steps 1** and **2** exactly as given in previous section (in Subheading 3.7).
2. After 1 h of incubation, take out dishes A1, B1, C1, and D1. Aspirate the existing medium, wash each dish twice with 2 ml 1× PBS, and add 2 ml of 2.5 % TCA in each dish. Then scrape the cells with a sterile scraper.
3. Transfer the contents of each dish (scraped cells in 2 ml TCA) to separate 15 ml centrifuge tubes, and centrifuge them at 13,000 × *g* for 2 min.

4. Transfer 500 μl of supernatant from each tube to fresh 15 ml centrifuge tubes. Also, to five separate 15 ml centrifuge tubes, add 500 μl of preformed MDA standards (0.1, 0.5, 1, 2, and 4 nmol/ml of PBS).
5. Add to each tube 400 μl of 15 % TCA and 800 μl of 0.67 % TBA/0.01 % BHT. Vortex the tubes for vigorous mixing of contents, and heat them for 20 min in a 95 °C water bath.
6. Allow the tubes to cool to room temperature. Then add 3 ml of butanol to each tube, mix gently, and transfer 200 μl of the butanol phase from each tube to individual wells in a 96-well plate.
7. Measure the fluorescent TBA adducts at their excitation (521 nm) and emission (552 nm) wavelengths on a microplate spectrophotometer. Determine the quantity of lipid peroxide in the sample wells by comparison to the MDA standard curve (*see* **Note 13**).
8. After 4 h of incubation from **step 1**, take out dishes A2, B2, C2, and D2, and repeat **steps 2** through 7.

4 Notes

1. We have chosen commercially available QDs with a core of cadmium selenide (CdSe) and a shell of zinc sulfide (ZnS), as they are the most common and efficient among all QDs. However, any commercial or laboratory-synthesized QD can be chosen for these studies.
2. As the presence of various functional groups on the QD surface plays a critical role in modulating the surface charge and directing the intracellular uptake of the QDs, we have selected two kinds of QDs displaying two different functional groups on their surface, namely, amino (QDa) and carboxyl (QDb).
3. We have selected Mia PaCa-2 cells as they are easy to grow and maintain in cell culture. However, any cell line can be chosen for these studies, provided they can be well maintained in tissue culture laboratory conditions.
4. The characterization studies are needed to be carried out to verify the size, surface charge (zeta potential), and optical properties (absorption and fluorescence spectra) of the QDs.
5. DLS gives the hydrodynamic size of nanoparticles (e.g., quantum dots), which combines the size of the nanoparticles with that of the surrounding hydration layer.
6. Avoid any air bubbles in the sample while DLS measurements (for both particle size and surface charge).
7. Since QDs have a broad absorption spectra, any wavelength within the 400–500 nm range can be used for excitation.

8. With confocal microscopy, we get a semiquantitative estimation of the uptake, as well as visualize the intracellular localization, of all fluorescent species (fluorescent nanoparticles in this case) in cells in culture. This uptake should increase with exposure time till a saturation point. In this overall protocol, this step serves as a preliminary step of verifying nanoparticle uptake of cells, prior to a quantitative estimation as described in the next section.
9. The centrifugation should not be carried at a speed higher than 704´*g*, as higher speed may lead to the settling down of some quantum dots, along with the cell debris, from the solution. This will introduce errors in the measurement.
10. NAC is a known ROS quencher. Therefore, it has been used to verify whether the apoptotic effect of QDs on cells is mediated by ROS or any other mechanism.
11. The reagent reacts with only live cells and produces water-soluble purple crystals that absorb at 490 nm. Therefore, the value of absorbance would directly correlate with the number of viable cells present in each well.
12. In addition to MTS assay, the toxic effects of QDs on treated cells will also be studied by flow cytometry via quantification of exposed membrane phosphatidylserine (PS) in apoptotic cells.
13. Both the QDs were selected in such a way that their emission maximum is at 655 nm, which should not interfere with the emission maximum of FITC-conjugated annexin-V (520 nm), or DCF (520 nm), or TBA (552 nm).
14. The levels of ROS within the cells will be determined by measuring the oxidative conversion of 20,70-dichlorofluorescin diacetate (DCFH-DA) to the fluorescent compound dichlorofluorescin (DCF).
15. GSH in samples is quantified by comparing the rates of 2-nitro-5-thiobenzoic acid production in the presence or absence of M4VP to that of a glutathione standard curve. The amount of reduced glutathione (GSH) in the cell lysates was determined by the kinetic method based on reduction of 5-5#-dithiobis(2-nitrobenzoic acid) (DTNB) by reduced glutathione (GSH) to form the colored product 2-nitro-5-thiobenzoic acid, which is measured at 415 nm. The resulting oxidized glutathione (GSSG) is then reduced by glutathione reductase to form GSH, allowing another cycle of DTNB reduction. Preincubation of the samples with the thiol-masking agent 1-methyl-4-vinyl-pyridinium (M4VP) prevents measurement of the GSH, so that only GSSG is measured.

16. Lipid peroxides in media were determined by the thiobarbituric reactive substances (TBARS) assay. The TBARS assay measures lipid hydroperoxides and aldehydes, such as malondialdehyde (MDA), in the cell lysate. MDA combines with thiobarbituric acid (TBA) in a 1:2 ratio to form a fluorescent adduct that is measured at its excitation (521 nm) and emission (552 nm) wavelengths with a microplate reader.

Acknowledgments

The authors would like to acknowledge the Start-up R&D grant and seed money provided by Delhi University, India.

References

1. Alivisatos AP (1996) Semiconductor clusters, nanocrystals, and quantum dots. Science 271:933
2. Prasad PN (2004) Nanophotonics. Wiley, New York
3. Alivisatos P (2004) The use of nanocrystals in biological detection. Nat Biotechnol 22:47–52
4. Gao XH, Yang LL, Petros JA, Marshall FF, Simons JW, Nie SM (2005) In vivo molecular and cellular imaging with quantum dots. Curr Opin Biotechnol 16:63–72
5. Michalet X, Pinaud FF, Bentolila LA et al (2005) Quantum dots for live cells, in vivo imaging, and diagnostics. Science 307:538–544
6. Yong KT, Roy I, Swihart MT, Prasad PN (2009) Multifunctional nanoparticles as biocompatible targeted probes for human cancer diagnosis and therapy. J Mater Chem 19:4655–4672
7. Peng XG, Wickham J, Alivisatos AP (1998) Kinetics of II-VI and III-V colloidal semiconductor nanocrystal growth: "Focusing" of size distributions. J Am Chem Soc 120:5343
8. Derfus AM, Chan WCW, Bhatia SN (2004) Probing the cytotoxicity of semiconductor quantum dots. Nano Lett 4:11–18
9. Kirchner C, Liedl T, Kudera S, Pellegrino T, Javier AM, Gaub HE et al (2005) Cytotoxicity of colloidal CdSe and CdSe/ZnS nanoparticles. Nano Lett 5:331–338
10. Clarke SJ, Hollmann CA, Zhang Z, Suffern D, Bradforth SE, Dimitrijevic NM et al (2006) Photophysics of dopamine-modified quantum dots and effects on biological systems. Nat Mater 5:409–417
11. Hsieh SC, Wang FF, Hung SC, Chen Y, Wang YJ (2006) The internalized CdSe/ZnS quantum dots impair the chondrogenesis of bone marrow mesenchymal stem cells. J Biomed Mater Res B Appl Biomater 79B:95–101
12. Lovric´ J, Bazzi HS, Cuie Y, Fortin GRA, Winnik FM, Maysinger D (2005) Differences in subcellular distribution and toxicity of green and red emitting CdTe quantum dots. J Mol Med 83:377–385
13. Chen FQ, Gerion D (2004) Fluorescent CdSe/ZnS nanocrystal-peptide conjugates for long-term, nontoxic imaging and nuclear targeting in living cells. Nano Lett 4:1827–1832
14. Finkel T, Holbrook NJ (2000) Oxidants, oxidative stress and the biology of ageing. Nature 408:239–247
15. Prestwich EG, Roy MD, Rego J, Kelley SO (2005) Oxidative DNA strand scission induced by peptides. Chem Biol 12:695–701
16. Noble M, Mayer-Proschel M, Proschel C (2005) Redox regulation of precursor cell function: insights and paradoxes. Antioxid Redox Signal 7:1456–1467

Chapter 18

Hybrid Polymeric Nanoparticles: Potential Candidate for Ophthalmic Delivery

Dhruba J. Bharali, Donald Armstrong, and Shaker A. Mousa

Abstract

Nanotechnology that deals with submicron particles has the potential to deeply impact ophthalmic delivery owing to its ability to transport incorporated active ingredient/drugs to the different compartments of the eye while protecting them from any external assault. Polymeric nanoparticles, in particular hybrid nanoparticles made up of different biodegradable and biocompatible polymers, has the potential to revolutionize ocular delivery carrier systems by taking advantage of their highly tunable physicochemical properties. Here we report a synthesis method of hybrid nanoparticles made up of chitosan-poly(lactide-co-glycolide) (PLGA, as a building block of the nanocarrier) as a carrier for ophthalmic delivery. The synthesis method, characterization, and uptake study in retinal epithelial cells by confocal imaging is described herein.

Key words Nanoparticles, Hybrid polymer, Chitosan, Ophthalmic delivery, RPE cell (retinal pigment epithelial) cell, Confocal imaging

1 Introduction

The advent of nanotechnology has had a profound effect in every route of drug delivery and in ocular delivery in particular [1]. This cutting-edge technology may provide an unprecedented opportunity for topical ophthalmic delivery because of its ability to transport incorporated active ingredient/drugs to the different compartments of the eye while protecting them from any external assault [2–6]. The major advantages of nanoparticulate carrier systems for ophthalmic delivery include: (a) sustained drug release, (b) protection of the drug from chemical and enzymatic assault, (c) P-gp circumvention – higher cellular permeability, (d) conjugation of a targeting moiety for site-specific delivery, (e) tunable size and surface charge, (d) mucoadhesive properties for prolonged residence time, and (f) robust stability of the formulation [4, 7–10].

Donald Armstrong and Dhruba J. Bharali (eds.), *Oxidative Stress and Nanotechnology: Methods and Protocols*, Methods in Molecular Biology, vol. 1028, DOI 10.1007/978-1-62703-475-3_18, © Springer Science+Business Media New York 2013

Chitosan, a natural polymer, is one of the most extensively used polymers for synthesizing nanoparticulate delivery carriers for ocular drug delivery. More than a decade's track record indicates that these chitosan nanoparticles have the ability to: (a) protect the drug from chemical and enzymatic assault, (b) enhance cellular permeability, (c) conjugate a targeting moiety for site-specific delivery, (d) be formed in variable sizes and acquire different surface charges according to the needs, (e) carry drugs to different compartments of the eye, and (f) adhere to surfaces (mucoadhesive properties) for prolonged residence time [1, 3, 11, 12]. But due to chitosan's hydrophilic nature, nanoparticles made up of chitosan are mainly restricted to hydrophilic drug delivery. Hybrid nanoparticles made up of a chitosan polymer conjugated to a hydrophobic polymer moiety like PLGA, poly(lactide-co-glycolide) result in an amphiphilic polymer that allows us to encapsulate lipophilic drugs [13–15].

Nanoparticles synthesized using the chitosan-PLGA conjugate to PLGA might have tremendous potential to address many of the current challenges in ocular delivery. Significant advantages of biodegradable PLGA polymers include their long track record of safety, proven biocompatibility, and the ability to control the time and rate of polymer degradation and hence the release of the drug incorporated in it [16–18]. Incorporation of an active molecule like an antioxidant in this submicron-sized particulate carrier system can be used to overcome ocular barriers and to prolong residence time in the target tissue. One of the major problems associated with many of the antioxidants is that they are poorly water soluble and are formulated with toxic solvents. This problem can be overcome by incorporating them into nanocarriers to get a solvent-free, aqueous formulation. This allows incorporation of lipophilic drugs without using any excipient or organic solvent.

Here we report nanoparticles made up of hybrid chitosan-PLGA as a building block of the nanocarrier that has the potential to be used as a carrier for ophthalmic delivery. The synthesis method, characterization, and uptake study in retinal epithelial cells by confocal imaging is described herein. Results of the preliminary experiments conducted in our laboratory demonstrate our ability to control the size and surface charge of the nanoparticles, both of which play a crucial role in retinal drug delivery. Additionally, nanoparticles as small as 200 nm can be synthesized using this hybrid polymer combining chitosan and PLGA. Since the smaller particles are better tolerated by patients than larger particles, nanoparticles of size ~200 nm have the potential to be used for prolonged action in the eye. Also as a proof of the concept, a method for encapsulation of astaxanthin (a commonly used antioxidant) in this chitosan-PLGA is described.

2 Materials

1. Alexa Fluor 488 sulfodichlorophenol ester (Alexa Fluor 488 5-SDP ester), Invitrogen Inc. (Grand Island, NY).
2. Poly(D-lactide-co-glycolide 70:30), Polysciences Inc. (Warrington, PA).
3. Polyvinyl alcohol, Sigma-Aldrich (Saint Louis, MO). One gram of this was solubilized in 100 ml of DI water for 1 % w/v solution.
4. EDC, *N*-(3-dimethylaminopropyl- *N*′-ethylcarbodiimide hydrochloride), Sigma-Aldrich (Saint Louis, MO).
5. Low-molecular-weight water-soluble chitosan (chitosan oligosaccharide), LSK BioPartners – distributors for Kittolife (S. Korea).
6. Human retinal pigment epithelium (RPE) cell line was purchased from ATCC (Manassas, VA) and other cell components from Invitrogen Inc. (Grand Island, NY).
7. Dimethyl sulfoxide, Sigma-Aldrich (Saint Louis, MO).
8. Dialysis membrane 12 kDa cutoff, Sigma-Aldrich (Saint Louis, MO). The preservatives were removed as per the manufacturer instructions (*see* **Note 1** below).

3 Methods

3.1 Synthesis of Chitosan-PLGA Nanoparticles (CHI-PLGA-NPs)

This method of synthesis of nanoparticles is aimed at establishing a scaffold for the nanoformulation and preliminary in vitro efficacy testing with confocal imaging. Briefly, in a typical experiment:

1. Prepare a solution of PLGA (poly-lactide-co-glycolide) solution in DMSO (40 mg/ml).
2. Add 200 μl of this 40 mg/ml PLGA solution in DMSO to 20 ml of a 1 % PVA (polyvinyl alcohol) solution and stir for 12 h at room temperature.
3. Dialyze the nanoparticles suspension for at least 8 h using 10–12 kDa membrane against water (*see* **Note 1** for preparation of dialysis membrane).
4. Lyophilize the water nanoparticles suspension to get the nanoparticles in powdered form.
5. Resuspend the nanoparticles in 18 ml MES buffer.
6. Add 1 ml of *N*-hydroxysuccinimide (NHS) solution (10 mg/ml) and 1 ml of 1-(3-dimethylaminopropyl)-3-3-ethylcarbodiimide hydrochloride (20 mg/ml) to the 18 ml of the nanoparticles suspension. Stir for 2 h.

7. Add 5 ml of chitosan solution (2 mg/ml) and stir for 3 h.
8. Dialyze the nanoparticles suspension for at least 8 h using 10–12 kDa membrane against DI water.
9. Lyophilize the suspension to get the nanoparticles in powdered form.

3.2 Conjugation of Alexa Fluor 488 Dye to the CHI-PLGA Nanoparticles

1. Resuspend the nanoparticles in 20 ml of PBS (pH 7.4).
2. Suspend 1 mg Alexa Fluor 488 5-SDP ester in 1 ml of DMSO.
3. Add 50 μl of Alexa Fluor 488 5-SDP ester to the 20 ml of CHI-PLGA nanoparticles suspension and stir for 4 h.
4. Dialyze the nanoparticles suspension for at least 8 h using 10–12 kDa membrane against DI water to remove unreacted dye.

3.3 Synthesis of Chitosan-PLGA Nanoparticles (CHI-PLGA-NPs) Encapsulating Astaxanthin (Antioxidant)

1. Prepare a solution of PLGA (poly-lactide-co-glycolide) solution in DMSO (80 mg/ml).
2. Prepare a solution of astaxanthin in DMSO (10 mg/ml).
3. Mix 100 μl of PLGA in DMSO and 100 μl of astaxanthin in DMSO.
4. Add this mixture of PLGA and astaxanthin (total 200 μl) to 20 ml of a 1 % PVA (polyvinyl alcohol) solution and stir for 12 h at room temperature.
5. Dialyze the nanoparticles suspension for at least 8 h using 10–12 kDa membrane against water.
6. Lyophilize the water nanoparticle suspension to get the nanoparticles in powdered form.

3.4 Size and Zeta Potential Measurement by Dynamic Light Scattering (DLS)

Size distribution and zeta potential of CHI-PLGA nanoparticles in aqueous dispersion were determined using a Malvern Zetasizer (Malvern Instrumentation Co., Westborough, MA).

1. For size measurement place 1 ml of CHI-PLGA nanoparticles in a four-sided clear plastic cuvette and analyze directly at 25 °C.
2. For zeta potential measurement place about 750 μl of the resuspended CHI-PLGA nanoparticles in a disposable capillary zeta potential cell (available for the Zetasizer Nano series) and do the measurement using the Malvern Zetasizer in zeta potential measurement mode.

The size distribution and zeta potential of CHI-PLGA nanoparticles in aqueous dispersion as determined by the above method found the average sizes of the nanoparticles to be < 200 nm in diameter (Fig. 1) and the zeta potential to be around +5 mV (Fig. 2). These results are within the acceptable range referenced in the literature of <500 nm and demonstrate that our proposed technique is satisfactory and that with refinements we can achieve our goal of preparing a nanoparticle delivery system. We also can manipulate the size of the nanoparticles by changing different

			Size (d.nm):	% Intensity	Width (d.nm):
Z-Average (d.nm):	170.8	Peak 1:	193.2	98.9	65.19
PdI:	0.258	Peak 2:	5384	1.1	320.0
Intercept:	0.961	Peak 3:	0.000	0.0	0.000
Result quality :	Good				

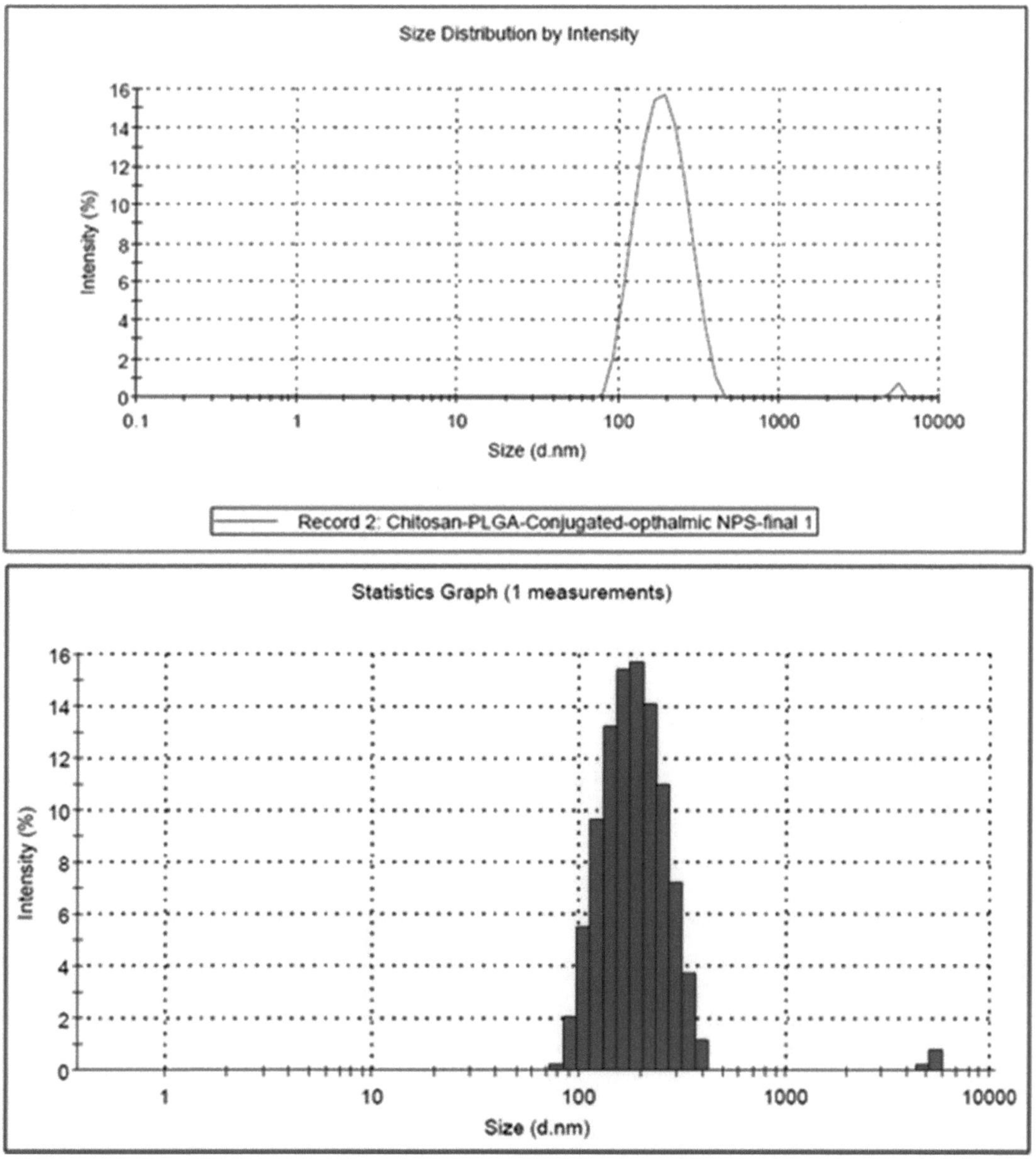

Fig. 1 Size measurement of chitosan-PLGA nanoparticles by dynamic light scattering (DLS) ($Z_{average}$ size = 170 nm)

parameters (refer to **Note 2** below). Also, the use of a suitable stabilizer increases the much-coveted stability of the nanoparticles/nanocarrier in the aqueous dispersion (Fig. 3).

			Mean (mV)	Area(%)	Width(mV)
Zeta Potential (mV):	5.03	**Peak 1:**	5.03	100.0	4.62
Zeta Deviation (mV):	4.62	**Peak 2:**	0.00	0.0	0.00
Conductivity (mS/cm):	1.48	**Peak 3:**	0.00	0.0	0.00
Result quality :	Good				

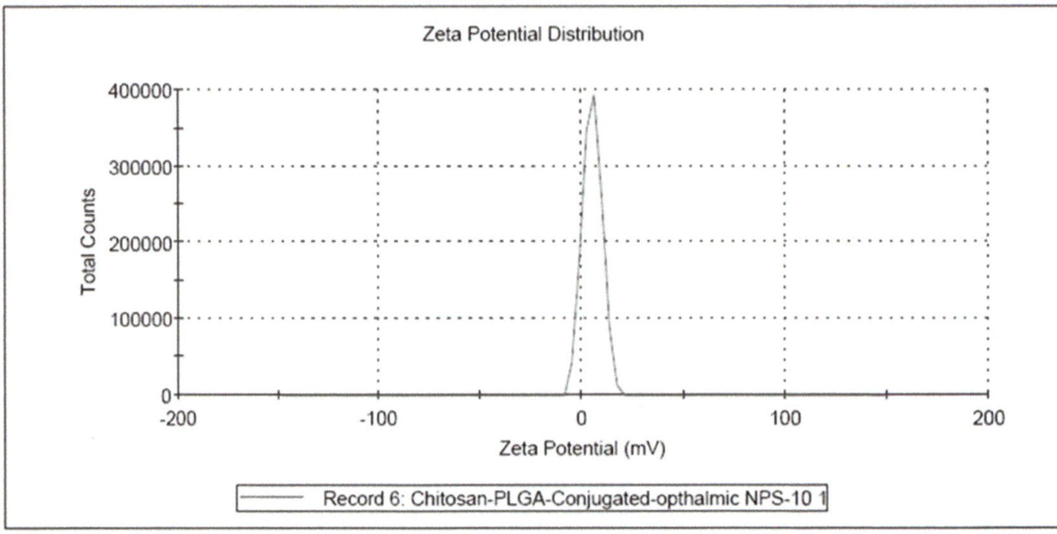

Fig. 2 Zeta potential measurement of chitosan-PLGA nanoparticles by dynamic light scattering (DLS) (zeta potential = +5 mV)

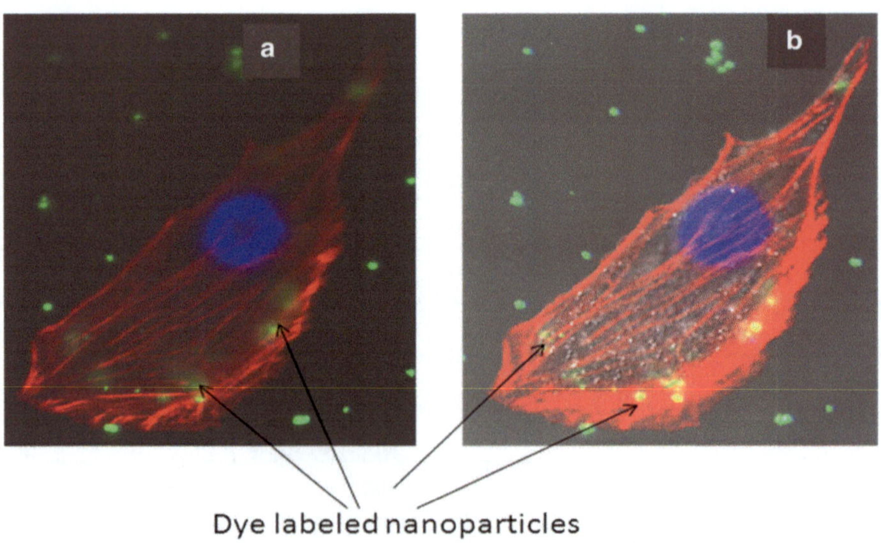

Fig. 3 Images showing the uptake of chitosan-PLGA nanoparticles labeled with Alexa Fluor 488 in RPE cell. (**a**) Confocal image and (**b**) confocal image superimposed with differential interface contrast (DIC). *Green* color indicates the fluorescently tagged nanoparticles, *blue* color indicates DAPI staining to the nucleus, and *red* color indicates the cell skeleton stained by rhodamine-phalloidin

3.5 Cell Culture

1. Maintain the RPE cells in DMEM supplemented with 10 % fetal bovine serum at 37 °C with 5 % CO_2.
2. Trypsinize the cells for about 3 min and centrifuge at around 670 × *g*; discard the supernatant. Resuspend the cell pellet in DMEM supplemented with 10 % fetal bovine serum at 37 °C with 5 % CO_2.
3. Transfer 1 ml of the suspension to 35 mm glass bottom culture dishes (MatTek Corp., Ashland, MA) and incubate for 24 h at 37 °C with 5 % CO_2 (Thermo Electron Corp., Forma Series II incubator; Marietta, OH).
4. Add 100 μl of the CHI-PLGA nanoparticles conjugated to dye to each plate and incubate for 2 h.
5. Wash the cells carefully with PBS 2–3 times.
6. Add ~1 ml of 1 % formaldehyde and incubate for 15–20 min. Wash with PBS again.
7. Stain the cells again with rhodamine-phalloidin and the nuclei with DAPI as per as the manufacturer instructions.
8. The cells are ready for confocal imaging (*see* **Note 3** below).

4 Notes

1. Removal of glycerol included as a humectant can be accomplished by washing the tubing in running water for 3–4 h. Removal of sulfur compounds can be accomplished by treating the tubing with a 0.3 % (w/v) solution of sodium sulfide at 80 °C for 1 min. Wash with hot water (60 °C) for 2 min, followed by acidification with a 0.2 % (v/v) solution of sulfuric acid, then rinse with hot water to remove the acid. This tubing will retain most molecules of molecular weight 12,000 or greater.
2. The size and surface charge of the nanoparticles can be manipulated by changing the ratios of amounts of the polymers chitosan and PLGA and the surfactant PVA.
3. Confocal imaging (using Leica TCS SPS; Buffalo Grove, IL): Nanoparticles labeled with Alexa Fluor 488 (green) were detected at excitation/emission 494 nm/518 nm. Rhodamine-phalloidin stain and DAPI were detected at excitation/emission 542 nm/565 nm and 345 nm/448 nm, respectively. Pictures were acquired at 63× magnification and 5× optical zoom.

References

1. de la Fuente M, Ravina M, Paolicelli P, Sanchez A, Seijo B, Alonso MJ (2010) Chitosan-based nanostructures: a delivery platform for ocular therapeutics. Adv Drug Deliv Rev 62:100–117
2. Bourges JL, Gautier SE, Delie F, Bejjani RA, Jeanny JC, Gurny R, BenEzra D, Behar-Cohen FF (2003) Ocular drug delivery targeting the retina and retinal pigment epithelium using polylactide nanoparticles. Invest Ophthalmol Vis Sci 44:3562–3569
3. de Campos AM, Diebold Y, Carvalho EL, Sanchez A, Alonso MJ (2004) Chitosan nanoparticles as new ocular drug delivery systems: in vitro stability, in vivo fate, and cellular toxicity. Pharm Res 21:803–810
4. Shah SS, Denham LV, Elison JR, Bhattacharjee PS, Clement C, Huq T, Hill JM (2010) Drug delivery to the posterior segment of the eye for pharmacologic therapy. Expert Rev Ophthalmol 5:75–93
5. Nagarwal RC, Kant S, Singh PN, Maiti P, Pandit JK (2009) Polymeric nanoparticulate system: a potential approach for ocular drug delivery. J Control Release 136:2–13
6. Wadhwa S, Paliwal R, Paliwal SR, Vyas SP (2009) Nanocarriers in ocular drug delivery: an update review. Curr Pharm Des 15:2724–2750
7. Mainardes RM, Urban MC, Cinto PO, Khalil NM, Chaud MV, Evangelista RC, Gremiao MP (2005) Colloidal carriers for ophthalmic drug delivery. Curr Drug Targets 6:363–371
8. Ludwig A (2005) The use of mucoadhesive polymers in ocular drug delivery. Adv Drug Deliv Rev 57:1595–1639
9. Li N, Zhuang C, Wang M, Sun X, Nie S, Pan W (2009) Liposome coated with low molecular weight chitosan and its potential use in ocular drug delivery. Int J Pharm 379:131–138
10. Eljarrat-Binstock E, Orucov F, Aldouby Y, Frucht-Pery J, Domb AJ (2008) Charged nanoparticles delivery to the eye using hydrogel iontophoresis. J Control Release 126:156–161
11. Paolicelli P, de la Fuente M, Sanchez A, Seijo B, Alonso MJ (2009) Chitosan nanoparticles for drug delivery to the eye. Expert Opin Drug Deliv 6:239–253
12. Wadhwa S, Paliwal R, Paliwal SR, Vyas SP (2009) Chitosan and its role in ocular therapeutics. Mini Rev Med Chem 9: 1639–1647
13. Park JH, Saravanakumar G, Kim K, Kwon IC (2010) Targeted delivery of low molecular drugs using chitosan and its derivatives. Adv Drug Deliv Rev 62:28–41
14. Sahu SK, Maiti S, Maiti TK, Ghosh SK, Pramanik P (2011) Hydrophobically modified carboxymethyl chitosan nanoparticles targeted delivery of paclitaxel. J Drug Target 19(2):104–113
15. Hwang HY, Kim IS, Kwon IC, Kim YH (2008) Tumor targetability and antitumor effect of docetaxel-loaded hydrophobically modified glycol chitosan nanoparticles. J Control Release 128:23–31
16. Budhian A, Siegel SJ, Winey KI (2008) Controlling the in vitro release profiles for a system of haloperidol-loaded PLGA nanoparticles. Int J Pharm 346:151–159
17. Vasir JK, Labhasetwar V (2007) Biodegradable nanoparticles for cytosolic delivery of therapeutics. Adv Drug Deliv Rev 59:718–728
18. Gupta H, Aqil M, Khar RK, Ali A, Bhatnagar A, Mittal G (2010) Sparfloxacin-loaded PLGA nanoparticles for sustained ocular drug delivery. Nanomedicine 6:324–333

Part III

Future Directions

Chapter 19

Relevance of Nanotechnology in Modulating Oxidative Stress: An Overview

Dhruba J. Bharali and Shaker A. Mousa

Abstract

The emergence of nanotechnology has already had a groundbreaking impact in biomedical research. In recent years, there has been significant interest by researchers to use nanotechnology to treat various oxidative stress-related ailments. We have seen a paradigm change in delivery methods of natural antioxidants for the treatment and prevention of various diseases. The use of nanotechnology to deliver natural antioxidants is highlighted, and current and future prospects of nanotechnology are briefly discussed.

Key words Nanotechnology, Oxidative stress, Antioxidant, Natural products, Half-life, Anticancer drugs

Antioxidants obtained from various natural sources are one of the most widely used entities to treat human diseases, and they have been used since ancient times. Castor oil, olive oil, beeswax, etc. were used widely in ancient times for different purposes. Today, the use of natural products is seen in a range of applications from appearance-enhancing cosmetics to life-threatening diseases like cancer. In recent years, numerous naturally derived products including aspirin, warfarin, heparin, and statins have been used to treat different diseases [1]. Additionally, natural antioxidants like flavonoids and isoflavones extracted from food sources like grapes, peaches, pomegranates, and green tea have been used to treat and prevent many diseases that are augmented by oxidative stress (directly or indirectly) [2]. Oxidative stress and the relationship between the various diseases are shown in Fig. 1. An example of one of the most important discoveries of an active compound derived from a natural product was taxol (derived from the bark of the Pacific yew tree and approved by the FDA); it is currently one of the major anticancer drugs used.

Donald Armstrong and Dhruba J. Bharali (eds.), *Oxidative Stress and Nanotechnology: Methods and Protocols*, Methods in Molecular Biology, vol. 1028, DOI 10.1007/978-1-62703-475-3_19, © Springer Science+Business Media New York 2013

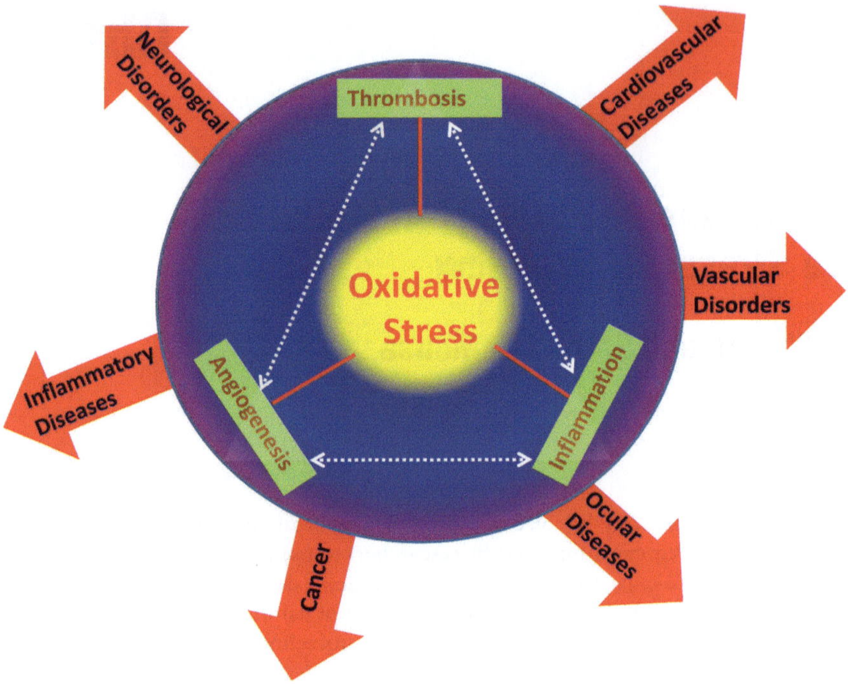

Fig. 1 A schematic diagram showing the role of oxidative stress as a pathological enhancer culminating in various disease types

The main obstacles that hinder the clinical application of naturally derived compounds are:

1. Poor solubility in aqueous media
2. Poor bioavailability and short half-life
3. Limited dose regimen

Most natural products are not soluble in water and need to be dissolved in organic solvent or in an excipient to make an injectable solution. This toxic organic solvent/external excipient may increase the cytotoxic effect of the natural product and may have its own side effects [3]. Additionally, the low molecular weight (meaning easily excreted) and the low therapeutic index of natural products are reasons for the typically higher dose administration (because they are poorly bioavailable and have a short half-life). There is the possibility that the higher doses of these therapeutically beneficial natural products will make them eventually more toxic. Also, when administrated alone, these drugs lack specificity and can cause significant damage to normal tissues. Thus, the conglomeration of nanotechnology and anticancer drugs might be the answer to many of the inherent problems associated with natural products/antioxidants.

The emergence of nanotechnology has already had a ground-breaking impact in biomedical research. In recent years, we have seen a paradigm change in delivery methods of natural antioxidants for the treatment and prevention of various diseases. One of the most important and useful methods is the utilization of nanotechnology as an efficient delivery system. The chemical structure, relatively safe history, and easy surface modulation of nanoparticles make them ideal carriers to deliver a multiple drug load to the desired site. Antioxidants like EGCG, curcumin, beta-carotene, and vitamin E (which is also known for its anticancer activity) already have been used in nanotechnology-mediated delivery to treat various cancers [4–8]. Apart from delivering antioxidants, there are some nanoparticles known to have antioxidant properties themselves, like silver nanoparticles, TiO_2, and cerium oxide nanoparticles [9–12]. These nanoparticles already have made their way to the market either as cosmetics or preservatives, and in the near future we may see them as antimicrobial agents.

With the human population approaching seven billion, the rapid emergence of nanotechnology can be considered as the technology of the future. In the last few years, this cutting-edge technology of minuscule format, comprising tiny nanomaterials, is making a sea change in every aspect of our daily life with or without our knowledge. The rapid growth of nanotechnology can be visualized by considering the fact that the allotted annual budget for the National Nanotechnology Initiative (NNI), a presidential research and development initiative in the USA, was increased from $650 million in 2005 to $1.8 billion in 2012, and more than $18 billion has been spent during 2001–2012 [13]. Furthermore, the enormous impact of nanotechnology, particularly in the biomedical field, is demonstrated by its growth of more than 17 % annually [14–16]. The National Science Foundation (NSF) of the USA estimates that half of all pharmaceutical industry products will eventually have some association with nanotechnology [17]. Within a very short time span, nanotechnology has made a deep impact: from MRI contrasting agents (supermagnetic iron oxide (SPIO)/USPIO nanoparticles) to cancer treatment (e.g., Abraxane, Doxil) to cosmetics (e.g., TiO_2 nanoparticles).

Multifunctional nanoparticles with imaging and drug/antioxidant delivering capacity will be reality in the near future. These all-in-one multivalent nanocarriers have the potential of detecting and delivering antioxidants as well as being used for live monitoring of various diseases caused by oxidative stress. The combination of antioxidants and nanotechnology might have the potential to treat different oxidative stress-related diseases, and thereby overall has the potential to increase the life expectancy and quality of life in the near future.

References

1. Wachtel-Galor S, Benzie IFF (2011) Herbal medicine: an introduction to its history, usage, regulation, current trends, and research needs. CRC Press, Boca Raton, FL
2. Pietta PG (2000) Flavonoids as antioxidants. J Nat Prod 63(7):1035–1042
3. Bawarski WE et al (2008) Emerging nanopharmaceuticals. Nanomedicine 4(4):273–282
4. Siddiqui IA et al (2009) Introducing nanochemoprevention as a novel approach for cancer control: proof of principle with green tea polyphenol epigallocatechin-3-gallate. Cancer Res 69(5):1712–1716
5. Ranjan AP et al (2012) Scale up, optimization and stability analysis of Curcumin C3 complex-loaded nanoparticles for cancer therapy. J Nanobiotechnol 10(1):38
6. Lim KJ et al (2011) A polymeric nanoparticle formulation of curcumin inhibits growth, clonogenicity and stem-like fraction in malignant brain tumors. Cancer Biol Ther 11(5): 464–473
7. Yusuf M et al (2012) Plausible antioxidant biomechanics and anticonvulsant pharmacological activity of brain-targeted beta-carotene nanoparticles. Int J Nanomedicine 7:4311–4322
8. Lian H et al (2011) Supramolecular micellar nanoaggregates based on a novel chitosan/vitamin E succinate copolymer for paclitaxel selective delivery. Int J Nanomedicine 6: 3323–3334
9. Celardo I, Traversa E, Ghibelli L (2011) Cerium oxide nanoparticles: a promise for applications in therapy. J Exp Ther Oncol 9(1): 47–51
10. Carlotti ME et al (2009) Specific effects of single antioxidants in the lipid peroxidation caused by nano-titania used in sunscreen lotions. J Photochem Photobiol B 96(2):130–135
11. Dimitrijevic NM, Rozhkova E, Rajh T (2009) Dynamics of localized charges in dopamine-modified TiO(2) and their effect on the formation of reactive oxygen species. J Am Chem Soc 131(8):2893–2899
12. Dowding JM et al (2012) Cerium oxide nanoparticles scavenge nitric oxide radical (NO). Chem Commun (Camb) 48(40):4896–4898
13. http://www.nano.gov/about-nni/what/funding. Accessed 16 Apr, 2013
14. Fako VE, Furgeson DY (2009) Zebrafish as a correlative and predictive model for assessing biomaterial nanotoxicity. Adv Drug Deliv Rev 61(6):478–486
15. The Freedonia Group, Inc. (2007) US demand for nanotechnology medical products to approach $53 billion in 2011 – Report. The Freedonia Group, Inc., Cleveland, OH. http://nanotechwire.com/news.asp?nid=4446. Accessed 16 Apr, 2013
16. The Freedonia Group, Inc. (2007) Nanotechnology in health care, US Industry Study with forecasts to 2011, 2016 & 2021. The Freedonia Group, Inc., Cleveland, OH, p 350
17. Lux Research (2006) The Nanotech Report™: investment overview and market research for nanotechnology. Lux Research, New York, NY

INDEX

Donald Armstrong and Dhruba J. Bharali (eds.), *Oxidative Stress and Nanotechnology: Methods and Protocols*, Methods in Molecular Biology, vol. 1028, DOI 10.1007/978-1-62703-475-3, © Springer Science+Business Media New York 2013

MIX
Papier aus verantwortungsvollen Quellen
Paper from responsible sources
FSC® C105338

If you have any concerns about our products,
you can contact us on
ProductSafety@springernature.com

In case Publisher is established outside the EU,
the EU authorized representative is:
Springer Nature Customer Service Center GmbH
Europaplatz 3, 69115 Heidelberg, Germany

Printed by Libri Plureos GmbH
in Hamburg, Germany